Advances in
Heterocyclic
Chemistry

Volume 20

Editorial Advisory Board

A. Albert
A. T. Balaban
J. Gut
J. M. Lagowski
J. H. Ridd
Yu. N. Sheinker
H. A. Staab
M. Tišler

Advances in
HETEROCYCLIC CHEMISTRY

Edited by

A. R. KATRITZKY

A. J. BOULTON

School of Chemical Sciences
University of East Anglia
Norwich, England

Volume 20

Academic Press · New York San Francisco London · 1976
A Subsidiary of Harcourt Brace Jovanovich, Publishers

COPYRIGHT © 1976, BY ACADEMIC PRESS, INC.
ALL RIGHTS RESERVED.
NO PART OF THIS PUBLICATION MAY BE REPRODUCED OR
TRANSMITTED IN ANY FORM OR BY ANY MEANS, ELECTRONIC
OR MECHANICAL, INCLUDING PHOTOCOPY, RECORDING, OR ANY
INFORMATION STORAGE AND RETRIEVAL SYSTEM, WITHOUT
PERMISSION IN WRITING FROM THE PUBLISHER.

ACADEMIC PRESS, INC.
111 Fifth Avenue, New York, New York 10003

United Kingdom Edition published by
ACADEMIC PRESS, INC. (LONDON) LTD.
24/28 Oval Road, London NW1

LIBRARY OF CONGRESS CATALOG CARD NUMBER: 62-13037

ISBN 0-12-020620-X

PRINTED IN THE UNITED STATES OF AMERICA

Contents

CONTRIBUTORS vii

PREFACE ix

Applications of the Hammett Equation to Heterocyclic Compounds

P. TOMASIK AND C. D. JOHNSON

I. Introduction 1
II. Six-Membered Rings 5
III. Five-Membered Rings 34
IV. Fused Five- and Six-Membered Rings 54
V. Three-Membered Rings 61

1,2,4-Oxadiazoles

LEALLYN B. CLAPP

I. Synthesis: General Methods 66
II. Synthesis: Elaboration 67
III. Synthesis of Oxadiazolines and Other Reduced Rings . . . 79
IV. Physical and Spectral Properties 85
V. Chemical Properties 93
VI. Uses 112

Covalent Hydration in Nitrogen Heterocycles

ADRIEN ALBERT

I. Introduction 117
II. Diagnosis and Location of Covalent Hydration 118
III. Time-Dependent Shifts in the Preferred Position of Hydration . 131
IV. Examples of Covalent Hydration in New Ring Systems . . 135
V. The Isolation of Stereoisomeric Hydrates 139
VI. Covalent Hydration in Nature 140

1,2,3,4-Thiatriazoles

Arne Holm

I. Introduction	145
II. Physicochemical Properties of 1,2,3,4-Thiatriazoles	146
III. Chemical Properties of 1,2,3,4-Thiatriazoles	149
IV. 1,2,3,4-Thiatriazoles Substituted with C-Radicals	162
V. 1,2,3,4-Thiatriazole-5-thiol and Its Derivatives	163
VI. 5-Alkoxy and 5-Aryloxy 1,2,3,4-Thiatriazoles	166
VII. N-Substituted 5-Amino-1,2,3,4-thiatriazoles	167
VIII. Synthesis and Chemical Properties of 1,2,3,4-Thiatriazolines and Alleged 1,5-Dihydro-1,2,3,4-thia(S^{IV})triazoles	168
Notes Added in Proof	173

The Nomenclature of Heterocycles

Alan D. McNaught

I. Introduction	176
II. Nomenclature of Heterocyclic Skeletons	180
III. Naming of Derivatives	227
IV. Miscellaneous Examples	243
Appendix: IUPAC Rules A-11 to A-56, B-1 to B-15, C-14, and C-15	247

Cumulative Index of Titles	321

Contributors

Numbers in parentheses indicate the pages on which the authors' contributions begin.

ADRIEN ALBERT,* *Research School of Chemistry, The Australian National University, Canberra, Australia* (117)

LEALLYN B. CLAPP, *Brown University, Providence, Rhode Island* (65)

ARNE HOLM, *The H. C. Ørsted Institute, Chemical Laboratory II (General and Organic Chemistry), University of Copenhagen, Copenhagen, Denmark* (145)

C. D. JOHNSON, *School of Chemical Sciences, University of East Anglia, Norwich, England* (1)

ALAN D. MCNAUGHT, *The Chemical Society, London, England* (175)

P. TOMASIK, *Pedagogical University, Zawadzkiego 13-15, Częstochowa, Poland* (1)

---

* Present address: Department of Pharmacological Sciences, Health Sciences Center, State University of New York at Stony Brook, Stony Brook, New York.

Preface

It is now 13 years since the appearance of Volume 1 of *Advances in Heterocyclic Chemistry,* and significant progress in many of the subjects covered in the early volumes of the series has been made since publication. We have therefore embarked on the policy of updating appropriate chapters, and three of the articles in the present volume are of this type. Adrien Albert has written an account of the progress of covalent hydration since its first appearance in Volume 4. P. Tomasik and C. D. Johnson have reported on progress in the application of the Hammett equation to heterocycles since the Jaffé and Lloyd Jones article in Volume 3, and A. Holm has similarly taken on the account of 1,2,3,4-thiatriazoles given by Jensen and Pedersen in the same volume.

Further updated articles are scheduled for the next volume of the series; however, sometimes a subject expands so considerably that updating is not possible in a single article. This was the case with the prototropic tautomerism of heterocycles, covered in four chapters in Volumes 1 and 2. A special supplementary volume to these Advances has now been published: J. Elguero, C. Marzin, A. R. Katritzky, and P. Linda cover the recent advances in this subject.

In the remaining two chapters in the present volume, L. B. Clapp deals with 1,2,4-oxadiazoles, and A. D. McNaught has tackled the daunting task of explaining the intricacies of heterocyclic nomenclature.

A. R. KATRITZKY
A. J. BOULTON

Applications of the Hammett Equation to Heterocyclic Compounds

P. TOMASIK

*Pedagogical University, Zawadzkiego 13–15,
Częstochowa, Poland*

AND C. D. JOHNSON

*School of Chemical Sciences, University of East Anglia
Norwich, England*

I. Introduction	1
II. Six-Membered Rings	5
A. Pyridine	5
1. The Ring Nitrogen (and Its Modified Forms) as a Substituent	5
2. The Ring Nitrogen as Reaction Site	19
B. Other Six-Membered Rings	25
1. Diazines and Triazines	25
2. Annelated Azines	27
3. Other Six-Membered Heterocycles Including Acyclic Compounds	31
III. Five-Membered Rings	34
A. Containing One Heteroatom	34
B. Containing Two or More Heteroatoms	48
IV. Fused Five- and Six-Membered Rings	54
V. Three-Membered Rings	61

I. Introduction

Over 10 years have elapsed since the first presentation of this topic,[1] and not surprisingly there has been a corresponding growth in pertinent literature which makes further appraisal profitable and interesting. This review covers contributions up to the end of 1974. Reiteration of references included in the first review is avoided as far as possible, and generally we have considered only papers from about 1962 onward, but even with these restrictions the amount of relevant work is so large that limitations of space restricts us in many cases to brief mention of studies otherwise worthy of detailed examination.

In the main the current situation reflects that of the Hammett equation's applicability to organic reactivity in general—much has been

---

[1] H. H. Jaffé and H. Lloyd Jones, this series. **3**, 209 (1964).

clarified and elucidated with overall the emergence of a fairly simple and encouraging picture, even though in a number of cases controversy and complexity remain. There is frequently, if not invariably, good correspondence and uniform behavior for thermodynamic (such as pK_a) and kinetic measurements, but often spectral measurements show anomalous or random patterns of behavior. At the same time it must be admitted that some of the most accurate and significant deductions have been made from measurements in this latter area. The inherent tendency of many heterocyclic systems to tautomerize or form zwitterions is also certainly one of the main factors leading to discrepancies and inaccuracies.

Work on specific heterocyclic systems may be grouped under two headings, both of which were employed in the previous article:[1] (1) the ring heteroatom as a substituent; (2) reactivity of heteroatoms, as affected by other substituents. Under (1), the heteroatom alone or the whole heterocyclic ring may be treated as a substituent. Additivity of effects can also be examined here. As one of us has noted previously[2] and as will be seen later in a number of examples, this is a more meaningful way of looking at the phenomena of "transmission through heterocyclic rings", a terminology which has been much used previously, but is difficult to define rigorously or express quantitatively. Since heteroatoms have often an extremely large effect on reactivity with concomitant propensity for solvent interactions, this is also a good testing ground for the evaluation of solvent effects on σ values. Under (2), the heteroatom alone may participate in reaction, as, for example, in nitrogen protonation or quaternization, or the whole ring may be involved, as when heteroatom protonation leads to subsequent nucleophilic attack and ring cleavage.

The applicability of Hammett-type parameters may be tested for heterocyclic compounds in a number of different ways. The simplest is to employ σ_m for substituents attached to positions not conjugated with the reaction site, and σ_p for those that are, or alternatively σ_p^+ or σ_p^- if there is a change in through-conjugation of the appropriate type. Such an approach assumes that the division between inductive and resonance effects is the same as for the defining benzenoid systems. Alternative procedures avoiding this obviously drastic and potentially inaccurate assumption involve the use of σ values to calculate basic inductive and resonance components of substituents and their subsequent incorporation in equations defining their relative ease of propagation. Thus in the Swain–Lupton derivation[3]

$$\sigma = f\mathscr{F} + r\mathscr{R} \tag{1}$$

[2] C. D. Johnson, "The Hammett Equation," p. 107. Cambridge Univ. Press, London and New York, 1973.
[3] C. G. Swain and E. C. Lupton, *J. Amer. Chem. Soc.* **90**, 4328 (1968).

where $\mathscr{F}$ and $\mathscr{R}$ are the inductive and resonance components and f and r are weighting factors differing for each different system. In this category belongs also the Dewar method[4]

$$\sigma = (F/r) + Mq \qquad (2)$$

where F and M are found for a given substituent by interpolation in the benzenoid system, r is the distance between substituent and reaction site, and q is a quantum-mechanical parameter defining propagation of the resonance effect. Subsequent treatment[5] involves consideration of the interaction between field and resonance components, the so-called FMMF method, and indeed in this area a whole variety of molecular orbital (MO) calculations may be made of different types and complexity whose description does not fall into the scope of this review. A third variant is the dual substituent parameter (DSP) procedure[6,7] of Taft and his co-workers:

$$\log k/k_0 \; (\log K/K_0) = \sigma_I \rho_I + \sigma_R \sigma_R \qquad (3)$$

$$\lambda = \rho_R/\rho_I \qquad (4)$$

where $\rho_R = \rho_I =$ unity for ionization of benzoic acids (H_2O, 25°).

Curiously enough, variation of λ in Eq. (4) and of the equivalent ratios in Eqs. (1) and (2) is quite clear-cut for passage of substituent effects across different aromatic and heteroaromatic systems, as will be emphasized in this review subsequently, but its deviation from unity in the case of reactivities of side chains of varying geometry is much less certain.[8–10] Alternative sets of σ_I and σ_R values to those of Taft have been offered by Exner.[11]

A further approach is represented by the Yukawa–Tsuno equations[12,13]

$$\log k/k_0 \; (\log K/K_0) = \rho\{\sigma + r(\sigma^+ - \sigma)\} \qquad (5)$$

$$\text{or } \log k/k_0 \; (\log K/K_0) = \rho\{\sigma + r(\sigma^- - \sigma)\} \qquad (6)$$

which accommodate the concept of a spread of σ values from σ^+ and

---

[4] M. J. S. Dewar and P. J. Grisdale, *J. Amer. Chem. Soc.* **84**, 3548 (1962).
[5] M. J. S. Dewar, R. Golden, and J. M. Harris, *J. Amer. Chem. Soc.* **93**, 4187 (1971).
[6] P. R. Wells, S. Ehrenson, and R. W. Taft, *Progr. Phys. Org. Chem.* **6**, 147 (1968).
[7] S. Ehrenson, R. T. C. Brownlee, and R. W. Taft, *Progr. Phys. Org. Chem.* **10**, 1 (1973).
[8] C. Eaborn, R. Eastmond, and D. R. M. Walton, *J. Chem. Soc. B,* 752, (1970).
[9] C. Eaborn, R. Eastmond, and D. R. M. Walton, *J. Chem. Soc. B,* 127 (1971).
[10] Ref. 2, pp. 85-86.
[11] O. Exner, *Collect. Czech. Chem. Commun.* **31**, 65 (1966).
[12] Y. Yukawa and Y. Tsuno, *Bull. Chem. Soc. Japan* **32**, 971 (1959).
[13] Y. Yukawa, Y. Tsuno, and M. Sawada, *Bull. Chem. Soc. Japan* **39**, 2274 (1966).

beyond to σ^- and beyond. σ^0 rather than σ may be employed in these equations.[13]

The correspondence of the Hammett equation and Eqs. (3) and (5) have been particularly well explored in the consideration of the pK_a values (MeCN, 25°) of 5H-dibenzo-1,3-diazepines (1),[14] and amidine structures 2 and 3,[15] which are generally well correlated by σ rather than σ^+. Relevant figures are shown in Table I, and reveal in particular that close correspondence to σ is equivalent to the use of Eq. (3) using σ^+ and putting ρ_I approximately twice ρ_R.

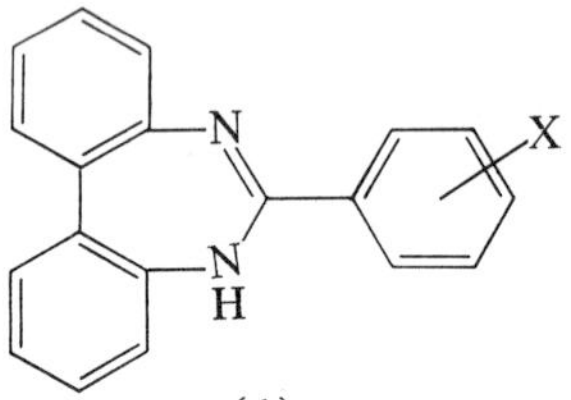

(1) 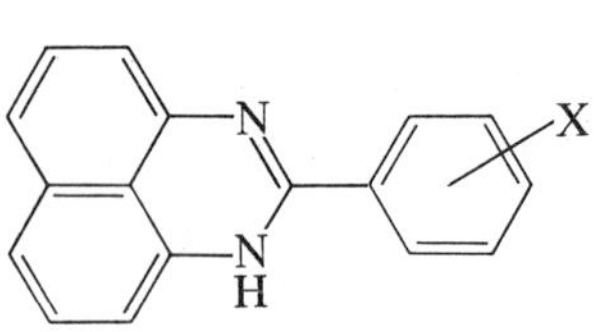 (2)

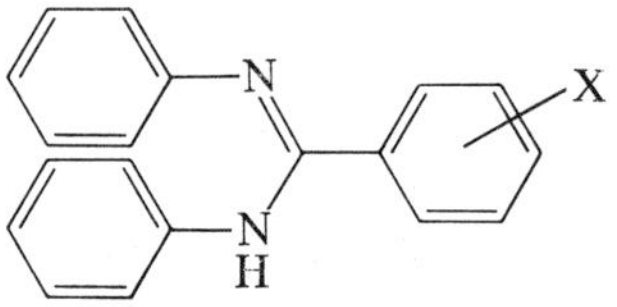

(3)

The question of the correspondence of all these approaches has also been taken up in a general review article,[16] but it must be stressed that the picture of developments in this area is still very much a dynamic one, and final conclusions have by no means yet been drawn.

As well as the previous chapter on the Hammett equation,[1] the importance of the approach has found expression in specific areas in other contributions to these volumes: side chain reactivities of thiophenes;[17] tautomerism;[18] reactions of azines with nucleophiles;[19,20] formation of benzofuroxans by decomposition of o-nitrophenyl azides;[21] acid dis-

---

[14] L. L. Popova, I. D. Sadekov, and V. I. Minkin, *Reakts. Sposobnost Org. Soedin.* **5**, 682 (1968).
[15] L. L. Popova, I. D. Sadekov, and V. I. Minkin, *Reakts. Sposobnost Org. Soedin.* **6**, 47 (1969).
[16] R. Zalewski, *Zesz. Nauk. Wyzsz. Szk. Ekon., Poznań Ser. 2* **51**. 3 (1973).
[17] S. Gronowitz, this Series **1**, 80-82 (1963).
[18] A. R. Katritzky and J. M. Lagowski, this Series **1**, 335-336, 354 (1963).
[19] G. Illuminati, this Series **3**, 305-306, 329-330 (1964).
[20] R. G. Shepherd and J. L. Fedrick, this Series **4**, 217 (1965).
[21] A. J. Boulton and P. B. Ghosh, this Series **10**, 14 (1969).

sociation constants for pyrrole carboxylic acids;[22] solvolysis of chloromethyl derivatives of selenophenes;[23] electrophilic substitution of five-membered rings;[24] base-catalyzed hydrogen exchange,[25] which includes in particular a survey of the important studies of Tupitsin and Zatsepina, and Zoltewicz; substituent effects in homolytic aromatic substitution;[26] reaction of styrenes with 9-substituted acridizinium ions;[27] pK_a values of 4-methyl- and 2,4-dimethyl-5-substituted oxazoles;[28] measurement of aromaticity by sensitivity to substituent effects in five-membered rings.[29]

TABLE I

CORRELATIONS FOR pK_a VALUES OF SYSTEMS 1, 2, and 3[a]

	$\log K/K_0 = \rho\sigma$	$\rho\sigma^+$	$\rho\{\sigma^0 + r(\sigma^+ - \sigma^0)\}$	$\rho_I\sigma_I + \rho_R\sigma_R^+$
1	$\rho = 2.12$ (0.984)	$\rho = 1.54$ (0.943)	$\rho = 2.14$ $r = 0.25$ (0.997)	$\rho_I = 2.08$ $\rho_R = 1.24$
2	$\rho = 1.92$ (0.994)	$\rho = 1.31$ (0.916)	$\rho = 1.93$ $r = 0.16$ (0.995)	$\rho_I = 2.00$ $\rho_R = 0.98$
3	$\rho = 1.87$ (0.988)	$\rho = 1.36$ (0.951)	$\rho = 1.84$ $r = 0.29$ (0.992)	$\rho_I = 1.92$ $\rho_R = 1.04$

[a] Correlation coefficients in parentheses.

These sections of work will only be considered briefly, with respect to more recent related studies, and insofar as they influence the generalized conclusions of the present chapter.

II. Six-Membered Rings

A. PYRIDINE

1. *The Ring Nitrogen Atom (and Its Modified Forms) as a Substituent*

In Table II are given recent values reported for pyridine σ constants. These values show a high degree of consistency and conformity, estimated within the usual accuracy of Hammett equation correlations. This

---

[22] R. A. Jones, this Series **11**, 415-417 (1970).
[23] N. N. Magdeseva, this Series **12**, 26 (1970).
[24] G. Marino, this Series **13**, 276-280, 298-313 (1971).
[25] J. A. Elvidge, J. R. Jones, E. A. Evans, and H. C. Sheppard, this Series **16**, 1 (1974).
[26] F. Minisci and O. Porta, this Series **16**, 157-158 (1974).
[27] C. K. Bradsher, this Series **16**, 320 (1974).
[28] R. Lakhan and B. Ternai, this Series **17**, 175 (1974).
[29] M. J. Cook, A. R. Katritzky, and P. Linda, this Series **17**, 287-288 (1974).

TABLE II
σ-CONSTANTS FOR THE PYRIDINE RING NITROGEN ATOM[a]

α	β	γ	Type	Solvent	Reaction	Reference footnotes
0.71	0.55	0.94	σ	—	Mean value from several reactions	30
0.73	0.51	0.99	σ	H_2O	Ionization of pyridine aldoximes	30
0.61	0.55	0.91	σ	H_2O	Ionization of acetyl pyridine ketoximes	30
0.75	0.65	0.96	σ	85% aq. MeOH	Alkaline hydrolysis of methyl benzoates	31
0.75	0.65	0.96	σ	55% aq. MeOH	Alkaline hydrolysis of alkyl benzoates (ethyl esters in acetone–water, otherwise methyl esters)	32
0.77	0.67	0.98	σ	56% aq. acetone		
0.75	0.67	0.99	σ	60% aq. dioxane		
0.60	0.59	0.92	σ	60% aq. DMSO		
0.64	0.63	0.93	σ	80% aq. DMSO		
0.88	0.74	1.11	σ		Mean value from several reactions	33, 34
1.00	0.60	0.80	σ	DMSO	PMR chemical shift of α-protons in β-pyridylacrylic acids	35
—	0.30	—	σ	DMF	$E_{1/2}$ for polarographic reduction of p-substituted nitrobenzenes and azobenzenes, respectively	36
—	0.40	—	σ	DMF		
0.68	0.33	0.66	σ	CCl_4	IR integral intensities of C—H stretching vibrations of substituted pyridines	37
0.88	0.74	1.11	σ	—	Diazodiphenylmethane with pyridine carboxylic acids	38
0.08	0.53	0.85	σ^0	H_2O	Ionization of pyridylacetic acids	30
1.0	0.5	0.6	σ^0	CCl_4	PMR of substituted pyridines	39
0.8	0.4	0.7	σ^0	CCl_4	IR of substituted pyridines	39
1.1	0.6	1.1	σ^-	MeOH	IR of substituted pyridines	39
1.00	0.59	1.17	σ^-	MeOH	Methanolysis of halogenopyridines	40, 41
0.80	0.30	0.87	σ^+	—	Pyrolysis of 1-arylethylacetates	42
0.75	0.54	1.16	σ^+	80% aq. EtOH	Solvolysis of 2-pyridyl-2-chloro propanes	43

[a] PMR, proton magnetic resonance; IR, infrared; DMSO, dimethyl sulfoxide; DMF, dimethylformamide.

is true even for the σ_α values (with the exceptions noted below), so that here steric effects are probably small. The averaged values of Blanch[30] are close to those of subsequent workers, and his concise study provides a good example of how complexities arising from tautomerism or zwitterion formation may be resolved. The values also appear to be sensibly independent of solvent, the results of the Australian workers[31,32] being particularly significant in this respect. Such a conclusion arises from consideration of general σ value compilations,[44,45] but it is remarkable in the case of the aza substitutent, where especially large differentials in solvent interaction would be expected. In other instances, however, there does appear to be some influence of solvent in the electronic effects of aza substituents. For example,[46] $E_{1/2}$ values for polarographic reduction of pyridylazobenzenes yield σ_p values for 4-pyridyl of 0.68 in 60% aq. EtOH, 0.56 in 75% aq. dioxane, 0.44 in dimethyl sulfoxide (DMSO), and 0.35 in dimethylformamide (DMF).

The similarity between the σ, σ^+, σ^-, σ, and σ^0 values also indicates the pyridine nucleus to be of low polarizability. The somewhat greater value of σ_γ^- than σ_γ could be ascribed to the extra stabilizing effect on negative charge in the former case (cf. p-NO$_2$, σ 0.78, σ^- 1.27), but this interpretation is clouded by the fact that σ_γ^+ shows the same trend for the S_N1 solvolysis of the pyridine analog of t-cumylchloride.[43] Charton has provided evidence for an "abnormal" electrical effect in α-substituted pyridines as well as quinolines and isoquinolines,[47] which he has

---

[30] J. H. Blanch, *J. Chem. Soc. B*, 337 (1966).
[31] A. D. Campbell, S. Y. Chooi, L. W. Deady, and R. A. Shanks, *Aust. J. Chem.* **23**, 203 (1970).
[32] L. W. Deady and R. A. Shanks, *Aust. J. Chem.* **25**, 2363 (1972).
[33] D. D. Perrin, *J. Chem. Soc.*, 5590 (1965).
[34] G. B. Barlin and D. D. Perrin, *Quart. Rev. Chem. Soc.* **20**, 75 (1966).
[35] A. R. Katritzky and F. J. Swinbourne, *J. Chem. Soc.*, 6707 (1965).
[36] P. Tomasik, *Pr. Nauk. Inst. Chem. Technol. Nafty Wegla Politech. Wroclaw.* **5**, 149 (1971).
[37] R. Joeckle, E. D. Schmid, and R. Mecke, *Z. Naturforsch. A* **21**, 1906 (1966); R. Joeckle, E. Lemperle, and R. Mecke, *Z. Naturforsch. A* **22**, 395, 403 (1967).
[38] D. M. Dimitrijević, Ž. D. Tadić, M. M. Mišić-Vuković, and M. Muškatrović, *J. Chem. Soc., Perkin Trans. II*, 1051 (1974).
[39] I. F. Tupitsin, N. N. Zatsepina, N. S. Kolodina, and A. A. Kane, *Reakts. Sposobnost Org. Soedin.* **5**, 931 (1968).
[40] M. Liveris and J. Miller, *J. Chem. Soc.*, 3486 (1963).
[41] J. Miller and W. Kai-Yan, *J. Chem. Soc.*, 3492 (1963).
[42] R. Taylor, *J. Chem. Soc.*, 4881 (1962).
[43] D. S. Noyce, J. A. Virgilio, and B. Bartman, *J. Org. Chem.* **38**, 2657 (1973).
[44] C. D. Johnson,[2] p. 20.
[45] C. D. Johnson, *Chem. Rev.*, **75**, 755 (1975).
[46] E. E. Pasternak and P. Tomasik, *Bull. Acad. Pol. Sci., Ser. Sci. Chim.*, in press.
[47] M. Charton, *J. Amer. Chem. Soc.* **86**, 2033 (1964).

attempted to quantify and which he ascribes, logically, to a short-range localized electrostatic interaction.

Another effect postulated for nitrogen in the α-position arises from the correlation of free energies of rotation of N,N-dimethylbenzamide and pyridinecarboxamides[48] with Jaffe's[1] σ values for the aza substituent, which differ only marginally from the mean values due to Blanch.[30] A large deviation is observed for N,N-dimethylpyridine-2-carboxyamide, which the authors tentatively ascribe to interaction between the nitrogen lone pair and a solvent $CDCl_3$ molecule increasing spatial requirements.

The low value for σ_α^{0} [30] undoubtedly arises from intramolecular hydrogen bonding stabilizing the neutral form of the acid (4), a circumstance also occurring in the pyridine-2-carboxylic and -2-hydroxamic acids as well as 2-hydroxypyridine.

(4)

The pK_a values of diazines have been used to calculate aza substituent constants, and from these, with a fair but not complete measure of success, the basicities of substituted pyridines and the strengths of heterocyclic acids can be predicted.[33,34]

Joeckle et al.[37] based their estimations of σ_I on the intensities of the C—H stretching vibration bands in the IR-absorption spectra of monosubstituted benzenes and pyridines. Tupitsin et al.[39] have disagreed, however, with this treatment and suggest that the bands in question can be perturbed by a mixing with overtones; it is also suggested that these intensities depend on solvent so that in CCl_4 the correlation is with σ^0 while CD_3OD/CD_3OK solvent produces a σ^- correlation. These workers have also demonstrated the applicability of the aza σ^0 values thus defined to the interpretation of methoxide-catalyzed hydrogen exchange in substituted pyridines.[25,49,50]

In Table III are given substituent constants for the modified aza substituents $=\overset{+}{N}H-$, $=\overset{+}{N}Me-$, $=\overset{+}{N}(O^-)-$, and $=\overset{+}{N}(OH)$.

Values for the N-oxide function are limited in extent, and there is a wide variation in the data that exist. Clearly there is further need for a

---

[48] F. G. Riddell and D. A. R. Williams, *J. Chem. Soc.*, 587 (1973).
[49] I. F. Tupitsin, N. N. Zatsepina, A. V. Kirowa, and J. M. Kapustin, *Reakts. Sposobnost Org. Soedin.* **5**, 601 (1968).
[50] I. F. Tupitsin, N. N. Zatsepina, A. V. Kirowa, Y L. Kaminskii, and A. G. Ivanenko, *Reakts, Sposobnost Org. Soedin.* **10**, 143 (1973).

TABLE III
σ-Constants for the Modified Nitrogen Atom in the Pyridine Ring[a]

α	β	γ	Type	Solvent	Reaction	Reference footnotes
				=$\overset{+}{N}H-$ (azonium)		
3.11	2.10	2.57	σ	—	Mean value from several reactions	30
3.45	2.06	2.86	σ	H_2O	Ionization of pyridine aldoximes	30
2.98	2.04	2.80	σ	H_2O	Ionization of acetylpyridine ketoximes	30
3.21	2.18	2.42	σ	—		33, 34
2.2	1.9	1.3	σ	DMSO, CF_3COOH	PMR chemical shift of α-protons in β-pyridylacrylic acids	35
2.20	—	1.50	$σ^0$	MeOH	Hydrogen exchange	51, 52
2.49	—	2.32	$σ^-$	MeOH	Hydrogen exchange	51, 52
—	2.10	4.00	$σ^-$	H_2O	Second pK_a of aminopyridines	1
—	2.02	4.06	$σ^-$	H_2O	Second pK_a of dimethylaminopyridines	53
				=$\overset{+}{N}Me-$ (methyl azonium)		
2.49	1.58	2.32	$σ^-$	MeOH	Methanolysis of halogenopyridinium	41
				=$\overset{+}{N}(O^-)-$		
—	1.48	1.35	σ	H_2O	pK_a of benzoic acids	33, 34
−0.39	1.31	1.14	σ	—	Diazodiphenylmethane with pyridine carboxylic acids	38
1.50	0.7	0.4	σ	DMSO, CF_3COOH	PMR chemical shift of α-protons in β-pyridylacrylic acids	35
1.0	0.7	0.5	$σ^0$	CCl_4	PMR of pyridines	39
1.0	0.8	1.1	$σ^0$	CCl_4	IR of pyridines	39
—	1.59	1.88	$σ^-$	H_2O	pK_a of phenols	1
1.50	1.18	1.53	$σ^-$	MeOH	Methanolysis of halogenopyridine-N-oxide	41
1.6	1.1	1.2	$σ^-$	MeOH	IR of pyridines	39
0.68	—	0.45	$σ^+$	80% aq. EtOH	Solvolysis of 2-pyridyl-2-chloropropanes	43
—	0.81	0.016	$σ^+$	—	Pyrolysis of 1-arylethylacetates	54
				=$\overset{+}{N}(OH)$		
—	2.3	3.9	$σ^-$	H_2O	Second pK_a of aminopyridine-1-oxides	1
—	2.25	3.49	$σ^-$	H_2O	Second pK_a of dimethylaminopyridine-1-oxides	53

[a] PMR, proton magnetic resonance; IR, infrared; DMSO, dimethyl sulfoxide.

systematic investigation of the electronic character of this group, but its amphoteric nature (ability to donate or accept electrons dependent on circumstances), the possibility of tautomerism, steric and proximity effects in the α-substituted compounds, and the intervention of differential solvent effects suggest that such an investigation is not a simple one.

The values quoted in Table III for the azonium substituent reveal it to be of high electron-withdrawing capacity, as expected. However, resonance electron withdrawal appears to be hardly enhanced over that of the aza substituent. This point is given further elucidation and confirmation by the IR intensity measurements on the ring stretching bands near 1600 cm^{-1} for determination of σ_R^0,[55] the values of which are given in Table IV, and which form part of an extensive and systematic investigation into such effects which has been found to have wide and reliable applicability.[56] It should be noted that in these studies the sign of σ_R^0 has to be arbitrarily assigned, the band intensity increasing with increasing resonance donation *or* acceptance. An analogous circumstance also arises for the influence of the inductive effect in a report[57] of a parabolic relationship for the dependence of the CH stretching mode bands on σ_I in a series of substituted benzenes, pyridines, pyridine N-oxides, and pyrazines and their deuterated derivatives. Additivity of effects for nuclear heteroatoms and substituents is observed to a high degree of accuracy.

Clearly therefore, Tables II, III, and IV indicate the large difference between the uncharged and positive pyridine nitrogen atom stems almost entirely from the inductive effect. The discrepancy between the σ_γ^- values reported for azonium in nucleophilic substitution[41] and from the second pK_a value of 4-aminopyridine[1] is large and hard to interpret; it has a bearing on the mode of correlation of the pK_a's of substituted pyridine derivatives with σ values and will be discussed in Section II, A, 2.

---

[51] N. N. Zatsepina and I. F. Tupitsin, *Isotopenpraxis* **3**, 103 (1967).
[52] N. N. Zatsepina, A. V. Kirova, and I. F. Tupitsin, *Reakts. Sposobnost Org. Soedin.* **5**, 70 (1968).
[53] P. Forsythe, R. Frampton, C. D. Johnson, and A. R. Katritzky, *J. Chem. Soc., Perkin Trans. II,* 671 (1972).
[54] R. Taylor, *J. Chem. Soc., Perkin Trans. II,* 277 (1975).
[55] A. R. Katritzky, C. R. Palmer, F. J. Swinbourne, T. T. Tidwell, and R. D. Topsom. *J. Amer. Chem. Soc.* **91**, 636 (1969).
[56] A. R. Katritzky and R. D. Topsom, *in* "Advances in Linear Free Energy Relationships" (N. B. Chapman and J. Shorter, eds.), Chapter 3. Plenum, New York, 1972.
[57] I. F. Tupitsin, N. N. Zatsepina, and N. S. Kolodina, *Reakts. Sposobnost Org. Soedin.* **6**, 11 (1969).

TABLE IV
RESONANCE EFFECTS FOR MODIFIED AZA-SUBSTITUENTS[a]

Substituent	N	=$\overset{+}{N}$H–	=$\overset{+}{N}$Me–	=$\overset{+}{N}$(OMe)–	=$\overset{+}{N}$(O^{t}Bu)–	=$\overset{+}{N}$(O$^-$)–	=$\overset{+}{N}$($\bar{O}$BF$_3$)–	=$\overset{+}{N}$($\bar{B}$H$_3$)–
σ_R^0	+0.24	~+0.35	+0.28	+0.17	+0.14	−0.21	+0.12	+0.21

[a] All measured in MeCN except =$\overset{+}{N}$(O$^-$)– and =$\overset{+}{N}$($\bar{B}$H$_3$)–, which were in CHCl$_3$.

The second protonation of dimethylaminopyridines and their N-oxides, all weak bases whose pK determination involves acidity function theory,[53] yield σ^- values in good agreement with previous values, affording thereby a check on the validity of the acidity function procedure.

Tomasik[36,58-63] has investigated the polarographic reduction of nitropyridines in DMF at 20°, and finds a good correlation of $E_{1/2}$ with σ for the 2-X,5-NO$_2$, 5-X,2-NO$_2$, 3-X,5NO$_2$, and 2-X,4-NO$_2$ systems, with ρ values of 0.35, 0.41, 0.46, and 0.42, respectively. The nitropyridine N-oxide system also afforded a correlation in this latter case ($\rho = 0.3$), but for the 2-substituted-3-nitropyridines there were only scattered points.[64] A series of 2-X,5Y-phenylazopyridine reductions also produced an $E_{1/2}$-σ correlation under the same conditions. For these 2X-5Y systems the defining equations were Eqs. (7) and (8) for the nitrobenzene and nitropyridine series and Eqs. (9) and (10) for the azobenzene and phenylazopyridine series, respectively.

$$E_{1/2} = 0.350\,\sigma_p - 1.034 \tag{7}$$

$$E_{1/2} = 0.350\,\sigma_p - 0.931 \tag{8}$$

$$E_{1/2} = 0.365\,\sigma_p - 1.301 \tag{9}$$

$$E_{1/2} = 0.365\,\sigma_p - 1.150 \tag{10}$$

These equations yield 0.30 and 0.40, respectively, for the σ_β aza substituent (see Table I) demonstrating accurate additivity of effects for both series. This is the basis of Stradins' concept,[65-67] which states that when the polarographic reductions of aromatic series (e.g., nitrobenzenes and nitropyridines) are run under identical conditions, the ρ value for both is the same. The generality of this conclusion, however, is under question,[46,47] and the role of solvent effects and applicability of approaches of the form of Eq. (3) are under scrutiny.

Noyce has also reported a very extensive series of investigations into the reactivity of substituted heteroaromatic compounds. Although the

---

[58] P. Tomasik, *Rocz. Chem.* **44**, 341 (1970).
[59] T. Batkowski, M. Kalinowski, and P. Tomasik, *Ann. Chim.* (*Rome*) **63**, 121 (1973).
[60] W. Drzeniek and P. Tomasik, *Ann. Chim.* (*Rome*) **63**, 135 (1973).
[61] P. Tomasik, *Abstr. Symp. Chem. Heterocycl. Compounds, 4th* 1972. p. B-2.
[62] M. K. Kalinowski, J. Skarzewski, Z. Skrowaczewska, and P. Tomasik, *Ann. Chim.* (*Rome*) **63**, 129 (1973).
[63] P. Tomasik, *Rocz. Chem.* **44**, 1211 (1970).
[64] P. Tomasik, unpublished work.
[65] J. P. Stradins, *Polarografia Organicheskikh Nitrosoedinerii*, Izdat. Akad. Nauk LSSR, Riga 1961, p. 118.
[66] M. K. Kalinowski, *Chem. Phys. Lett.* **8**, 378 (1971).
[67] H. Kryszczynska and M. K. Kalinowski, *Rocz. Chem.* **47**, 1747 (1971).

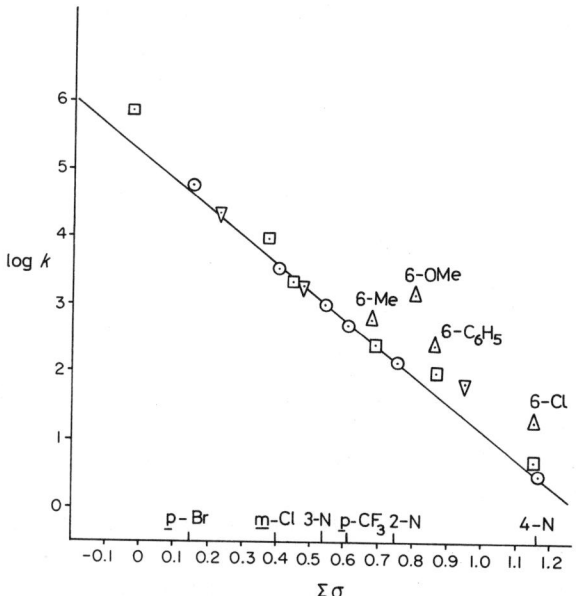

Fig. 1. Log rates of solvolysis of substituted 2-(pyridyl or phenyl)-2-chloropropanes vs. $\Sigma\sigma^+$ ($\rho = -4.0$). ⊙, Monosubstituted benzenes or 2-, 3-, and 4-pyridines; ▫, 2-pyridyl systems; ▽, 3-pyridyl systems; △, 2-pyridyl systems (6-substituted).

general title is "Transmission of Substituent Effects in Heterocyclic Systems," which might be considered slightly misleading (see Section I), the approach taken seems to us to be a most profitable one—that of consideration of additivity between heteroatom and substituent in their effect on the reactivity of a side chain in which positive charge is developed in conjugation with the ring (Scheme 1). Figure 1 shows the

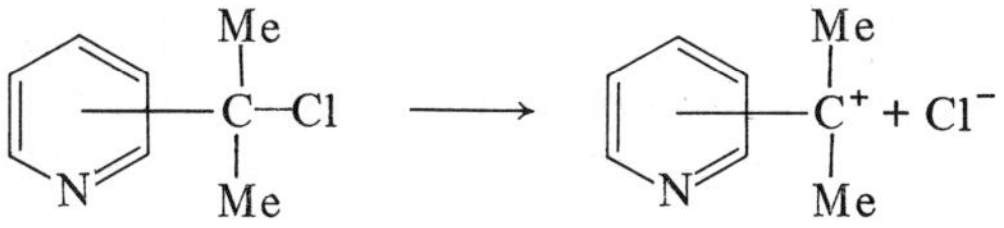

SCHEME 1

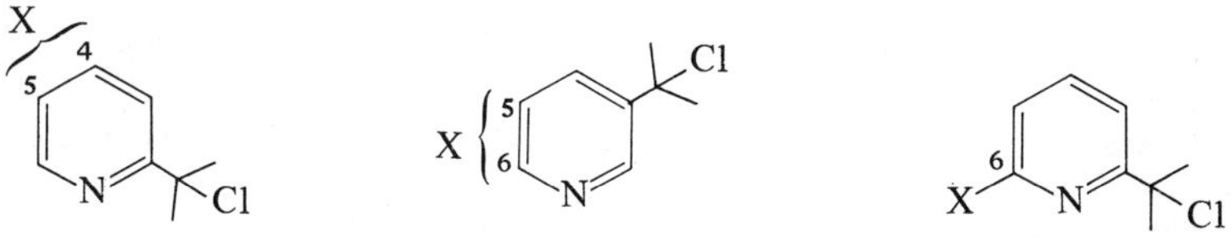

(5)   (6)   (7)

results obtained using the σ^+ values for the aza substituent reported in Table II[43] for systems **5**, **6**, and **7**[68] (80% aq. EtOH, 75°). Additivity is quite accurate, the chief deviations arising from the data for **7**, where resonance donors stand ortho to the aza substituent and can interact as in **8**.

(8)

Such a situation could arise in **5**, but unfortunately here the molecule **5** (4X = OMe), for which this effect would be most marked, has not been examined.

Surprisingly similar conclusions have been reached in a further series,[69-71] where the side chain reaction is that of alkaline methylcarboxylate hydrolysis in aq. MeOH at 25°. For methyl 5-X-picolinates (**9**), methyl 4-X-picolinates (**10**), and methyl 5-X-nicotinates (**11**), accurate additivity of effects were found in a σ correlation with $\rho = 2.00$, confirming earlier results for alkaline hydrolysis of 5-substituted ethyl nicotinates in aq. EtOH at temperatures from 0° to 25°.[72]

(9) (10) (11)

For series **12**, **13**, and **14**, deviations from additivity were observed where X represents a strong electron donor, e.g., OMe. Figure 2 shows the results obtained graphically. The upper line is the correlation for the hydrolysis of methylbenzoates and diazine carboxylates[73] under identical conditions with the exception of temperature, 10° in this case. The coincidence of the three pyridine carboxylates on both lines makes it

---

[68] D. S. Noyce and J. A. Virgilio, *J. Org. Chem.* **38**, 2660 (1973).

[69] A. D. Campbell, S. Y. Chooi, L. W. Deady, and R. S. Stranks, *J. Chem. Soc. B*, 1063 (1970).

[70] A. D. Campbell, E. Chan, S. Y. Chooi, L. W. Deady, and R. A. Shanks, *J. Chem. Soc. B*, 1065 (1970).

[71] A. D. Campbell, E. Chan, S. Y. Chooi, L. W. Deady, and R. A. Shanks, *J. Chem. Soc. B*, 1068 (1970).

[72] Y. Ueno and E. Imoto, *J. Chem. Soc. Jap.* **88**, 1210 (1967).

[73] L. W. Deady, D. J. Foskey, and R. A. Shanks, *J. Chem. Soc. B*, 1962 (1971).

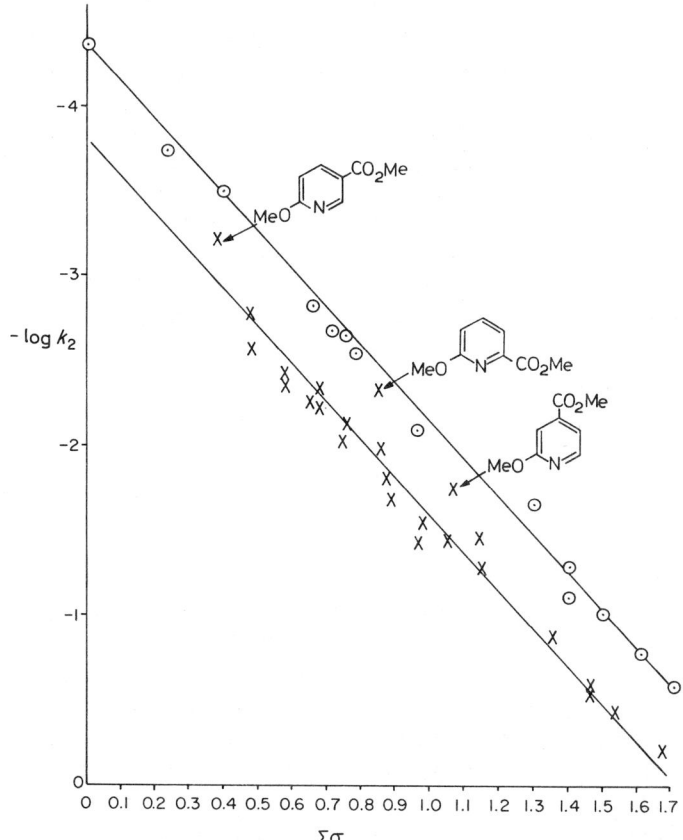

FIG. 2. Log rates of alkaline hydrolysis of substituted methyl pyridinecarboxylates vs. Σσ. ⊙, 10°; ×, 25°.

certain that a single correlation line would be produced for the same temperature, the only significant deviations being produced by ortho interactions of the type shown in **8**.

Another interesting set of results relating to additivity of substituent effects in heterocyclic systems is provided by the work of Katritzky and co-workers on electrophilic substitution in acidic media for heteroaromatic systems. These studies, together with those by Ridd and Schofield,[74-76] add considerably to our knowledge of reactivity in such species, emphasizing particularly methods for distinguishing between reaction on tautomeric forms, and free bases or conjugate acids; nevertheless the applicability of the Hammett equation, while leading to a degree of consistency, has not always been successful. This must be due in part, if not entirely, to the very high degree of substitution in these compounds with concomitant steric and proximity effects, and to through conjugation between one substituent and a second standing ortho or para to it, as illustrated in **8**, for example.

(15)

Data for the bromination[77] and H-exchange[78] of 5-substituted 2-aminopyridines (**15**) in the 3-position in aqueous sulfuric acid media enabled ρ values of -4.4 (107°) and -3.5 (25°), respectively, to be calculated, from whence by consideration of the rates for the analogous reactions of 4-substituted anilines σ_m^+ values could be found for the heteroatom. For the H-exchange results (**15**, R = H) reaction was shown to occur through the monoprotonated form, the resultant σ_m^+ value ($= \overset{+}{N}H-$) being 1.82 and, from the N-oxide, σ_m^+ ($= \overset{+}{N}(OH)-$) 1.99. The bromination experiments were in more dilute sulfuric acid, and this reaction occurred on the free base, giving σ_m^+ ($= N-$) 0.65. For the dimethylamino compounds (**15**, R = Me), however, there was a very considerable rate enhancement indicating a value of 0.06 only for σ_m^+; this is most logically explained by release of steric hindrance by a negative buttressing effect on the dimethylamino group by substitution of N for CH.

Further H-exchange results, on substituted aminopyrimidines,[79] gave values of σ_m^+ for the aza substituent of 0.60 and for the azonium substituent of 1.95, and on both free base and conjugate acid forms of 4-

---

[74] A. R. Katritzky and C. D. Johnson, *Angew. Chem., Int. Ed. Engl.* **6**, 608 (1967).
[75] J. H. Ridd, *J. Chem.* **8**, 201 (1968).
[76] J. G. Hoggett, R. B. Moodie, J. R. Penton, and K. Schofield. "Nitration and Aromatic Reactivity," pp. 192, 206-220. Cambridge Univ. Press, London and New York, 1971.
[77] P. J. Brignell, P. E. Jones, and A. R. Katritzky, *J. Chem. Soc. B*, 117 (1970).
[78] G. P. Bean and A. R. Katritzky, *J. Chem. Soc. B*, 864 (1968).
[79] A. R. Katritzky, M. Kingsland, and O. S. Tee, *J. Chem. Soc. B*, 1484 (1968).

pyridone and 1-hydroxy-4-pyridone,[80] gave values of 2.2 for $=\overset{+}{N}(OH)-$ and 0.8 for $=\overset{+}{N}(O^-)-$. All these values are in good agreement with those in Table III, given that σ_m^+ is approximately equal to σ_m.

All this H-exchange work has been summarized[81] by use of the equation

$$\log f = p\Sigma\sigma^+ + p\Sigma\pi\sigma^+ \tag{11}$$

where f is a partial rate factor defined for comparison with benzene reactivity at $H_0 = 0$ and 100°, $\Sigma\sigma^+$ is the summation term for the substituents present, and $\Sigma\pi\sigma^+$ incorporates all cross products, taken in pairs, triplets, or higher terms. The best fit to the points is given by

$$\log f = -6.3\ \Sigma\sigma^+ - 2.1\ \Sigma\pi\sigma^+ + 1.6 \tag{12}$$

an empirical equation, which, encouragingly, has considerable predictive power, although the calculations frequently involve very lengthy temperature or acidity extrapolations. Nevertheless, the theoretical significance of Eq. (12) is dubious. Moreover, no corresponding correlation is produced by use of the approach for nitration rates of both five- and six-membered systems containing one or more heteroatoms.[82,83]

The estimates of σ_γ^+ for $=\overset{+}{N}(O^-)-$ by Noyce[43] and Taylor[54] (Table III) differ considerably, the latter result indicating an electrophilic reactivity of the 4-position of pyridine 1-oxide similar to that of benzene. No estimate for σ_α^+ can be made in this case, because the pyrolysis takes an anomalous path, involving direct intervention of the N-oxide function (Scheme 2). The alternative estimate[43] of 0.45 for σ_α^+ is in keeping with

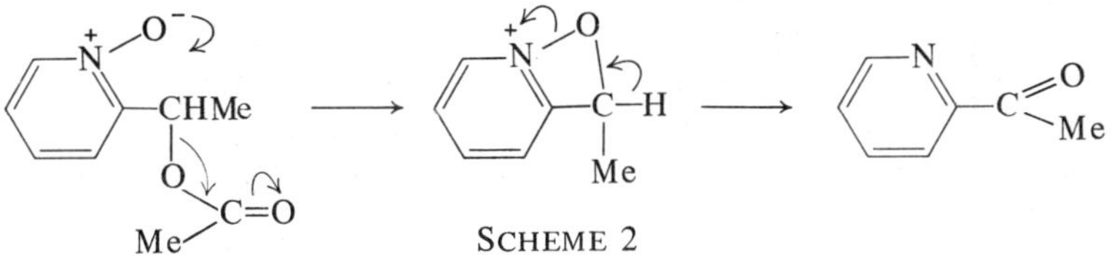

SCHEME 2

the nitration results, however, the reaction being known to proceed via the free base,[76] not via the conjugate acid as previously,[84] and now more recently,[54] claimed.

---

[80] P. Bellingham, C. D. Johnson, and A. R. Katritzky, *J. Chem. Soc. B*, 866 (1968).
[81] S. Clementi, C. D. Johnson, and A. R. Katritzky, *J. Chem. Soc., Perkin Trans. II*, 1294 (1974).
[82] S. Clementi, P. P. Forsythe, C. D. Johnson, A. R. Katritzky, and B. Terem, *J. Chem. Soc., Perkin Trans. II*, 399 (1974).
[83] S. Clementi, A. R. Katritzky, and H. O. Tarhan, *Tetrahedron Lett.*, 1395 (1975).
[84] J. Cleghorn, R. B. Moodie, K. Schofield, and M. J. Williamson, *J. Chem. Soc. B*, 870 (1966).
[85] J. Kavalek, J. Polanský and V. Štěrba, *Collect. Czech. Chem. Commun.* **39**, 1049 (1974).

By the same token that aza substituents retard electrophilic substitution, so they accelerate nucleophilic substitution,[19,20,40,41] particularly when positively charged. In an interesting study based on this type of reactivity, the equilibrium **16** = **17** has been investigated,[85] and this and the rate of subsequent ring opening leading to substituted anils of glutaconic aldehyde found to correlate with σ^-.

(16)   (17)

The spectral properties of substituted pyridines and the correlation with Hammett parameters have been investigated in a number of cases, but with only a limited measure of success. The intensity of the C—H stretching vibrations in the IR-absorption spectra of a large number of monosubstituted benzenes, pyridines, and pyridine 1-oxides are found to correlate with σ_I in the form of a parabola.[57,86,87] Other efforts to correlate the position of C—H deformation frequencies in 2-(p-phenylacyl)pyridines,[88] stretching vibrations of the nitro group in 2-substituted-5-nitropyridines,[89] and N=N stretching vibrations in 2-substituted-5-phenylazopyridines[90] met with no success; either peak position was insensitive to substituent change, or all types of σ values produced scatter. As noted previously,[55] the intensity of the IR bands in the 1600–1580 cm^{-1} region for pyridines, pyridine 1-oxides, and pyridinium compounds afford a sound basis for establishment of reliable σ_R^0 values.

A recently estimated set of spectral σ constants applicable to the correlation of substituent effects on K-band shifts in the UV spectra of p-disubstituted benzenes[91–93] has been applied to monosubstituted and

---

[86] E. D. Schmid and R. Joeckle, *Spectrochim. Acta* **22**, 1645 (1966).
[87] R. Joeckle and R. Mecke, *Ber. Bunsenges. Phys. Chem.* **71**, 165 (1967).
[88] R. F. Brauch, *Nature (London)* **177**, 671 (1956).
[89] P. Tomasik, *Rocz. Chem.* **47**, 1747 (1971).
[90] P. Tomasik, *Rocz. Chem.* **44**, 1369 (1970).
[91] P. Tomasik and M. K. Krygowski, *Bull. Acad. Polon. Sci., Ser. Sci. Chim.* **22**, 443 (1974).
[92] P. Tomasik and M. K. Krygowski, *Bull. Acad. Polon. Sci., Ser. Sci. Chim.* **22**, 877 (1974).
[93] P. Tomasik, M. K. Krygowski, and T. Chellathurai, *Bull. Acad. Polon. Sci., Ser. Sci. Chim.* **22**, 1065 (1974).

2,5-disubstituted pyridine systems and the effects found to parallel those of the benzenoid analogs, with the exception of compounds that can tautomerize.[94–96] σ Values also correlate substituent effects on the magnetic shielding of the ^{13}C atom in the 5-position of substituted pyridines;[97] but other workers[98,99] have found no correlation of the chemical shift of protons with the electron effect of substituents in the pyridine ring, and it has been suggested that this is due to the varying assymetry of the ring current in the particular series.[100] The relation between proton shifts in 2-substituted pyridines and pyrazines and σ has also received attention.[101]

2. The Ring Nitrogen as Reaction Site

The most-studied reaction at pyridine nitrogen is protonation, and correlation of pK_a values with Hammett parameters has received extensive investigation.[7,47,102–114] Quite clearly, electronic interactions of α-substituents are tempered by general steric effects,[47,102] and there is also doubt as to the mode by which electronic effects are conveyed from the β- and γ-positions. Groups in the 4-position of potential (-I-R) character

---

[94] P. Tomasik and Z. Skrowaczewska, *Rocz. Chem.* **42**, 759 (1968).
[95] P. Tomasik, *Rocz. Chem.* **42**, 2037 (1968).
[96] P. Tomasik and A. Zakowicz, *Abstr. Symp. Chem Heterocycl. Compounds, 5th,* 1975 p. 126.
[97] L. L. Retcofsky and F. R. McDonald, *Tetrahedron Lett.,* 2575 (1968).
[98] H. H. Perkampus and U. Krüger. *Chem. Ber.* **100**, 1105 (1967).
[99] W. Brügel, *Z. Electrochem.* **66**, 159 (1962).
[100] P. Tomasik and A. Zabza, unpublished results.
[101] G. P. Syrova and Y. N. Sheinker, *Khim. Geterotsikl. Soedin.,* 345 (1972).
[102] P. Tomasik and R. Zalewski, *Abstr. Symp. Chem Heterocycl. Compounds, 5th,* 1975 p. 127.
[103] D. Wegmann and W. Simon, *Helv. Chim. Acta* **45**, 962 (1962).
[104] A. Fischer, W. J. Galloway, and J. Vaughan, *J. Chem. Soc.,* 3591 (1964).
[105] C. S. Tsai, *Can. J. Chem.* **45**, 2862 (1967).
[106] J. C. Doty, J. L. R. Williams, and P. J. Grisdale, *Can. J. Chem.* **47**, 2355 (1969).
[107] G. B. Ellam and C. D. Johnson, *J. Org. Chem.* **36**, 2284 (1971).
[108] R. T. C. Brownlee and R. D. Topsom, *Tetrahedron Lett.,* 5187 (1972).
[109] J. M. Essery and K. Schofield, *J. Chem. Soc.,* 2225 (1963).
[109] J. M. Essery and K. Schofield, *J. Chem Soc.,* 2225 (1963).
[110] G. Gauzzo, G. Galiazzo, U. Mazzucato, and N. Mongiat, *Tetrahedron* **22**, 589 (1966).
[111] I. R. Bellobono and G. Favini, *J. Chem. Soc. B,* 2034 (1971).
[112] L. Sucha, Z. Urner, and M. Suchanek, *Sb. Vys. Sk. Chem.-Technol. Praze, Anal. Chem.* **H7**, 103 (1971) [*CA* **79**, 114958e (1973)].
[113] M. R. Chakrabarty, C. S. Handloser, and M. W. Mosher, *J. Chem. Soc., Perkin Trans, II,* 938 (1973).
[114] C. L. Liotta, E. M. Perdue, and H. P. Hopkins, *J. Amer. Chem. Soc.* **96**, 7308 (1974).

influence the pK_a values by induction alone,[104,109] and for other groups the correlation with σ is good. Equation (13) for the latter and (14) for the former are therefore appropriate:

$$\log (K/K_0) = \rho\sigma \qquad (13)$$

$$\log (K/K_0) = \rho\sigma_I \qquad (14)$$

alternatively, the dual-parameter Eq. (15) may be used.[7,108,115]

$$\log (K/K_0) = 5.15\,\sigma_I + 2.69\,\sigma_R^+ \qquad (15)$$

In fact this is a specific example of a general contention[116] that in all heterocyclic systems Eq. (3) rather than the straightforward Hammett equation is the more correct. Equation (15), however, gives no indication that electron acceptors influence pK_a values of pyridinium ions by induction alone, although this circumstance has been carefully authenticated,[104,109] but it does give voice to the contention that a correlation with σ^+ would be more realistic, to allow for the presence of canonical **19** in y-substituted pyridines for groups of +R character. Evidence

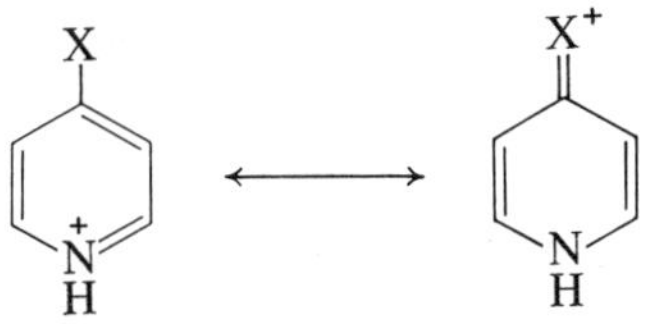

(18)   (19)

could be provided by consideration of the σ and σ^- values of the $=\overset{+}{\text{N}}\text{R}-$ group, but the results are ambiguous;[107] σ reveals an exaltation for the second pK_a of 4-aminopyridine (R = H),[1,107,111] but for nucleophilic reaction on 4-chloropyridium (R = Me) no enhancement occurs[41] (see Table III).

Grob and his co-workers[117] and others[118] have measured pK_a values of 4-substituted quinuclidines (**20**) in water[117] and 5% and 50% aq. EtOH and 50% aq. methyl Cellosolve,[118] as well as quaternization

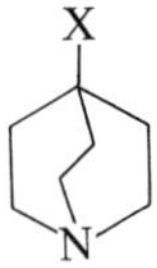

(20)

[115] C. A. Grob and R. W. Taft, *J. Amer. Chem. Soc.* **96**, 1236 (1974).
[116] L. E. Kholodov, *Reakts. Sposobnost Org. Soedin.* **5**, 246 (1968).
[117] E. Ceppi, W. Eckhardt, and C. A. Grob, *Tetrahedron Lett.*, 3627 (1973).
[118] J. Paleček and J. Hlavatý, *Collect. Czech. Chem. Commun.* **38**, 1985 (1973).

rates.[119] with MeI (MeOH, 10°). These quinuclidine structures may be taken as a model for the definition of inductive effects in pyridines,[115] the pK_a values in water obeying Eq. (16). The ratio of transmission through the pyridine to that through the quinuclidine nucleus is 1.14, and these results suggest the efficacy of Eqs. (13) and (14).

$$\log (K/K_0) = 5.15\,\sigma_I \tag{16}$$

Some experiments have also been reported on pyridine pK_a's in the excited state, a part of a comprehensive investigation in which pK_a^*'s are evaluated by the Forster cycle.[120] Correlation is found to be best with exalted parameters (σ^+ or σ^-), the degree of correlation often depending on the wavelength for investigation. Similar conclusions have been reached for 2- and 4-substituted styrylpyridines,[106] where in addition it is observed that the correlation of pK_a^* values for cis-2-styrylpyridines ($\rho = 12.07$) is better than for the pK_a values. Widely variant, but often very large, positive ρ values are typical of this type of reactivity, suggesting that in the excited state the rings become planar with consequent increased interaction.[106] MO calculations on the changes in charge densities of substituted pyridines on elevation to the first excited state show they correspond with σ values.[121]

The $\pi \rightarrow \pi^*$ transition energies for various pyridinium iodides yield a ρ value of 18.3, the transition energies being regarded as the spectroscopic equivalent of activation energies.[122] Stability constants of charge transfer complexes of pyridines[123,124] and 3- and 4-styrylpyridines[125,126] with iodine and other halogens[127] follow the Hammett equation although there are some discrepancies in the latter case, probably owing to interaction of iodine with the phenyl ring.

The hydrogen-bonding abilities of the ring nitrogen atom have also been the subject of investigation using IR spectroscopy, with proton donors such as phenol, pyrrole, and acetylenes in dilute carbon tetrachloride solution. Good correlations are observed;[128–131] thus Eq.

---

[119] C. A. Grob and M. C. Schlageter, Helv. Chim. Acta **57**, 509 (1974).
[120] H. H. Jaffé and H. Lloyd Jones, J. Org. Chem. **30**, 964 (1965).
[121] J. Del Bene and H. H. Jaffé, J. Chem. Phys. **49**, 1221 (1968).
[122] M. Kosower and J. A. Skorcz, Advan. Mol. Spectrosc., 413 (1962).
[123] W. J. McKinney, M. J. Wong, and A. J. Popov, Inorg. Chem. **7**, 1001 (1968).
[124] G. Aloisi, G. Cauzzo, and U. Mazzucato, Trans. Faraday Soc. **63**, 1858 (1968).
[125] G. Aloisi, G. Cauzzo, G. Giacometti, and U. Mazzucato, Trans. Faraday Soc. **61**, 1406 (1965).
[126] U. Mazzucato, G. Aloisi, and G. Cauzzo, Trans. Faraday Soc. **62**, 2685 (1966).
[127] G. Aloisi, G. Beggiato, and U. Mazzucato, Trans. Faraday Soc. **66**, 3075 (1970).
[128] J. Rubin, B. Z. Senkowski, and G. S. Panson, J. Phys. Chem. **68**, 1601 (1964).
[129] J. Rubin and G. S. Panson, J. Phys. Chem. **69**, 3089 (1965).
[130] C. Laurence and B. Wojtkowiak, C. R. Acad. Sci. **264**, 1216 (1967).
[131] S. Ghersetti, S. Giorgianni, A. Mangini, and G. Spunta, Spectrosc. Lett. **5**, 111 (1972).

(17) describes the correlation for 3- and 4-substituted pyridines with phenol, where ν is the OH stretch.[130] In other IR measurements the stretching vibrations of 3-substituted bipyridyl rings at 1565 cm^{-1} were

$$\nu_X - \nu_H = -140\,\sigma^0 \qquad (17)$$

shown to follow Eq. (18).[105]

$$\nu = 1590 - 30\,\sigma_p - 33\,\sigma_m \qquad (18)$$

pK_T values calculated from the nonlinear dependence of pK_a on σ^0 for the tautomeric equilibrium constants of an extensive range of 2-arylsulfonamidopyridines [Eq. (19)] in 80% dioxane and 80% alcohol have been shown to correlate excellently with σ^0, the ρ values being positive, indicating that electron-attracting substituents shift the equilibrium toward the zwitterionic form.[132]

$$\text{(structure)} \quad \underset{K_T}{\rightleftarrows} \quad \text{(structure)} \qquad (19)$$

The nucleophilic properties of pyridine nitrogen have been assessed in studies on pyridine-catalyzed Schotten-Baumann reactions[133] and aromatic sulfonyl chloride hydrolyses,[134] from which highly accurate Brønsted and Hammett treatments arise. There is some doubt as to the mechanism of this reaction.[45] Equations (13) and (14) accurately express the reactivity of 3- and 4-substituted pyridines toward ethyl iodide (using rate for equilibrium constants),[135] while rates of N-methylation of 2-substituted pyridines have been used to estimate both steric and electronic effects.[136]

A situation in which the pyridine nitrogen is not strictly the reaction site, but where the inductive stabilization of negative charge by the azonium site is the key role leading to reactivity is that of pyridine ylide formation, by deprotonation of the 2- and 6-positions of 3-substituted 1-methylpyridinium ions in D$_2$O buffer solution catalyzed by OD$^-$. A reasonable correlation is produced using σ_I ($\rho = 8.9$) for the reactivity of

---

[132] T. A. Mastrukova, Y. N. Sheinker, I. K. Kuznetzova, E. M. Peresleni, T. B. Sakharova, and T. M. Kabachnik, *Tetrahedron* **19**, 357 (1963).
[133] L. M. Litvinenko and A. I. Kirichenko, *Dokl. Akad. Nauk SSSR* **176**, 97 (1967).
[134] O. Rogne, *J. Chem. Soc. B*, 727 (1970); 489 (1972).
[135] A. Fischer, W. J. Galloway, and J. Vaughan, *J. Chem. Soc.* 3596 (1964).
[136] L. W. Deady and J. A. Zoltewicz, *J. Org. Chem.*, **37**, 603 (1972).

the 2-position, but there is evidence for a small resonance influence, while the reactivity of the 6-position accords with σ_p^0 ($\rho = 3.7$).[137] In neutral solution at 218°, the reactivity of the 2- and 6-position is hardly influenced at all by 3-substituents,[138] possibly owing to their compensatory effect on the dual-stage process shown in Scheme 3. The base-catalyzed exchange of the methyl group in 3- and 4-substituted 1-methylpyridinium iodides in both KOH/H$_2$O and MeOH with morpholine as catalyst has also been investigated,[139] and here the correlation is with σ_R^0.

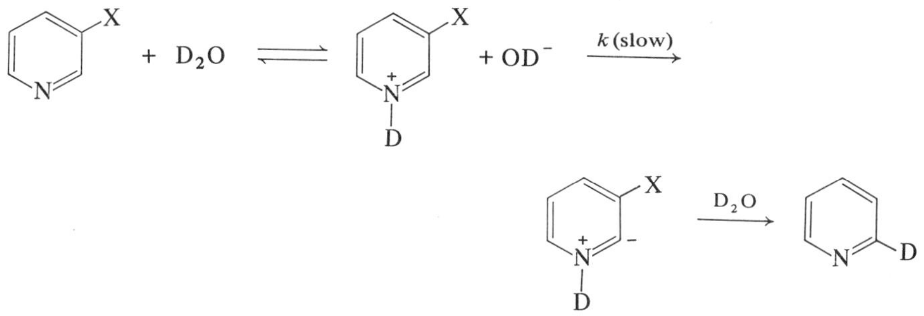

SCHEME 3

Pyridines are also well known as ligands in transition metal complexes, and if the equilibrium constants for the formation of such complexes can be related to base strength, it is expected that such constants would follow the Hammett equation. The problem has been reviewed,[140] and a parameter S_f formulated which is a measure of the contribution of the additional stabilization produced by bond formation to the stabilization constants of complexes expressed in terms of σ.[141] The Hammett equation has also been applied to pyridine 1:1 complexation with Zn(II), Cd(II), and Hg(II) $\alpha,\beta,\gamma,\delta$-tetraphenylporphins,[142,143] the σ values being taken as measures of cation polarizing ability. Variation of the enthalpy of complexation for adducts of bis(2,4-pentanediono)-Cu(II) with pyridines plotted against σ, however, exhibited a curved relationship.[144]

Decarboxylation of 6-substituted 2-pyridinecarboxylic acids in 3-nitrotoluene[145] are considered to involve transition state **21**. The sub-

[137] J. A. Zoltewicz and R. E. Cross, *J. Chem. Soc., Perkin Trans. II*, 1363 (1974).
[138] J. A. Zoltewicz and R. E. Cross, *J. Chem. Soc., Perkin Trans. II*, 1368 (1974).
[139] I. F. Tupitsin. N. N. Zatsepina, and A. V. Kirova, *Reakts. Sposobnost Org. Soedin.* **9**, 223 (1972).
[140] J. J. R. F. da Silva and J. G. Calado, *J. Inorg. Nucl. Chem.* **28**, 125 (1966).
[141] R. Irving and J. J. R. F. da Silva, *Proc. Chem. Soc.*, 250 (1962).
[142] C. H. Kirksey, P. Hambright, and C. B. Storm, *Inorg. Chem.* **8**, 2141 (1969).
[143] C. H. Kirksey and P. Hambright, *Inorg. Chem.* **9**, 958 (1970).
[144] W. R. May and M. M. Jones, *J. Inorg. Nucl. Chem.* **25**, 507 (1963).
[145] E. V. Brown and R. J. Moser, *J. Org. Chem.* **36**, 454 (1911).

stituents exert their effect again predominantly through σ_I, revealing the

(21)

strong influence of short-range induction, and the ρ value of -1.92 indicates that N—H bond formation precedes C—C bond cleavage. Resonance donors fall off the line, reacting much more slowly than predicted, probably owing to steric inhibition of resonance. In related studies[146,147] on the decarboxylation of some heterocyclic acetic acids, evidence from Hammett plots and from other sources indicates reaction via the zwitterionic form. One interesting[45] feature of the correlations is curvature explained[147] on the basis of being indicative of transition state drift along the reaction coordinate with increasing electronegativity of the heterocyclic ring.

The reactions of pyridine 1-oxides are considered in this section, although reaction occurs at O rather than N; indeed the pK_a's of the oxides yield a ρ value essentially equivalent to that for the iso electronic phenol dissociations. This ionization has been investigated extensively[1,53,148–150] and shown to follow a "sliding σ scale" using σ^+ for resonance donors and σ^- for resonance acceptors in the 4-position. A self-consistent set of σ values applying specifically to these compounds has been suggested,[148] but this tends to mask the expression of the amphoteric character of the N-oxide function which typifies its behavior, excellently summarized in the σ^+–σ–σ^- type of relationship. Excited state pK_a's show an enormous range, over 22 pK_a units,[120] but the correlation is poor, probably owing to high polarizability and charge transfer.

Other properties of pyridine 1-oxides have been investigated with regard to applicability of Hammett parameters. The type of parameter required and the degree of correlation achieved varies, but again the need for a "sliding" σ scale is generally indicated. Such measurements

[146] P. J. Taylor, *J. Chem. Soc. B,* 1077 (1972).
[147] R. G. Button and P. J. Taylor, *J. Chem. Soc., Perkin Trans. II,* 557 (1973).
[148] J. H. Nelson, R. G. Garvey, and R. O. Ragsdale, *J. Heterocycl. Chem.* **4**, 591 (1967).
[149] C. D. Johnson, A. R. Katritzky, B. J. Ridgwell, N. Shakir, and A. M. White, *Tetrahedron* **21**, 1055 (1965).
[150] C. Klotufar, S. Paljk, and B. Barlič, *Spectrochim. Acta, Part A* **29**, 1069 (1973).

involve: stretching vibrations of the N—O group in pyridine 1-oxides and *trans*-ethylene (pyridine 1-oxides) Cl_2 Pt(II);[151] energy differences between singlet and triplet states of pyridine 1-oxide Cu(II) halide complexes;[152] N—O stretching frequencies in phenol complexes and in complexes of the type |Ni(4-X-pyridine 1-oxide)|$^{2+}$ 2Y$^-$, where Y is ClO_4^- and BF_4^-;[153] N—O and Ti—O stretching vibrations in complexes with TiF_4;[154] selected IR band energies for complexes |VO(4-X-pyridine 1-oxide)$_2$|$^{2+}$ 2Y$^-$, where Y is Cl or Br,[155] and for pyridine 1-oxide adducts of bis(2,4-pentanedionato)oxovanadium(IV);[156] ^{19}F chemical shift in complexes of pyridine 1-oxides with TiF_4[157] and $TiF_4 \cdot CH_3CON(CH_3)_2$, 4-X-pyridine 1-oxide;[158] K_{eq} for 4-X-pyridine 1-oxide/I_2 charge transfer complexes;[159] intensity of stretching vibrations of N—O and metal —O groups in pyridine 1-oxide complexes with V(IV), Ti(IV) and Zr(IV);[160] half-wave oxidation potentials and electronic spectra of heteroaromatic amine *N*-oxides.[161]

B. OTHER SIX-MEMBERED RINGS

1. *Diazines and Triazines*

Correlations of pK_a values with the Hammett or dual-parameter equations have been reported for 5-substituted 4-amino-2-methylpyrimidines, which are shown to exist entirely in the amino form rather than in the imino form, and 4,5-disubstituted 2-methylpyrimidines,[162,163] 2- and 4-diaminopyrimidines,[164] 5- and 6-substituted 2,4-diaminopyrimidines,[165] substituted pyrazines and pyridazines,[102,166]

---

[151] S. I. Shupack and M. Orchin, *J. Amer. Chem. Soc.* **85**, 902 (1963).
[152] W. E. Hatfield and J. S. Pascal, *J. Amer. Chem. Soc.* **86**, 3888 (1964).
[153] D. W. Herlocker, R. S. Drago, and V. I. Meek, *Inorg. Chem.* **5**, 2009 (1966).
[154] F. E. Dickson, E. W. Gowling, and F. F. Bentley, *Inorg. Chem.* **6**, 1099 (1967).
[155] R. G. Garvey and R. O. Ragsdale, *J. Inorg. Nucl. Chem.* **29**, 745 (1967).
[156] R. G. Garvey and R. O. Ragsdale, *Inorg. Chim. Acta* **2**, 191 (1968).
[157] D. S. Dyer, and R. O. Ragsdale, *Inorg. Chem.* **6**, 8 (1967).
[158] C. E. Michelson, D. S. Dyer, and R. O. Ragsdale, *J. Chem. Soc. A*, 2296 (1970).
[159] R. C. Gardner and R. O. Ragsdale, *Inorg. Chim. Acta* **2**, 139 (1968).
[160] F. E. Dickson, E. W. Baker, and F. F. Bentley, *Inorg. Nucl. Chem. Lett.* **5**, 825 (1969).
[161] M. Miyazaki, T. Kubota, and M. Yamakava, *Bull. Chem. Soc. Jap.* **45**, 780 (1972).
[162] S. Mizukami and E. Hirai, *J. Org. Chem.* **31**, 1199 (1966).
[163] S. Mizukami and E. Hirai, *Chem. Pharm. Bull.* **14**, 1321 (1966).
[164] N. V. Khromov-Borisov, *Dokl Akad. Nauk SSSR. Khim.* **180**, 1129 (1968).
[165] B. Roth and J. Z. Strelitz, *J. Org. Chem.* **34**, 821 (1969).
[166] R. F. Cookson and G. W. H. Cheeseman, *J. Chem. Soc., Perkin Trans. II*, 392 (1972).

and pyridazones,[166,167] and s-triazines.[168,169] In such studies, the use of acidity function theory may be needed, and a distinction between H_A and H_0 for pyridazone protonation in one case.[166] The generally excellent Hammett correlations achieved again provide reassuring evidence for the authenticity and accuracy of this methodology.

Correlation of spectral studies for these classes of compounds involve K-band shifts in the UV spectra of 6-substituted 2,4-diaminopyrimidines,[165] NMR chemical shifts of the two amide protons and the 6-proton in 5-substituted uracils[170] and of protons in pyrazines,[171] MeO $^{13}C-H$ coupling constants in methoxy-s-triazines,[172] and ^{35}Cl NQR in chloropyrimidines.[173]

Such compounds are highly susceptible to nucleophilic attack, owing to stabilization of negative charge by both the inductive and resonance effect of the nuclear nitrogens, and there is a profusion of studies relating to the applicability of the Hammett treatment in these cases. Tupitsin and Zatsepina and their co-workers have examined hydrogen exchange of nuclear positions of various diazines[49] and diazine N-oxides.[174] Among reactions proceeding via Meisenheimer-type adducts with subsequent expulsion of halogen on which substituent studies have been made may be included substitution of a wide variety of 2- and 4-substituted chloropyrimidines,[175] substituted 2-halogenopyrimidines with piperidine in various media,[176,177] 2,4-dichloro-6-substituted s-triazines with isopropylamine,[178] alkaline[179] and acidic[180] hydrolyses of 2-chloro-

---

[167] I. V. Turovskii, V. T. Glezer, O. Avota, J. P. Stradins, and S. Hillers, *Khim. Geterotsikl. Soedin.*, 993 (1973).

[168] R. Bacaloglu, E. Fliegl, and G. Ostrogovitch, *Rev. Roum. Chim.* **16**, 1447 (1971).

[169] H. Matsui and Y. Hashida, *Kagaku No Ruoiki* **26**, 584 (1972) [*CA* **77**, 23897y (1972)].

[170] J. P. Kokko, L. Mandel, and J. H. Goldstein, *J. Amer. Chem. Soc.* **84**, 1042 (1962).

[171] G. S. Marx and P. E. Spoerri, *J. Org. Chem.* **37**, 111 (1972).

[172] Y. Fukushima, Y. Hashida, and K. Matsui, *Nippon Kagaku Kaishi*, 629 (1972) [*CA* **76**, 153087e (1972)].

[173] G. K. Semin, T. A. Babushkina, V. P. Mamaev, and V. P. Krivopalov, *Izv. Sib. Otd. Akad. Nauk SSSR, Ser. Khim Nauk*, 82 (1971) [*CA* **76**, 39820m (1972)].

[174] I. F. Tupitsin, N. N. Zatsepina, J. M. Kapustin, and A. W. Kirowa, *Reakts. Sposobnost Org. Soedin.* **5**, 601 (1968).

[175] V. P. Mamaev and O. A. Zagulyaeva, *Zh. Org. Khim.* **8**, 583 (1972) [*CA* **77**, 19002p (1972)].

[176] S. M. Shein, O. A. Zagulyaeva, A. J. Shvets, and V. P. Mamaev, *Reakts. Sposobnost Org. Soedin.* **9**, 890 (1973).

[177] S. M. Shein, V. P. Mamaev, O. A. Zagulyaeva, and A. J. Shvets, *Reakts. Sposobnost Org. Soedin.* **9**, 897 (1973).

[178] D. H. Sim, S. Y. Jo and Y. J. Li, *Hwahak Kwa Hwahak Kongop*, 252 (1971) [*CA* **76**, 153080v (1972)].

[179] T. N. Bykhovskaya and O. N. Vlasov, *Reakts. Sposobnost Org. Soedin.* **4**, 510 (1967).

[180] T. N. Bykhovskaya, O. N. Vlasov, J. A. Malnikova, and N. N. Melnikov, *Reakts. Sposobnost Org. Soedin.* **9**, 1149 (1973).

4,6-disubstituted s-triazines, 2-substituted 4-amino-6-chloro-s-triazines with polyvinyl alcohol,[181] successive replacement of chlorine by substituted anilines in 2,4,6-s-triazines,[182] and 2,4-disubstituted 6-chloro-s-triazines with diethylamine.[183]

Toxicity effects in triazines have also been treated by the Hammett approach.[184,185]

2. Annelated Azines

The σ values of rings annelated to pyridine in systems such as quinoline, isoquinoline, benzoquinolines, acridine, and phenanthridine and determined by the basicity of the ring nitrogen atom are predicted[186] and found to be small.[33,34] This is an important point stressed and elucidated previously,[187] and demonstrating the validity of the basic assumptions of the Hammett equation.

The influence of substituents, including aza, on reactivity of the nitrogen atom in such systems have been considered in detail[1,6,7,47,102,188,189] and may be correlated with Dewar–Grisdale calculations [Eq. (2)],[6,33,190–192] or dual-substituent parameter [Eqs. (3) and (4)].[6,7,47,188–190] Thus, analysis[6] of pK_a's (H_2O, 25°) of substituted quinolinium and isoquinolinium cations according to Eq. (20) gave the results shown in Table V, where they are compared with those for substituted naphthoic acids (aq. EtOH, 25°).

$$\log (K/K_0) = \sigma_I \rho_I + \sigma_R^0 \rho_R \tag{20}$$

A systematic difference between λ (1) and λ (2) might have been expected, since one reaction site is in the ring and the other is well removed from it, but Table V shows that for the α site the values are almost con-

---

[181] T. Goto and M. Nagano, *Nippon Kagaku Zasshi*, 1514 (1972) [*CA* **77**, 165133y (1972)].
[182] T. M. Chinh, J. Kaválek, and M. Večeřa, *Collect. Czech Chem. Commun.* **37**, 832 (1972).
[183] A. S. Estrin, E. G. Sochilin, and L. M. Dolgopolskii, *Reakts. Sposobnost Org. Soedin.* **3**, 111 (1966).
[184] O. Andrysova, V. Rambousek, J. Jirasek, V. Zverina, M. Matrka, and J. Marhold, *Physiol. Bohemoslov.* **21**, 63 (1972) [*CA* **77**, 135918r (1972)].
[185] G. F. Kolar and R. Preussmann, *Z. Naturforsch. B* **26**, 950 (1971).
[186] A. Streitwieser, "Molecular Orbital Theory," pp. 419-420. Wiley, New York, 1961.
[187] C. D. Johnson,[2] pp. 114 and 156.
[188] P. Tomasik, R. Zalewski, and W. Drzeniek, *Rocz. Chem.*, in press.
[189] P. Tomasik, R. Zalewski, and J. Chodzinski, *Rocz. Chem.*, in press.
[190] M. Charton, *J. Org. Chem.* **30**, 3341 (1965).
[191] C. D. Johnson,[2] p. 132.
[192] C. W. Donaldson and M. M. Joullié, *J. Org. Chem.* **33**, 1504 (1968).

TABLE V

ρ_R/ρ_I (λ) Values for Naphthoic Acid (1) and Quinolinium and Isoquinolinium (2) pK_a Values

	1-Naphthoic acids and quinolinium cations Substituent position					2-Naphthoic acids and isoquinolium ions Substituent position				
	3	4[a]	5[a]	6	7[a]	4	5	6[a]	7	8[a]
λ (1)	0.47	1.51	0.62	0.87	1.17	0.61	0.62	1.06	0.73	1.46
λ (2)	0.27	1.39	0.61	0.76	1.05	0.46	0.47	1.43	0.91	0.81

[a] Conjugation between substituent and reacting side chain position.

stant even for the conjugated positions where both σ_I and σ_R are large, while for the β site such differences as do exist are random.

The various factors that influence reactivity in these types of molecules are clearly illustrated in a study[193] of their reactions with methyl iodide and p-nitrophenyl acetate, giving rise in certain cases to deviations from Hammett-type plots. Thus, rates for isoquinoline, pyrimidine, and pyridazine fit reasonably well on to the $pK_a - \log K_{rel}$ (MeI) plot (Fig. 3) and thus conform to the Hammett reaction constant

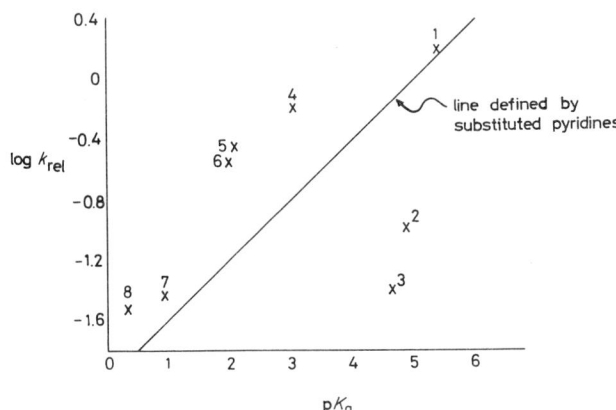

FIG. 3. Reactivity effects in diazines and annelated pyridines: 1, isoquinoline; 2, quinoline; 3, 1,10-phenanthroline; 4, phthalazine; 5, cinnoline; 6, pyridazine; 7, pyrimidine; 8, pyrazine.

[193] J. A. Zoltewicz and L. W. Deady, *J. Amer. Chem. Soc.* **94**, 2765 (1972).

−2.30 for methylation in DMSO. Quinoline and 1,10-phenanthroline react more slowly than predicted, owing to greater steric hindrance in the methylation than the protonation reaction, while phthalazine, pyridazine, and cinnoline react faster, a kinetic effect discussed in terms of electron-pair repulsions, the so-called α-effect.

Nucleophilic substitution reactions in this area have been described previously,[19,20] as have free-radical reactions.[26] It is of interest that in this latter field homolytic acetylation of 4-substituted protonated quinolines in the 2-position give only approximate correlation with σ_m, OMe being rate retarding instead of accelerating as expected from its positive σ_m value; however, relative rates log k_{CN}/k_{Cl}, where CN and Cl refer to the substituents in the 4-position, did correlate with the σ values for 3- and 4-substituted benzoyl radicals, giving $p = -0.49$.[194] σ^+-Values for quinoline have been defined from pyrolysis rates of 1-quinolylethyl acetates,[195] with the results shown in structure **22**, but agreement with electrophilic substitution rates is uncertain because either the reaction goes via the protonated form or else product yields are very low.

(22)

The influence of substituents in mono- and disubstituted 2,3-dimethylquinoxalines and their conjugate acids on the UV and NMR spectra have been correlated with σ,[196] while stretching frequencies of the carbonyl group of 6-substituted quinazolinediones and their half-neutralization potentials have been related to σ_p.[197]

A number of studies have been directed specifically at the acridine nucleus. The effect of substituents in the 9-position in acridine N-oxides on R_f values for thin-layer chromatography (TLC) is measured by σ_p,[198] and there is also a relation between this parameter and the N—O stretching frequencies. The effect of substituents on the basicity of acridine and the possibility of amine–ketimine tautomerism in 9-aminoacridines have been evaluated, taking quinoline as a model

[194] T. Caronna, G. Fronza, F. Minisci, O. Porta, and G. P. Gardini, *J. Chem. Soc., Perkin Trans. II*, 1477 (1972).
[195] R. Taylor, *J. Chem. Soc. B*, 2382 (1971).
[196] P. Vetesnik, V. Bekarek, V. Beranek, and J. Kavalek, *Sci. Pap. Univ. Chem. Technol. Pardubice II*, 37 (1966).
[197] E. Wulfert, P. Bolla, and J. Mathieu, *J. Chim. Ther.* **4**, 257 (1969).
[198] M. Ionescu, I. Goia and H. Mantsch, *Rev. Roum. Chem.* **11**, 243 (1966).

system: the 1-, 2-, 7-, and 8-positions of acridine have been assigned to the meta class and the 3- and 6-positions to the para class.[102]

Rates of addition of styrene to 9-substituted acridizinium ions have been measured,[199] and the work was extended via substituted styrenes to a multiple structure–reactivity relationship.[200]

Substituent effects in 1,10-phenanthrolines have been comprehensively investigated[201,202] from their pK_a's, stability constants in Fe(III) and Cu(II) complexes, and redox potentials; the Ni(II) complexes have also been examined,[203] as have other metal complexes.[204] ρ-Values have been obtained for the three successive proton losses (1, 2, 3, respectively) from 4-substituted dications of 10-hydroxy-1,7-phenanthrolines (**23**).[205]

(23)

The half-wave potentials for polarographic reduction of 2- and 3-substituted phenazines and phenazine-N-oxides follow a Hammett correlation only for the latter substituents,[206] although independent investigations on phenazines afford a general correspondence.[207–210]

---

[199] I. J. Westerman and C. K. Bradsher, *J. Org. Chem.* **36**, 969 (1971).
[200] N. A. Porter, I. J. Westerman, T. G. Wallis, and C. K. Bradsher, *J. Amer. Chem. Soc.* **96**, 5104 (1974).
[201] M. Charton, *J. Org. Chem.* **31**, 3799 (1966).
[202] O. Enea, G. Berthon, and V. Bokra, *Termochim. Acta* **4**, 449 (1972).
[203] R. K. Steinhaus and D. W. Margerum, *J. Amer. Chem. Soc.* **88**, 441 (1966).
[204] Y. Bokra and G. Berthon, *J. Chim. Phys. Physiocochim. Biol.* **69**, 1547 (1972) [*CA* **78**, 8512z (1972)].
[205] G. P. Bean, M. J. Cook, T. M. Dand, A. R. Katritzky, and J. R. Lea, *J. Chem. Soc. B*, 2339 (1971).
[206] V. Riganti, M. Mazza, and S. Locchi, *Farmaco Ed. Sci.* **23**, 778 (1968) [*CA* **69**, 86192y (1968)].
[207] G. Canallini, I. Degani, R. Focchi, and G. Spunta, *Ann. Chim. (Rome)* **57**, 1045 (1967).
[208] L. L. Gordenko and Y. S. Rozum, *Elektrokhimya* **7**, 1834 (1971) [*CA* **76**, 153072u (1972)].
[209] S. Nakamura and T. Yoshida, *Denki Kagaku* **40**, 714 (1972) [*CA* **78**, 91765c (1973)].
[210] L. L. Gordenko, *Elektrokhimya* **1**, 1497 (1965) [*CA* **64**, 9243 (1966)].

Sec. II.B]   HAMMETT EQUATION AND HETEROCYCLES   31

3. *Other Six-Membered Heterocycles Including Alicyclic Compounds*

A series of studies have been reported on the basicities and nucleophilicities of 4-substituted and 4,4'-disubstituted piperidines, together with related work on decahydroquinolines.[211-215] Substituent effects on pK_a values determined in methanol of 4,4'-disubstituted N-methylpiperidines[216,217] and 1,2,5-trimethylpiperidines (**25**)[218] were found to be dependent on σ^*; i.e., as expected, resonance effects are absent, the hydroxyl substituent where present taking the axial conformation (**24**),[216] owing to a bridging molecule of methanol.

(24)   (25)

This dependence of pK_a on inductive effects was also found in nitrobenzene[218] and aqueous methanol.[219] Anomalies were found, however, particularly where the predominant conformer had an axial OH group.[215,217,218] These were explained in terms of interaction between inductive effects of geminal groups, or of steric effects,[215,218,219] termed by the authors a "stereopolar effect." These studies led to the conclusion that the inductive effect of substituents in such systems is registered at

---

[211] G. S. Litvinenko, V. I. Artyukin, A. A. Andrushenko, D. V. Sokolova, and K. I. Khludneva, *Reakts. Sposobnost Org. Soedin.* **5**, 263 (1968).
[212] D. V. Sokolova, G. S. Litvinenko, V. I. Artyukin, and A. A. Andrushenko, *Izv. Akad. Nauk. SSSR, Ser. Khim.* **4**, 73 (1965).
[213] T. D. Sokolova, S. V. Bogatkov, Y. F. Malina, and B. V. Unkovskii, *Reakts. Sposobnost Org. Soedin.* **7**, 626 (1970).
[214] G. S. Litvinenko, V. I. Artyukin, A. A. Andrushenko, D. V. Sokolov, V. V. Sosnova, and M. M. Akimova, *Reakts. Sposobnost Org. Soedin.* **7**, 960 (1970).
[215] K. I. Romanova, S. V. Bogatkov, T. D. Sokolova, Y. F. Malina, and B. V. Unkovsky, *Reakts. Sposobnost Org. Soedin.* **9**, 93 (1972).
[216] T. D. Sokolova, S. V. Bogatkov, J. F. Malina, B. V. Unkovskii, and E. M. Cherkasova, *Reakts. Sposobnost Org. Soedin.* **4**, 68 (1967).
[217] T. D. Sokolova, S. V. Bogatkov, J. F. Malina, B. V. Unkovskii, and E. M. Cherkasova, *Reakts. Sposobnost Org. Soedin.* **5**, 160 (1968).
[218] T. D. Sokolova, S. V. Bogatkov, J. F. Malina, B. V. Unkovskii, and E. M. Cherkasova, *Reakts. Sposobnost Org. Soedin.* **4**, 445 (1967).
[219] T. D. Sokolova, S. V. Bogatkov, Y. F. Malina, B. V. Unkovskii, and E. M. Cherkasova, *Reakts. Sposobnost Org. Soedin.* **6**, 610 (1969).

the reacting N atom predominantly by σ-bond transmission, and yet its magnitude and accuracy of correlation with σ* *does* depend on the spatial orientation of the substituent.[214,215] These conclusions apply not only for different solvent systems, but also to the *trans*-decahydroquinoline derivatives,[211] and also to reactivities other than protonation. Thus the catalytic activity of 4,4'-disubstituted-1,2,5-trimethylpiperidines (24) on hydrolysis of benzoyl chloride in 98% aqueous dioxane gave an accurate Brønsted plot of slope 0.65 using the pK_a's previously determined[220] (see Table VI).

TABLE VI

pK_a Values and Benzoylation Rates for 25 (25°)

25			
X	X'	pK_a	$k_2 \times 10^2$
H	H	8.19	3.7
Et	OH	7.82	2.2
Ph	OH	7.69	2.2
OH	Et	7.32	0.82
OH	C≡CH	6.69	0.37
C≡CH	OH	7.18	0.72
C≡N	OH	6.18	0.21

A satisfactory correlation of half-wave potentials for polarographic reduction of 2,5-diaryl-4-pyrones,[221] as well as Δλ for the electronic absorption spectra maxima of 3-phenylazojulolidines (26) and their

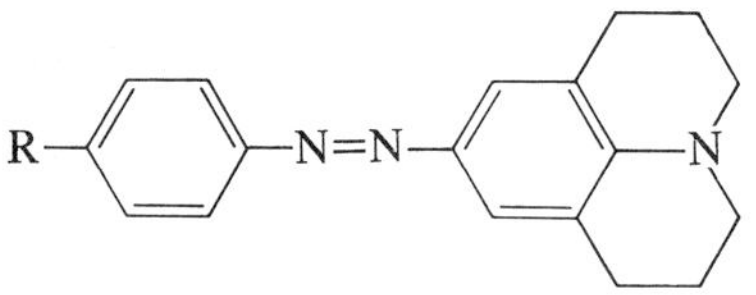

(26)

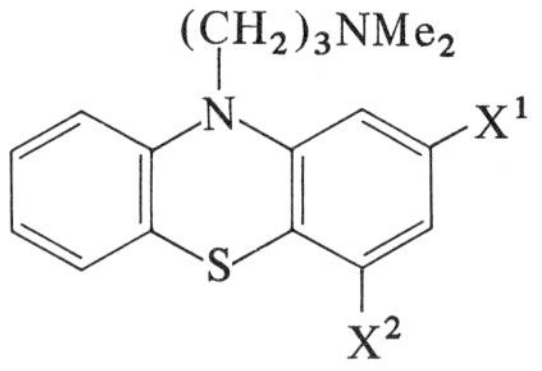

(27)

azonium salts,[222] with σ has been obtained. The reactivity of several six-membered heterocyclic thiones toward alkylation has also been con-

---

[220] T. D. Sokolova, S. V. Bogatkov, V. M. Manyashkina, Y. F. Malina, and B. V. Unkovskii, *Reakts. Sposobnost Org. Soedin.* **8**, 363 (1971).
[221] H. C. Smitherman and L. N. Ferguson, *Tetrahedron* **24**, 923 (1967).
[222] R. W. Castelino and G. Hallas, *J. Chem. Soc. B,* 793 (1971).

TABLE VII
Substituent Constants for Six-Membered Rings

Substituent X	σ	Type	Reaction	Reference footnotes
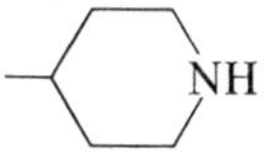 piperidin-1-yl (−NH)	−0.12	σ_p	Ionization constants of amidines p-Ph·C$_6$H$_4$(=NH)NHX (50% aq. EtOH)	229
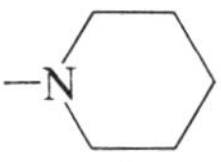 piperidino (−N⟨⟩)	−2.43	σ_p^+	Dipole moments of 5-substituted furfurals	230
	−0.57	σ_p	Polarography of 6-substituted 2,3-dimethyl-5,8-quinoxalinediones	231
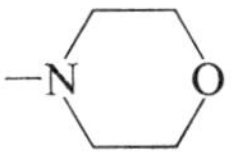 morpholino (−N⟨O⟩)	−1.61	σ_p^+	Dipole moments of 5-substituted furfurals	230
	−0.50	σ_p	Polarography of 6-substituted 2,3-dimethyl-5,8-quinoxalinediones	231
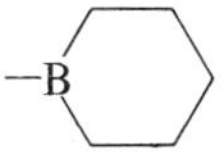 −B⟨⟩	0.46	σ_p	Proton magnetic resonance of substituted propylenes	232
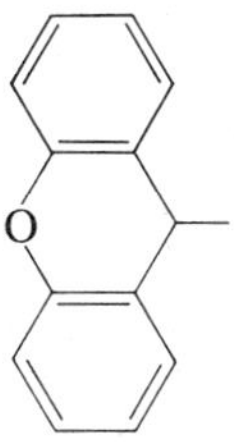 xanthenyl	0.35	σ^*	Ionization constants of XCH(NO$_2$)COOEt	233
	−2.0	E_s^0		
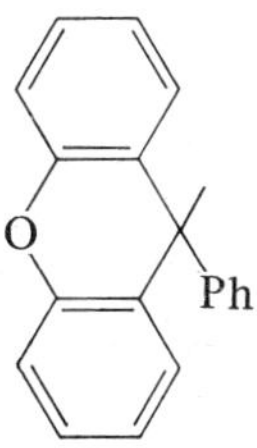 9-phenylxanthenyl	0.58	σ^*		
	−4.0	E_s^0		

sidered,[223] as has the complexation by H-bonding on the exocyclic N, of phenothiazine (promazine) derivatives (**27**) with alcohols,[224] and their

---

[223] G. J. Ovechkina, L. A. Ignatov, and B. V. Unkovskii, *Khim. Technol. Tr. Yubileinoi Konf. Posvyashch. 70—Letiyu Inst.*, 158 (1970) [*CA* **81**, 24867m (1974)].

[224] B. Tilguin, G. Tilleux, and T. Zeegers-Huykens, *Ann. Soc. Sci. Bruxelles, Ser. 1* **87**, 448 (1973) [*CA* **80**, 81822w (1974)].

psychotropic activity.[225] More unusual σ correlations are reported in the effect of ac frequency on the dissolution of iron in the presence of pyrylium salts,[226] the inhibition efficiency of iron corrosion by piperidines,[227] and the passivation of titanium in hydrochloric acid solutions containing pyrones.[228]

From the pK_a values of 2,6-dimethyl-4-pyrone and -4-thiapyrone and the ρ value of -1.75 for the correlation of 4-substituted 3,5-dimethylphenols, σ_p^- values of 5.8 and 5.4 were derived for the heteroatoms O^+ and S^+, respectively, while values of σ_m^+ of 3.0 and 3.2, respectively, similarly uniquely large, were calculated from hydrogen exchange data.[80] Table VII gives substituent parameters for the total effect of some six-membered rings.

III. Five-Membered Rings

A. Containing One Heteroatom

A large number of studies, some reviewed previously[1] and others considered subsequently here, have sought to apply the Hammett equation to reactions of five-membered rings by considering the 2-4 and 2-5 relationships as meta- and para-like, respectively, and thus using σ_m and σ_p values. In such cases, it is of interest to compare the resultant ρ values with those for the corresponding benzenoid side-chain reactions. As pointed out in the Introduction, differences between such ρ values may be discussed in terms of "intensity of transmission of effects," but it appears to be a more illuminating and meaningful approach to consider such differences as representing nonadditivity of substituent effects, the heteroatom being taken as a substituent replacing $-CH=CH-$ in a benzenoid ring and able on occasion to interact via proximity effects with the side chain, altering its susceptibility to substituent influences.

Since these five-membered rings, however, differ markedly from benzene in their geometry a more rigorous approach, as for substituent effects in polycyclic systems, would appear to be calculation of appropriate σ values by approaches of the Dewar–Grisdale equation [Eq. (2)] or multiparameter form [Eq. (3)]; in the former case, the ρ values are taken as equivalent to that in benzenoid compounds.

[225] M. J. Mercier and P. A. Dumont, *J. Pharm. Pharmacol.* **24**, 706 (1972).
[226] V. P. Grigonev and A. V. Nikolaev, *Issled. Obl. Korroz. Zashch. Met.*, 39 (1971) [*CA* **78**, 78924j (1973)].
[227] K. Aramaki, *Denki Kagaku* **41**, 321 (1973) [*CA* **79**, 86718f (1973)].
[228] V. P. Grigonev and J. M. Gershanova, *Zh. Prikl. Khim.* **42**, 2135 (1969).

Corollaries to these considerations are the designation of σ values for heteroatoms in such rings, in cases where steric interaction with the reaction site appears absent, and also for such entire ring systems as substituents meta or para to a reacting side chain in a benzenoid system. Such an approach, however, would certainly leave no basis for the contention that the sizes of ρ values are indicative of aromaticity.[29,234,235]

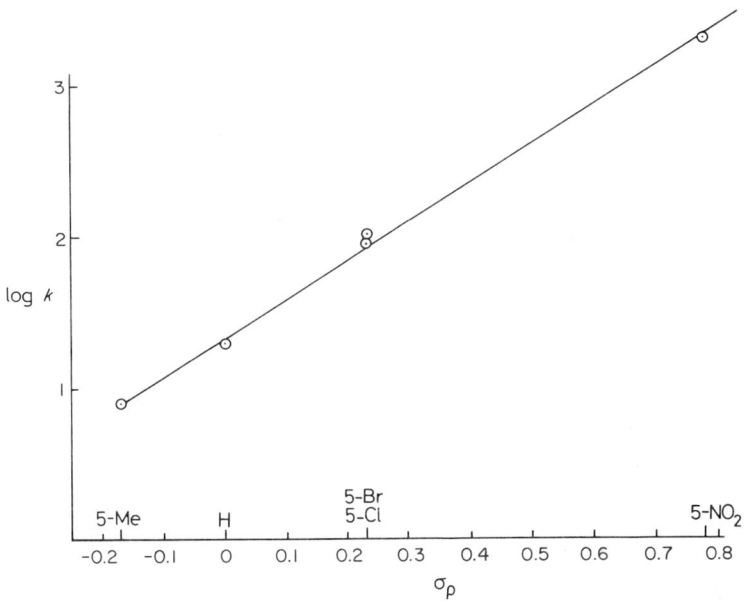

FIG. 4. Log rates of basic hydrolysis of 5-X-thiophene-2-ethyl carboxylates vs. σ_p (see Table VIII).

We first take up the question of substituted five-membered heterocycles, σ_m and σ_p values being used as described above. Table VIII gives relevant data including information on ρ values.

---

[229] J. Cymerman-Craig, M. J. Parker, and P. Woodhouse, *J. Chem. Soc.*, 3050 (1953).
[230] M. G. Koghan, V. S. Pustovarov, Y. V. Kolodyazhnyi, Z. N. Nazarova, and O. A. Osipov, *Zh. Org. Khim.* **4**, 2216 (1968).
[231] W. F. Gum and M. M. Joullié, *J. Org. Chem.* **32**, 53 (1967).
[232] M. P. Simonnin and J. Braun, *Bull Soc. Chim. Fr.*, 4918 (1968).
[233] H. Timotheus, R. Tampere, and R. Hiob, *Reakts. Sposobnost Org. Soedin.* **8**, 109 (1971).
[234] O. Exner and W. Simon, *Collect. Czech. Chem. Commun.* **29**, 2016 (1964).
[235] W. K. Kwok, R. A. More O'Ferrall, and S. I. Miller, *Tetrahedron* **20**, 1913 (1964).

TABLE VIII
ρ-VALUES FOR SIDE-CHAIN REACTIVITIES OF FIVE-MEMBERED HETEROCYCLES

Reaction		Reference footnotes
Ionization of 4- and 5-X-pyrrole-2-carboxylic acids (H_2O, 25°)	1.65 (1.00)[a]	236
Ionization of 4-X-3,5-dimethylpyrrole-2-carboxylic acids (50% aq. EtOH, 25°)	2.18 (1.51)	237
Ionization of 5-X-2,4-dimethylpyrrole-3-carboxylic acids (50% aq. EtOH, 25°)	1.60 (1.51)	237
Ionization of 5-X-2-furoic acids (H_2O, 25°)	1.40 (1.00)	236
Ionization of 5-X-2-furoic acids (methyl cellosolve, 25°)	2.20 (1.68)	234
Esterification of 5-X-2-furoic acids (diphenyldiazomethane, EtOH, 25°)	0.98 (0.84)[b]	235
Ionization of 4- and 5-X-thiophene-2-carboxylic acids (H_2O, 25°)	1.20 (1.00)	238, 239
(50% aq. EtOH, 25°)	1.58 (1.58)	
Basic hydrolysis of 5-X-thiophene-2-ethylcarboxylates (52% aq. acetone)[c]	2.50 (2.23)	240
Ionization of selenophene-2-carboxylic acids (H_2O), 25°)	1.23 (1.00)	241
Ionization of tellurophene-2-carboxylic acids (H_2O, 25°)	1.20 (1.00)	242

[a] Values from benzenoid reaction are given in parentheses.
[b] In MeOH.
[c] See Fig. 4.

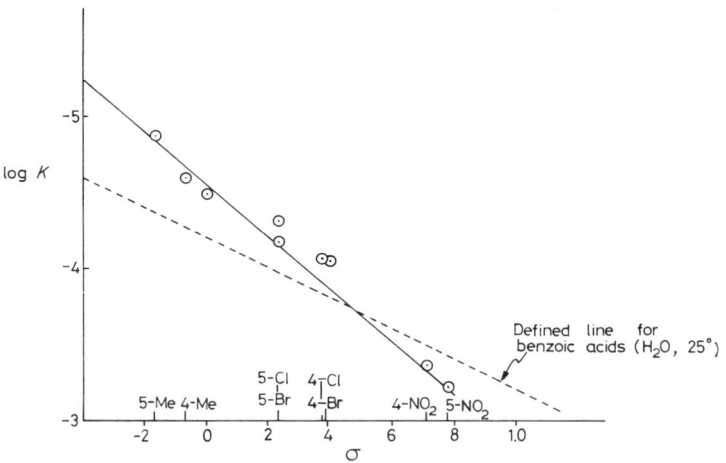

FIG. 5. Ionization of pyrrole-2-carboxylic acids. Log $K = \rho(\sigma + \sigma_N) + \log K_0$; ∴ for 5-Me, $-4.88 = 1.00 (-0.17 + \sigma_N) -4.20$ and $\sigma_N = -0.51$, for 5-NO_2, $-3.22 = 1.00 (0.78 + \sigma_N) -4.20$, and $\sigma_N = 0.20$.

Figure 5 shows the results graphically for pK_a values of pyrrole-2-carboxylic acids.[236] The apparent σ_N value is clearly variable, changing from -0.51 to $+0.20$ over the substituent range examined; here there is the possibility of H-bonding interactions **28** and **29** in the neutral form or in the anionic form.[235] From Fig. 6 on the other hand, it can be noted that the effect of α_s, from the pK_a values of thiophene-2-carboxylic acids, is almost constant at 0.50.[234] Clearly there is minimal interaction

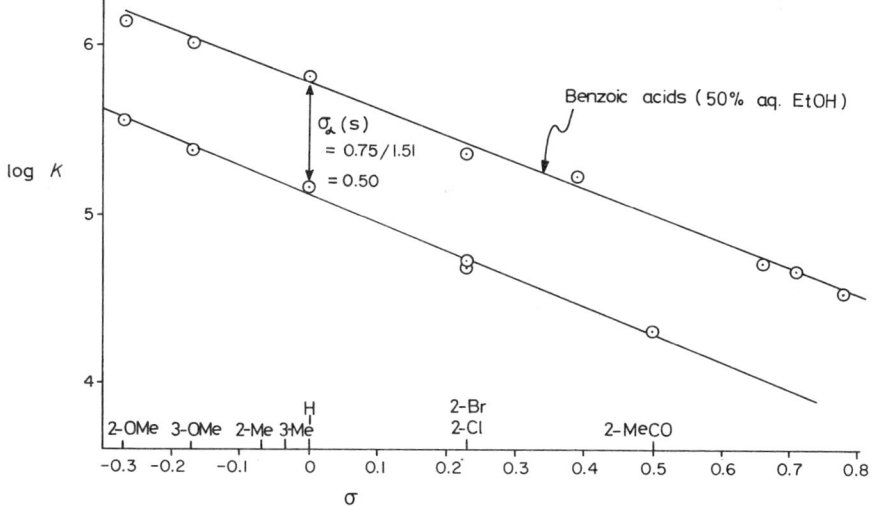

FIG. 6. Ionization of thiophene-2-carboxylic acids.

here with the reaction site. Similarly, close correspondence is observed between ρ for the benzenoid reaction and that for the ionization of 5-X-2,4-dimethylpyrrole-3-carboxylic acids where the 3-position of the carboxyl group and the interposing Me group make interactions of form **28** and **29** impossible.

(28)   (29)

[236] F. Fringuelli, G. Marino, and G. Savelli, *Tetrahedron* **25**, 5815 (1969).
[237] T. A. Melenteva, L. V. Kazanskaya, and V. M. Berezovskii, *Dokl. Akad. Nauk. SSSR* **175**, 354 (1967).
[238] F. Freeman, *J. Chem. Educ.* **47**, 140 (1970).

Butler[239] has used the Dewar–Grisdale approach [Eq. (2)] to estimate σ values in the thiophene series. Obviously the good correlation of the 4- and 5-substituent effects with σ_m and σ_p, respectively, together with the correspondence of the ρ values in the thiophene and benzenoid systems, means that the calculated values differ little from those in benzene. An analysis of NMR effects for both substituted thiophenes and furans leads to similar conclusions.[243,244]

The possibility of proximity effects make σ values for heteroatoms, particularly pyrrole, of doubtful general validity. Those gathered in

TABLE IX

σ-VALUES FOR HETEROATOMS EVALUATED FROM pK_a MEASUREMENTS OF APPROPRIATE CARBOXYLIC ACIDS

Heteroatom	α	β	Reference footnotes
O	1.10	—	234
	1.04	0.25	242
S	0.66	—	234
	0.67	0.12	239, 240
	0.50	—	239
Se	0.60	—	242
Te	0.23	—	242
NH	−0.15	−0.75	245

Table IX should therefore be viewed with some caution, although certainly the strong electron withdrawal they indicate by induction increasing in the order Te < Se < S < O is realistic. However, it has been reported[246,247] that these values are completely inappropriate, and from early references are quoted[247] values of σ_α (O) 0.32, σ_β (O) 0.04, σ_α (S)

[239] A. R. Butler, *J. Chem. Soc.*, 867 (1970).
[240] P. A. Ten Thije and M. J. Janssen, *Rec. Trav. Chim. Pays-Bas* **84**, 1169 (1965).
[241] D. Spinelli, G. Guanti, and C. Dell'Erba, *Ric. Sci.* **38**, 1048 (1968).
[242] F. Fringuelli, G. Marino, and A. Tattichi, *J. Chem. Soc., Perkin Trans. II*, 1738 (1972).
[243] R. A. Gavars and J. P. Stradins, *Khim. Geterotsikl. Soedin.* 15 (1965) [*CA* **63**, 5509 (1965)].
[244] R. A. Gavars and J. P. Stradins, *Reakts. Sposobnost Org. Soedin.* **2**, 22 (1965).
[245] M. K. A. Khan and K. J. Morgan, *Tetrahedron* **21**, 2197 (1965).
[246] D. S. Noyce, and R. L. Castenson, *J. Amer. Chem. Soc.* **95**, 1247 (1973).
[247] G. T. Bruce, A. R. Cooksey, and K. J. Morgan, *J. Chem. Soc., Perkin Trans. II*, 551 (1975).

0.03, σ_β (S) 0.04, σ_α (NH) -0.58, and σ_β (NH) -0.94. These are commented on later.

These five-membered rings are also highly susceptible to electrophilic attack as a consequence of their high polarizability, a topic previously reviewed in these pages.[24] Under this heading may also be considered S_N1 hydrolyses and solvolyses, where a carbonium ion is generated in conjugation with the ring, both types of reactivity requiring correlation by σ^+ constants. Marino and his co-workers have demonstrated in elegant studies[24] that such reactions are best treated by the Extended Selectivity Relationship, in which log rate factors relative to the equivalent benzenoid reaction are plotted against the ρ values appropriate for the reaction and estimated from the benzenoid reaction series. These reactions include not only electrophilic substitution reactions, but also side chain carbonium ion reactions, where the positive charge is conjugated with the aromatic nucleus.[248-250]

A resultant curved plot indicates a variable σ^+ value, and the α- and β-positions of thiophene and the α-positions of furan and selenophene yield good straight lines indicating constant σ^+ values of -0.79, -0.52, -0.93, and -0.88, respectively. This treatment appears to constitute a direct denial of the practical validity of the Yukawa–Tsuno equation [Eq. (5)] in general, despite its theoretical appeal. Less extensive results gave σ^+ values of -0.44, -1.53, and -1.89 for the β-position of furan and the α-position of pyrrole and N-methylpyrrole, respectively. Hydrogen exchange results indicate σ^+ (NMe) to be -1.4 for both the α- and β-positions of N-methylpyrrole,[251,252] somewhat less negative for the α position than the previous result suggests. Also for pyrrole, the carbonyl stretch of the appropriate esters indicates values of σ_α^+ of -2.0 and σ_β^+ of -1.5.[245] A more extensive compilation[247] involving solvolysis of 1-arylethyl chlorides (95% aq. acetone, 45°) gives σ_α^+ (O) -0.85, σ_β^+ (O) -0.44, σ_α^+ (S) -0.76, and σ_β^+ (S) -0.44. Values of σ_α^+ (NH) -1.61 and σ_β^+ (NH) -1.20 were derived from the application of the Yukawa–Tsuno equation [Eq. (5)] to borohydride reduction rates of arylmethyl ketones.[247]

From a series of studies of the reaction of arylsulfonyl chlorides with aniline in methanol at 25°, substituents in the aryl moiety appear to exert an effect registered best by σ^+. The plot incorporating results by both

---

[248] D. S. Noyce, C. A. Lipinski, and G. M. Loudon, *J. Org. Chem.* **35**, 1718 (1970).
[249] E. A. Hill, M. L. Gross, M. Stasiewicz, and M. Manion, *J. Amer. Chem. Soc.* **91**, 7381 (1969).
[250] R. Taylor, *J. Chem. Soc. B*, 1397 (1968).
[251] G. P. Bean, *Chem. Commun.*, 421 (1971).
[252] S. Clementi, P. P. Forsythe, C. D. Johnson, and A. R. Katritzky, *J. Chem. Soc., Perkin Trans. II*, 1675 (1973).

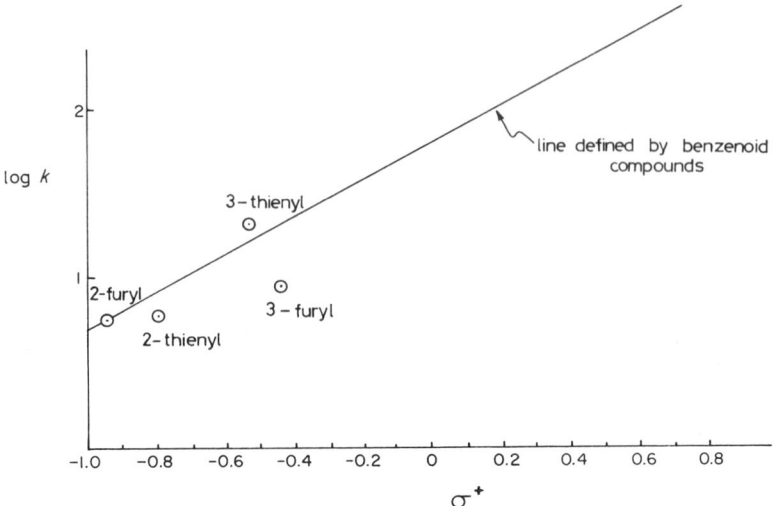

FIG. 7. Log rates of aminolysis of arylsulfonyl chlorides with aniline vs. σ^+.

Rogne[253] and Arcoria and his co-workers[254-256] (Fig. 7) shows only 3-furyl deviating markedly; certainly using σ values for the five-membered rings would predict reaction acceleration compared with phenyl. This approximate correlation indicates that the rate-determining step is addition of the nucleophile (Scheme 4) rather than synchronous attack of amine and displacement of chloride.

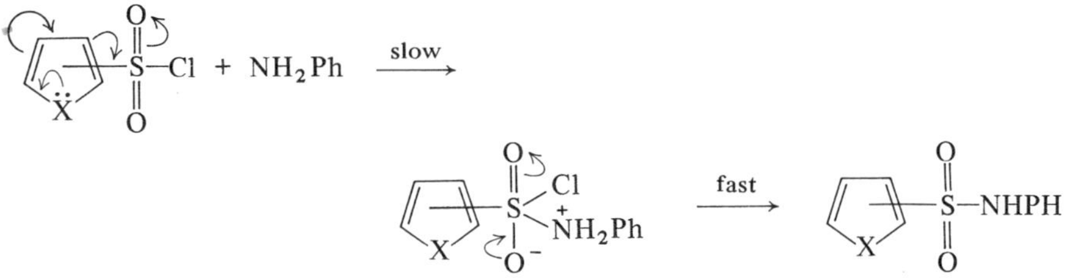

SCHEME 4

Another important study is that of the participation of heterocyclic moieties in the solvolytic rearrangement of β-arylethyl tosylates,[246] in-

---

[253] O. Rogne, *J. Chem. Soc. B*, 1855 (1971).
[254] A. Arcoria, E. Maccarone, G. Musumarra, and G. A. Tomaselli, *J. Org. Chem.* **38**, 2457 (1973).
[255] A. Arcoria, E. Maccarone, and G. A. Tomaselli, *J. Org. Chem.* **39**, 1689 (1974).
[256] A. Arcoria, E. Maccarone, G. Musumarra, and G. A. Tomaselli, *J. Org. Chem.* **39**, 3595 (1974).

volving a transition state (**30**) leading to the symmetrical intermediate **31**.

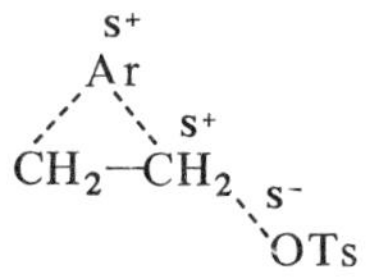

$$
\begin{array}{cc}
\text{(30)} & \text{(31)} \\
\end{array}
$$

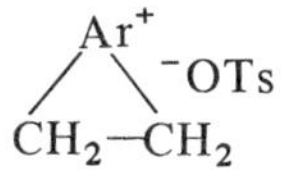

The Yukawa–Tsuno equation [Eq. (5)] appears best to correlate the results, but as commented previously the σ values used for the five-membered rings, 2- and 3-thienyl and 2-furyl, appear hard to rationalize although other workers have also argued for their authenticity.[247]

It is subsequently interesting to enquire the degree of additivity of effects in electrophilic substitution and carbonium ion reactions of substituted five-membered rings, using σ_m^+ for 4-substituents and σ_p^+ for 5-substituents. As reasoned previously, the ρ values for such correlations compared with those for the equivalent reactions in benzene indicate the status of the σ^+ values for the heteroatoms. If the ρ values are the same

TABLE X

COMPARISON OF ρ VALUES FOR BENZENOID AND THIOPHENOID REACTIVITY

Reaction	Reactivity		Reference footnotes
	Benzene	Thiophene	
Bromination	−12.1	−10.0	256, 258, 259
Chlorination	−9.6	−9.4	257
Protodetritiation	−8.2	−7.2	259, 260
Acetylation	−9.1	−5.7	259, 261, 262
Mercuration	−4.0	−5.3	259, 263

or similar, it indicates the absence of proximity interactions between these heteroatoms and the reaction site and the σ^+ value of the heteroatom is constant. Unfortunately, the only nucleus for which sufficient experimental data have been provided is that of thiophene. Here studies[45,257] indicate, for example, that for the chlorination of 2-substituted thiophenes in acetic acid the ρ value is very similar to that of benzenoid compounds. Other values are compared in Table X.

[257] R. N. McDonald and J. M. Richmond, *Chem. Commun.*, 333 (1974).

The difference in ρ values for equivalent reactions is probably not significant, particularly as there are often differences in reaction conditions and interpretations. Thus the kinetic form of some of the thiophene brominations are in doubt,[258] there is a variation in acidic media used for the protodetritiation,[260] and a different catalyst, $SnCl_4$, is used for the thiophene acetylations;[261,262] cf that for the benzenoid reaction, $AlCl_3$. In general, therefore, and in particular for the chlorination reaction,[257] there appears to be a constant σ_α^+ value for thiophene with additivity of substituents effects, but there is certainly room in this area for further experimentation and debate.

For the trifluoroacetylation of 2-substituted thiophenes, furans, and pyrroles in $C_2H_4Cl_2$, 75°, the ρ values are -7.4, -10.3, and ca. -4.5, respectively.[259] The value for substituted benzenes is not known. In the gas phase ionization of substituted furans, thiophenes, selenophenes, and pyrroles,[264] a reaction proceeding through a positively charged molecular ion taken to be analogous to the Wheland intermediate for electrophilic substitution, the ρ values are reported to be -20.2, -16.5, ~ -16.5, and 18.2, respectively. The ρ value for the benzenoid reaction is reported as -14.7 "or larger." The authors' conclusion, that equating such differences as exist between these values with relative ring aromaticity must be viewed with caution, is a very valid one.

The rates of solvolysis of 4- and 5-substituted 1-(2-thienyl)ethyl p-nitrobenzoates in 80% EtOH at 25° are correlated by a ρ value of -6.79,[265] to be compared with -5.7 for the benzenoid reaction. Using σ^+ values calculated by the Dewar–Grisdale approach [Eq. (2)] incorporating q parameters derived from CNDO/2 or INDO molecular orbital procedures (which give closely similar results) gives a value of ρ of -7.14. For the equivalent 4-substituted 1-(2-furyl)ethyl systems,[266,267] however, the correlation with σ_m^+ is poor, ρ being -8.5. In this case Eq. (2) yields σ^+ values providing a close fit to experimental determinations with a ρ value that is appropriate for the benzenoid reaction.

The general rationale behind the use of the Dewar–Grisdale approach[268,269] and the best methods to employ in the calculation of the

[258] A. R. Butler and J. B. Hendry, *J. Chem. Soc. B*, 848 (1970).
[259] S. Clementi and G. Marino, *J. Chem. Soc., Perkin Trans. II*, 71 (1972).
[260] A. R. Butler and C. Eaborn, *J. Chem. Soc. B*, 370 (1968).
[261] P. Linda and G. Marino, *Tetrahedron* **23**, 1739 (1967).
[262] S. Clementi, P. Linda, and M. Vergoni, *Tetrahedron* **27**, 4667 (1971).
[263] R. Motoyama, J. Ogaiwa, and E. Imoto, *Nippon Kagaku Zasshi* **78**, 962 (1957).
[264] P. Linda, G. Marino, and S. Pignataro, *J. Chem. Soc. B*, 1585 (1971).
[265] D. S. Noyce, C. A. Lipinski, and R. W. Nichols, *J. Org. Chem.* **37**, 2615 (1972).
[266] D. S. Noyce and H. J. Pavez, *J. Org. Chem.* **37**, 2620 (1972).
[267] D. S. Noyce and H. J. Pavez, *J. Org. Chem.* **37**, 2623 (1972).
[268] D. S. Noyce and R. W. Nichols, *Tetrahedron Lett.*, 3889 (1972).
[269] D. A. Forsyth and D. S. Noyce, *Tetrahedron Lett.*, 3893 (1972).

constituent resonance and inductive parameters have been carefully discussed.[270] This approach is clearly essential for substituent effects in annelated pyridines as discussed previously, and also in benzofurans, benzothiophenes, and indoles as discussed later, but as the work on furan[266,267] suggests the differences in geometry and conjugation effects between benzenoid and five-membered heteroaromatic systems makes the Dewar–Grisdale approach relevant, and spotlights the inaccuracies, not large but sometimes significant, in simply assuming that one system is generated from the other by replacement of —CH=CH— with a heteroatom X.

TABLE XI

NUCLEOPHILIC REACTIVITY IN BROMOTHIOPHENES

Series	Conditions	ρ	Type	Reference footnotes
32	Sodium thiophenoxide (EtOH, 25°)	3.21	σ_p^-	273–275
33	Sodium thiophenoxide (MeOH, 20°)	4.51	σ_m	276
34	Sodium thiophenoxide (MeOH, 20°)	3.96	σ_p	277
35	Sodium thiophenoxide (MeOH, 20°)	8.18	σ_p^-	277
32	Piperidine (EtOH, 20°)	3.21	σ_p^-	278
32	Piperidine (MeOH, 20°)	3.18	σ_p^-	279
36	Piperidine (MeOH, 20°)	4.02	σ_p^-	280
37	Piperidine (MeOH, 20°)	3.24	σ_p^-	279

The reactivity of these five-membered systems with nucleophiles has received note. Zatsepina and Tupitsin and their co-workers recorded the use of deuterium exchange rates in methyl groups of five-membered heterocycles in ethanolic KOEt as a source of σ^- values, with a ρ value of 4.7.[271] Spinelli and his co-workers revealed that even for tri- and tetrasubstituted five-membered heterocycles the Hammett equation employing σ^- values where appropriate gives good correlations. The systems studied are **32–37**, in all cases the slow step

[270] D. A. Forsyth, *J. Amer. Chem. Soc.* **95**, 3594 (1973).
[271] N. N. Zatsepina, I. F. Tupitsin, Y. L. Kaminskii, and N. S. Kolodina, *Reakts. Sposobnost. Org. Soedin.* **6**, 766 (1969).

being taken as the formation of the Meisenheimer complex.[272] The results obtained are reported in Table XI.

(35)   (36)   (37)

The very large ρ value for **35** obtained using σ_p^- reveals the strong conjugation between positions 2 and 3, which is not present between 3 and 4 (**34**) yielding a σ_p correlation, and which is much reduced between 2 and 5 (**32**). These results also show that ortho-steric interactions in five-membered rings are generally much lower than for six-membered rings, a point stressed in other work.[275,281–283] This is illustrated in Fig. 8, where the plots for **32** and **37** show that in the latter no twisting of the nitro group by the adjacent methyl occurs. The degree of correlation is not so good as claimed in the original work,[279] where the origin of some of the σ_p^- values used is obscure, although there is a very high degree of consistency of such values within this general series of reactions.

The successful application of the Hammett equation to methanolysis of 2-chloro-5-X-methylselenophenes,[284] and debromination by sodium thiophenoxide of 2-bromo-5-X-3-nitroselenophenes (EtOH, 20°)[275] is also reported. In the latter case ρ is 3.15; cf. 3.21 for the equivalent thiophene reaction.

Additivity of effects is not observed in homolytic reactivity, however, where the phenylation of 2- and 3-substituted thiophenes reveals only a

[272] G. Guanti, C. Dell'Erba, and P. Macera, *J. Heterocycl. Chem.* **8**, 537 (1971).
[273] C. Dell'Erba and D. Spinelli, *Tetrahedron* **21**, 1061 (1965).
[274] L. Chierici, C. Dell'Erba, A. Guareschi, and D. Spinelli, *Ann. Chim.* (*Rome*) **52**, 632 (1967).
[275] C. Dell'Erba, A. Guareschi, and D. Spinelli, *J. Heterocycl. Chim.* **4**, 438 (1967).
[276] G. Guanti, C. Dell'Erba, and D. Spinelli, *J. Heterocycl. Chem.* **7**, 1333 (1970).
[277] D. Spinelli, G. Guanti, and C. Dell'Erba, *J. Chem. Soc., Perkin Trans. II*, 441 (1972).
[278] C. Dell'Erba and D. Spinelli, *Tetrahedron* **21**, 1061 (1965).
[279] D. Spinelli, G. Consiglio, and A. Corrao, *J. Chem. Soc., Perkin Trans. II*, 1866 (1972).
[280] D. Spinelli, D. Consiglio, R. Noto, and A. Corrao, *J. Chem. Soc., Perkin Trans. II*, 620 (1975).
[281] D. Spinelli, G. Consiglio, R. Noto, and A. Corrao, *J. Chem. Soc., Perkin Trans. II*, 1632 (1974).
[282] S. Clementi, P. Linda, and M. Vergoni, *Tetrahedron Lett.*, 611 (1971).
[283] R. G. Gallo, M. Chanon, H. Lund, and J. Metzger, *Tetrahedron Lett.*, 3857 (1972).
[284] Y. K. Yurev, M. A. Galbershtam, and A. F. Prokofeva, *Izv. Vysshikh Uchebn. Zaved. Khim. Technol.* **8**, 421 (1965) [*CA* **63**, 16151 (1965)].

very small effect of structure on reactivity due to through conjugation between heteroatom and substituents.[285,286]

Substituent effects in the dye-sensitized (Rose Bengal or methylene blue) singlet oxygen photooxidation of a wide variety of furans have been investigated.[287] Solvents effects on the value of ρ, which became less negative as polarity of the medium was increased was indicative of

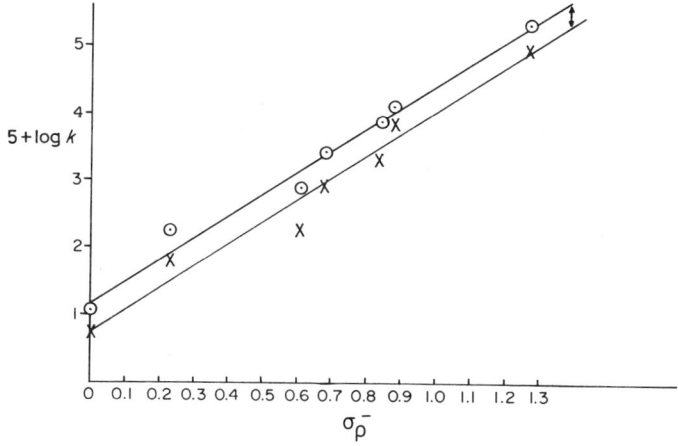

FIG. 8. Log rates of nucleophilic displacement from bromothiophenes by thiophenoxide. ⊙, 32; ×, 37. ↕ : $\sigma_{m-Me} = 0.29/3.21 = 0.09$.

polar character in the transition state leading to adduct **38** or the formation of a charge transfer complex (**39**).

(38)   (39)

Also in the furan series Stradins and his co-workers have demonstrated[288–290] that ρ for the correlation of polarographic reduction

---

[285] C. M. Camaggi, G. De Luca, and A. Tundo, *J. Chem. Soc., Perkin Trans. II*, 412 (1972).

[286] C. M. Camaggi, G. De Luca, and A. Tundo, *J. Chem. Soc., Perkin Trans. II*, 1594 (1972).

[287] R. H. Young, R. L. Martin, N. Chinh, C. Mallon, and R. H. Kayser, *Can. J. Chem.* **50**, 932 (1970).

[288] J. P. Stradins, S. Hillers, and Y. K. Yurev, *Dokl. Akad. Nauk. SSSR* **129**, 816 (1963).

[289] J. P. Stradins and S. Hillers, *Nitro Compounds, Proc. Int. Symp., 1963*, 409 (1964).

[290] J. P. Stradins and G. O. Reikhmanis, *Elektrokhimya* **3**, 178 (1967).

potentials for 5-substituted 2-nitro compounds in aqueous ethanol mixtures depends on pH, being given by (0.11 + 0.02 pH). Correlations of polarography of substituted bromothiophenes[291] and of other thiophene[292] and selenophene[293] series, the latter in solutions of variable pH, are also available.

Dipole moments in 5-substituted furfurals follow:

$$\mu = 3.60 - 0.76\, \sigma \tag{21}$$

σ^+ or σ^- being utilized for σ where substituent resonance donation or acceptance is possible,[293] while correlations have also been made with dipole moments of furan analogs of chalcones.[294]

Electron spin resonance (ESR) spectroscopy reveals that an increase in the electron-donating ability of substituents in the 2-position of the furan ring produces an increase in the spin density at the 3-position and a decrease at the 4-position; the coupling constants in the ESR spectra of anion radicals generated by electrochemical reduction of 2-substituted 5-nitrofurans equate to σ_p values.[295]

Katritzky and Topsom and co-workers have extended their studies of IR intensities of ring stretching bands[55,56] to include 2-substituted furans and thiophenes and demonstrate a correlation with σ_R^0.[296] The marked contrast between σ and σ^+ values for α and β NH, S, and O (see Table IX and pp. 38 and 39), the former values indicating the strong electron withdrawal of these heteroatoms, and the latter their high resonance polarizability, have been emphasized by IR intensity studies on 2- and 3-substituted cyanopyrroles, -thiophenes, and -furans.[297] The intensities of CH stretching vibrations of substituted thiophenes and furans are parabolically dependent on σ_I constants,[298] as noted previously for the six-membered congeners.[57,86,87] The relevant equation is

$$\text{Intensity} = 0.196\, \Sigma\sigma_I^2 - 0.283\, \Sigma\sigma_I + 0.117 \tag{22}$$

whence σ_I values of 0.9, 0.4, 0.7, and 0.6 are calculated for α and β S

---

[291] M. Person and R. Mora, *Bull. Soc. Chim. Fr.* 521 (1973).

[292] F. M. Stoyanovich, S. G. Mairanovskii, Y. L. Goldfarb, and I. A. Dyachenko, *Izv. Akad. Nauk. SSSR, Ser. Khim.*, 1439 (1971).

[293] M. G. Koghan, V. S. Pustovarov, Y. V. Kolodyazhnyi, Z. N. Nazarova, and O. A. Osipov. *Zh. Org. Khim.* **4**, 2216 (1968).

[294] S. V. Tsukerman and V. M. Nikitchenko, *Khim. Geterotsikl. Soedin.*, 586 (1972).

[295] R. A. Gavars, J. P. Stradins, and S. A. Hillers, *Dokl. Akad. Nauk. SSSR* **157**, 1424 (1964).

[296] J. M. Angelelli, A. R. Katritzky, R. F. Pinzelli, and R. D. Topsom, *Tetrahedron* **28**, 2037 (1972).

[297] L. W. Deady, R. A. Shanks, and R. D. Topsom, *Tetrahedron Lett.*, 1881 (1973).

[298] I. F. Tupitsin, N. N. Zatsepina, N. S. Kolodina, and Y. L. Kaminskii. *Reakts. Sposobnost Org. Soedin.* **6**, 548 (1969).

and O, respectively. Carbonyl and NH stretching frequencies in substituted 2- and 4-carboethoxypyrroles can be correlated with σ_m or σ_p,[299] while means of symmetrical and asymmetrical C=O stretches for 2-(5-phenyl-2-furfurylidene)-1,3-indandiones (**40**) measured in both $CHCl_3$ and CCl_4 plot linearly with σ^+.[300, 301]

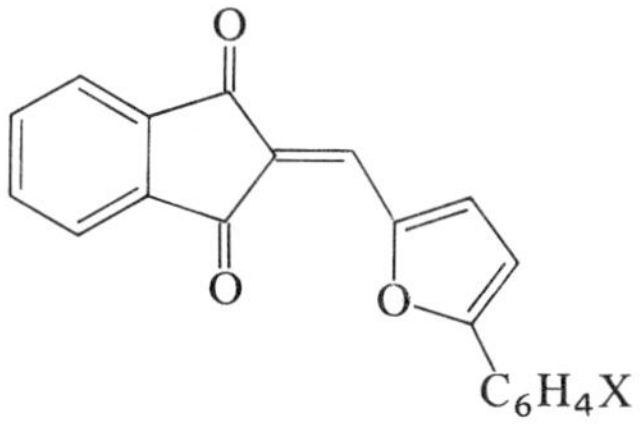

(40)

NMR studies have been reported on thiophenes, which coupled with MO calculations[302,303] indicate that effects of substituents are transmitted through the hydrocarbon portion of the molecule rather than via the heteroatom, an interesting study which thereby justifies the use of σ_m

TABLE XII

σ Values for Five-Membered Heteroaromatic Rings as Substitutes

2-Furyl	2-Thienyl	3-Thienyl	Type		Reference footnotes
0.06	0.09	0.03	σ_m	Ionization of benzoic acids (aq. EtOH, 25°)	304, 305
0.02	0.05	−0.02	σ_p		
0.11	0.11	0.07	σ_m^-	Ionization of phenols (aq. EtOH, 25°)	304, 305
0.21	0.19	0.13	σ_p^-		
—	0.16	0.08	σ_m^+	Solvolysis of phenyldimethyl carbinyl chlorides (aq. acetone 25°)	306
—	−0.43	−0.38	σ_p^+		
0.10	0.15	—	σ_m^+	Solvolysis of 1-arylethyl acetates (aq. EtOH, 80°, 100°, or 120°)	305, 307
−0.39	−0.33	—	σ_p^-		
0.10	0.15	—	σ_m^+	IR carbonyl stretching frequencies of substituted acetophenones (CCl_4)	307
−0.45	−0.38	—	σ_p^+		

[299] L. V. Kazanskaya, T. A. Melenteva, V. M. Berezovskii, *Zh. Obshch. Khim.* **38**, 2020 (1968).
[300] A. Perjéssy, P. Hrnčiar, and A. Krutošiková, *Tetrahedron* **28**, 1025 (1972).
[301] B. Kamienski and T. M. Krygowski, *Tetrahedron Lett.*, in press.
[302] W. Kemula and T. M. Krygowski, *Tetrahedron Lett.*, 5135 (1968).
[303] W. Kemula and T. M. Krygowski, *Bull. Acad. Pol. Sci., Ser. Sci. Chim.* **15**, 479 (1967).

and σ_p values to approximate substituent effects on reacting side chains in such systems.

Italian workers have reported extensive studies on the evaluation of σ values for thienyl and furyl groups. These are collected in Table XII. The σ_m, σ_m^+, and σ_m^- values reveal the electron-withdrawing inductive effect of the rings, while as expected from the previous discussion of this section, these substituents fall into that small category whose σ_p, σ_p^+, and σ_p^- values are all different, revealing an ability to both accept and donate electrons by resonance under the appropriate incentives.

B. Containing Two or More Heteroatoms

Not a great deal of work has been done in this area, but what has been done shows that the Hammett equation can be applied with some degree of success to five-membered rings containing several heteroatoms, often together with other substituents and/or reacting side chains. This illustrates the efficacy of the equation and suggests that steric and proximity effects in such systems frequently remain constant or are smaller than expected.

A detailed analysis of substituent effects on the pK_a values of imidazoles and tetrazoles as well as benzimidazoles and naphthimidazoles has been made.[308] The ortho effect is shown to parallel that of 2-substituted pyridines and quinolines[47,190] and application of the Hammett equation to the tautomerism of these systems is also considered. The equation also satisfies the effect of substituents on the basicity of the nitrogen in the 1-position for 1-pyrazolines.[309]

The pK_a's of substituted 2-, 4-, and 5-substituted carboxythiazoles and rates of basic hydrolysis of substituted ethyl 2-, 4-, and 5-thiazolocarboxylates, in water and aqueous ethanol, correlate approximately with the Hammett equation,[310,311] but the ρ values are much different to their benzenoid counterparts demonstrating the presence of proximity effects such as hydrogen bonding, as well as possibly tautomeric equilibria. Other correlations of acid-base properties in this area are for

---

[304] F. Fringuelli, G. Marino, and A. Taticchi, *J. Chem. Soc. B*, 1595 (1970).
[305] F. Fringuelli, G. Marino, and A. Taticchi, *J. Chem. Soc. B*, 2304 (1971).
[306] F. Fringuelli, G. Marino, and A. Taticchi, *J. Chem. Soc. B*, 2302 (1971).
[307] F. Fringuelli, G. Marino, and A. Taticchi, *J. Chem. Soc., Perkin Trans. II*, 158 (1972).
[308] M. Charton, *J. Org. Chem.* **30**, 3346 (1965).
[309] J. Elquero, E. Gonzalez, and R. Jacquier, *Bull. Soc. Chim. Fr.*, 2054 (1969).
[310] Y. Otsuji, T. Kimura, Y. Sugimoto, E. Imoto, Y. Omori, and T. Okawara, *Nippon Kagaku Zasshi* **80**, 1024 (1959) [*CA* **55**, 5467 (1961)].
[311] A. Benko, I. Zsako, and P. Nagy, *Chem. Ber.* **100**, 2178 (1967).

2-arylimidazoles,[312,313] where an equation of the form of Eq. (3) using σ_R^+ has been used for interpretation,[312] 5-substituted 3-bromo-1,2,4-triazoles,[314] and 3-nitro-1,2,4-triazoles,[315] 3-substituted sydnonimines,[316–318] 5-benzylidene hydatoins,[318] 2-aminothiazoles and 2-aminooxazoles,[319] 1,2-diaryl-2-imidazolines,[320] and 1,5-disubstituted tetrazoles.[321] Enthalpies of ionization[322] of alkyl-substituted thiazoles have been correlated with σ^*.

The correlation of the ratio of solvolysis of substituted 1-(5-thiazolyl)ethyl chlorides (41) in 80% ethanol at 25° with σ_p^+ is of such high precision that it permits a determination of -0.34 for σ_p^+ for the coplanar phenyl substituent,[323] reflecting the smaller steric interactions in these five-membered ring systems already commented on (see Section III, A). The ρ value here is 6.14 (6.2 at 45°)[324] close to 5.82 and 6.05,[269] the values for the substituted benzene and pyridine systems, respectively, under identical conditions, revealing that the correlation given in Fig. 1 for six-membered ring systems can be extended to include these five-membered congeners. Solvolysis data for the three isomeric 1-thiazolyl, -0.01 for 4-thiazolyl, and $+0.26$ for 2-thiazolyl system are taken[325] to reflect the instability of the canonical form of the latter where electron deficiency is placed directly on N (see 42 and 43).

The rates of solvolyses of 1-(2-substituted 4-thiazolyl) ethyl chlorides (44) do not, however, correlate with σ_m.[326] There is nevertheless a correlation with reactivities for 6-substituted 2-pyridyl[68] systems, which also contain the reaction site and substituents flanking the N. Analysis of the results in terms of the multiparameter Eqs. (1) and (3) reveals the predominance of resonance effects. The same conclusions apply to the

---

[312] V. A. Bren, V. I. Minkin, A. D. Garnovsky, E. V. Botkina, and B. S. Tenaysechuk. *Reakts. Sposobnost Org. Soedin.* 5, 651 (1968).
[313] N. Blazevic, F. Kajfez, and V. Sunjic, *J. Heterocycl. Chem.* 7, 227 (1970).
[314] W. Freiberg and C. F. Kroeger, *Chimia* 21, 159 (1967).
[315] L. I. Baghal and M. S. Pevzner. *Khim. Geterotsikl. Soedin.* 558 (1970).
[316] V. G. Yashunskii, O. I. Samoilova, and L. E. Kholodov, *Zh. Obsch. Khim.* 34, 2050 (1964).
[317] L. E. Kholodov and V. G. Yashunskii, *Khim. Geterotsikl. Soedin.* 1, 328 (1965).
[318] B. A. Ivin, G. V. Rutkovskii, V. A. Kirillova, and E. C. Sochilin, *Reakts. Sposobnost Org. Soedin.* 6, 1055 (1969).
[319] U. Strauss, H. P. Haerter, and O. Schindler, *Chimia* 27, 99 (1973).
[320] B. Fernandez, I. Perillo, and S. Lamdan, *J. Chem. Soc., Perkin. Trans. II*, 1371 (1973).
[321] N. D. Agibalov, A. S. Enin, G. I. Koldobskii, B. V. Gidaspov, and T. N. Tomofeeva, *Zh. Org. Khim.* 8, 2414 (1972).
[322] P. Gourstot and I. Wadso, *Acta Chem. Scand.* 20, 1314 (1966).
[323] D. S. Noyce and S. A. Fike, *J. Org. Chem.* 38, 2433 (1973).
[324] D. S. Noyce and S. A. Fike, *J. Org. Chem.* 38, 3318 (1973).
[325] D. S. Noyce and S. A. Fike, *J. Org Chem.* 38, 3316 (1973).
[326] D. S. Noyce and S. A. Fike, *J. Org. Chem.* 38, 3321 (1973).

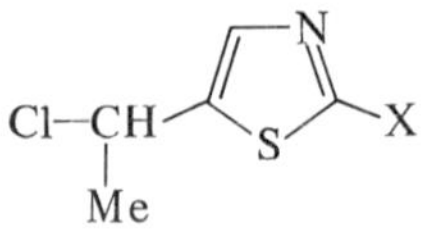

4-substituted 1-(2-thiazolyl) ethyl chlorides (**45**), as well as to the rates of solvolysis of 4-substituted 1-(1-methylimidazolyl) ethyl *p*-nitrobenzoates (**46**),[327] which solvolyzed much more rapidly than predicted by σ^+ and the ρ value of -5.6 established for the 5-substituted series, and from substituted benzenoid compounds. Values of -0.82, -1.01 and -1.02 have been calculated for the σ^+ values of 1-methyl-2-imidazoyl, 1-methyl-4-imidazoyl, and 1-methyl-5-imidazoyl systems, respectively.

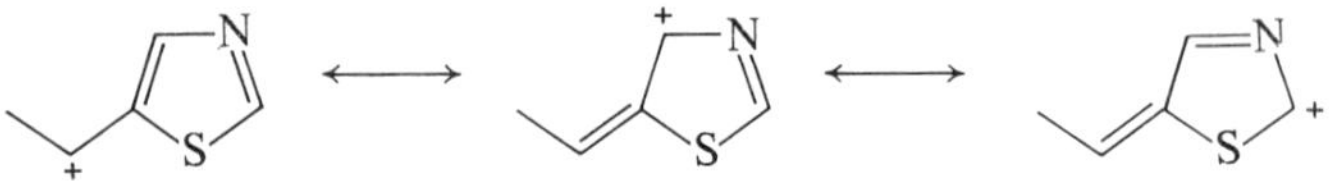

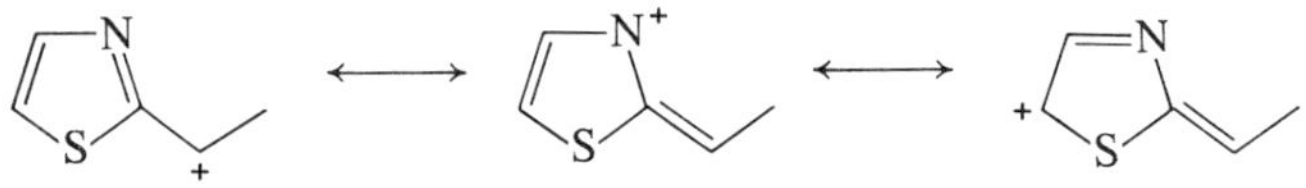

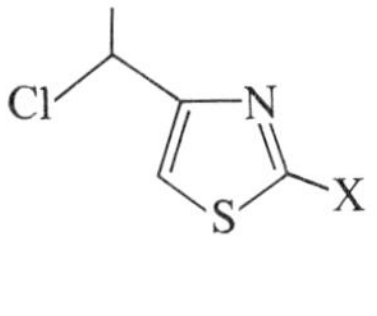

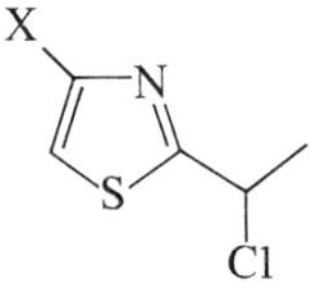

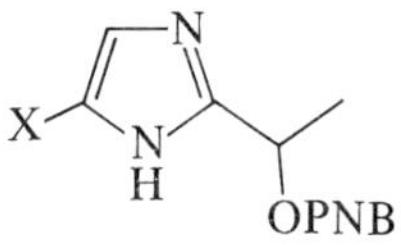

Extensive studies on hydrogen exchange in aqueous sulfuric acid on 1-methylpyrazole,[252] isoxazole,[328] and isothiazole[328] and their methylated derivatives[252,328] have been reported, but calculated partial rate factors derived from treatment of the form of Eq. (12) do not tally with those observed. This emphasizes the contribution from interactions between donor and acceptor groups or solvent effects which also lead to breakdown in the interpretation of nitration results.[83]

The hydrolysis of 4-substituted 1-acetylpyrazoles can be correlated with σ_p values;[329] the fact that the reacting group is attached to the

---

[327] D. S. Noyce and G. T. Stowe, *J. Org. Chem.* **38**, 3762 (1973).
[328] S. Clementi, P. P. Forsythe, C. D. Johnson, A. R. Katritzky, and B. Terem, *J. Chem. Soc. Perkin Trans. II*, 399 (1974).
[329] R. Hüttel and I. Kratzer, *Chem. Ber.* **92**, 2014 (1959).

heteroatom does not appear to complicate the transmission of substituent effects through the ring.

The thermodynamics and kinetics of conversion of dyes with pyrazolone or antipyrine rings into carbinol compounds have been extensively studied. For compounds of the type **47**, log K and log k correlate best with the Yukawa–Tsuno equation [Eq. (5)], with $r = 1.10$ and 2.05 using σ^+,[330,331] indicating in contrast to the correlation for triphenyl carbinols where ring twisting occurs[332] that conjugation is about as efficient as in the defining system for σ^+. This does serve again to emphasize the reduced steric effects in the five-membered ring

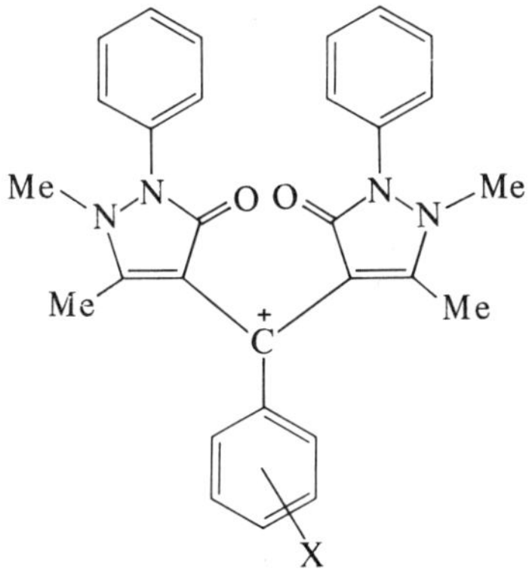

(47) 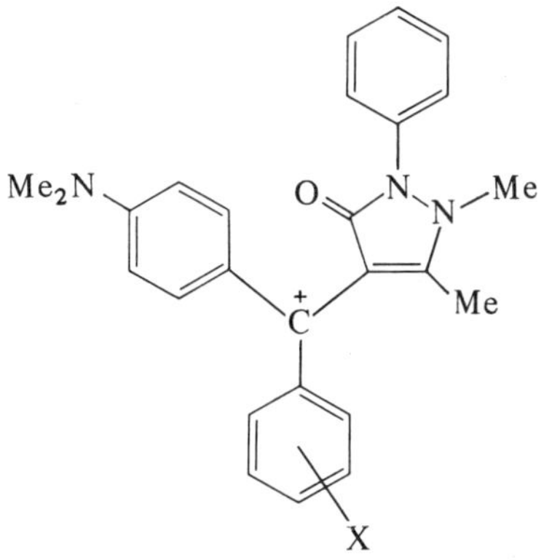 (48)

system.[275,281-283,323] The dependence of the ρ parameter for the rate correlation on temperature is given by

$$\rho = 3.65\,(1-223/T) \qquad (23)$$

233°K being the isokinetic temperature where $\rho = 0$.[333]

Similar conclusions have been reached for the triarylmethane monoantipyrine dyes (**48**) where here r in the Yukawa–Tsuno equation is 1.50[334] and the isokinetic temperature is 740°K.[335]

[330] O. F. Ginzburg, E. I. Kvyat, and G. S. Idlis, *Zh. Obshch. Khim.* **32**, 2633 (1962).
[331] V. V. Sinov, O. F. Ginzburg, and E. I. Kvyat. *Reakts. Sposobnost Org. Soedin.* **2**, 84 (1965).
[332] M. J. Cook, N. L. Dassanayake, C. D. Johnson, A. R. Katritzky, and T. W. Toone, *J. Amer. Chem. Soc.* **97**, 760 (1975).
[333] V. V. Sinov, O. F. Ginzburg, and E. I. Kvyat, *Reakts. Sposobnost Org. Soedin.* **2**, 90 (1965).
[334] V. V. Sinov and E. P. Shepel. *Reakts. Sposobnost Org. Soedin.* **7**, 43 (1970).
[335] V. V. Sinov, E. P. Shepel and O. F. Ginzburg, *Reakts. Sposobnost Org. Soedin.* **7**, 119 (1970).

These multiheteroatom systems are also highly susceptible to nucleophilic attack and to ring-opening reactions. Nucleophilic substitution of 1-methyl-3-nitro-5-halotriazoles by substituted amines in aqueous ethanol has been investigated.[336] The acid hydrolysis of *m*- and *p*-substituted 1-aryl-3-methyl-5-pyrazolones[337] yields a σ correlation with a negative ρ value of -1.11 indicating that substituent effects on the catalyzing protonation equilibrium are the important ones; alternatively, the ρ values for both acid and base hydrolysis of *N*-acylimidazoles,[338,339] sydnonimine derivatives,[116,340] and 2-amino-3-aryl-4-oxathiazolidines[341] **49** → **50** are positive, indicating the influence of substituents on the nucleophilic ring attack.

(49) → (50)

The thermal rearrangement of 2,4,4-triaryl-5-methylthio-4*H*-imidazoles[342] to yield 2,4,5-triarylimidazoles gave no correlation between substituents in the migrating aryl-group and σ^+, but that of 5,5-diaryl-2-phenylimino-Δ^3-1,3,4-oxadiazolines[343] to diaryldiazomethane and phenylisocyanate via a retro-1,3-dipolar cycloaddition did relate to the Yukawa–Tsuno equation [Eq. (5)] with $r = 0.55$, and a Hammett correlation was afforded by the rates of dissociation of dimers of triarylimidazolyl radicals.[344] Hegarty and Scott have employed the Hammett equation extensively for mechanistic studies within this area: bromination of 5-(arylmethylenehydrazino)tetrazoles in aqueous acetic acid,[345] cyclization of *N*-imidazolidin-2-ylidenehydrazonyl chlorides (**51**) to 6,7-

---

[336] N. N. Melnikova, M. S. Pevzner, and L. I. Baghal, *Reakts. Sposobnost Org. Soedin.* **9**, 563 (1973).
[337] T. Y. Chepurnaya and O. F. Ginzburg, *Reakts. Sposobnost Org. Soedin.* **2**, 38 (1965).
[338] M. Caplow and W. P. Jencks, *Biochemistry* **1**, 883 (1962).
[339] J. A. Fee and T. H. Fife. *J. Org. Chem.* **31**, 2343 (1966).
[340] L. E. Kholodov and V. G. Yashunskii, *Reakts. Sposobnost Org. Soedin.* **1**, 77 (1964).
[341] Y. V. Svetkin, S. A. Vasileva, and A. N. Minlibaeva, *Reakts. Sposobnost Org. Soedin.* **4**, 705 (1967).
[342] J. Nyitrai, K. Lempert, and T. Cserfalvi, *Chem. Ber.* **107**, 1645 (1974).
[343] P. R. West and J. Warkentin, *J. Org. Chem.* **34**, 3233 (1969).
[344] B. S. Tanaseychuk and L. G. Rezepova, *Zh. Org. Khim.* **6**, 1065 (1970).
[345] J. C. Tobin, A. F. Hegarty, and F. L. Scott, *J. Chem. Soc. B*, 2198 (1971).

dihydro-3-aryl-5H-imidazo[2,1-c]-s-triazoles (**52**) in aq dioxane,[346] and hydrolysis of 1-(N'-aroylamidino)-3,5-dimethylpyrazoles.[347]

$$XC_6H_4\overset{Cl}{\underset{}{C}}=N-N=\underset{H}{\overset{H}{\underset{N}{\overset{N}{\diagup}}}} \qquad XC_6H_4-\underset{}{\overset{N-N}{\underset{N}{\diagup}}}\underset{}{\overset{}{\diagdown}}NH$$

(51)   (52)

Zuman[348-350] and others[316,351] have conducted extensive studies on the polarographic reduction of sydnone derivatives. Substituted aryl rings in the 3-position have only a small effect, correlated by σ. Others have examined correlations in the case of polarographic reduction of 1,2-dimethyl-3-aryl pyrazolium salts,[352] and polarographic oxidation of triarylimidazoles,[353] as part of a general study of substituent effects in triarylimidazolyl radicals.[354]

IR studies of the carbonyl stretching band of 3-methyl-4-arylazoisoxazol-5-ones reveal a positive ρ value for p-substituents in the aryl nucleus, but a negative value for m-substituents.[355] Other azo compounds investigated are hydroxyphenylazopyrazolones, for which λ_{max} in the visible spectra correlates with σ for phenyl substituents, as does ν_{OH} and $\nu_{C=O}$ in the IR spectra,[356] and 1-phenyl-3-methyl-4-arylazopyrazol-5-ones for which $\nu_{C=O}$ also corresponds to σ.[357]

A good linear relationship with σ constants has been found for the relative pulse height as well as for the fluorescent intensity of a series of

[346] A. F. Hegarty, J. O'Driscol, J. K. O'Halloran and F. L. Scott, *J. Chem. Soc. B*, 1887 (1972).
[347] A. F. Hegarty, C. N. Hegarty, and F. L. Scott, *J. Chem. Soc., Perkin Trans. II*, 2054 (1973).
[348] P. Zuman, *Collect. Czech. Chem. Commun* **27**, 630 (1962).
[349] P. Zuman, *Z. Phys. Chem.*, 246 (1968).
[350] P. Zuman, *Collect. Czech. Chem. Commun.* (in press).
[351] L. E. Kholodov and V. G. Yashunskii, *Reakts. Sposobnost Org. Soedin.* **1**, 77 (1964).
[352] Z. N. Timofeeva, N. M. Omar, L. S. Tikhonova, and A. V. Eltsov, *Zh. Obshch. Khim.* **40**, 2072 (1970).
[353] B. S. Tenaseychuk, L. G. Tikhonova, and A. A. Bardina, *Khim. Geterotsikl. Soedin.*, 387 (1973).
[354] B. S. Tenaseychuk, L. G. Tikhonova, and A. P. Dydykina, *Zh. Org. Khim.* **9**, 1273 (1973).
[355] G. Cum, G. Lo Vecchio, M. C. Aversa, and M. Crisafulli, *Gazz. Chim. Ital.* **97**, 346 (1967).
[356] S. Toda and Y. Kojima, *Nippon Kagaku Zasshi* **88**, 18 (1967) [*CA* **66**, 120417a (1967).].
[357] S. Toda, *Nippon Kagaku Zasshi* **80**, 402 (1959) [*CA* **55**, 4150 (1961)].

1,3,5-triphenyl-2-pyrazolines substituted in the 3-phenyl ring,[358] but the linear shift of the fluorescence emission wavelength corresponds with σ^+. Chemiluminescent intensities of derivatives of imidazoles, lophines, bearing substituents in the 2-phenyl group show a σ relationship over a wide series.[359] This involves the oxidation of the lophine anion to the dimer 1,1'-bi(2,4,5-triphenylimidazolyl) which dissociates into radicals and then forms a peroxide with O_2, which in turn decomposes with emission of light.

The 4-proton chemical shifts in the NMR spectra of 1,3,5-trisubstituted pyrazoles correlate with $\Sigma\sigma$, indicating that such effects are electronic rather than anisotropic,[360] while correlations for the PMR spectra of antiparasitic 2-(substituted styryl)-5-nitro-N-substituted imidazoles have also received attention.[361]

UV measurements and their correlation have been investigated for 3-pyrazolindiones,[362] γ-arylazopentamethine cyanine dyes,[363] and nitro derivatives of 1,2,4-triazoles.[364]

IV. Fused Five- and Six-Membered Rings

In Table XIII are gathered effective σ values of various types for aryl groups as a whole, in which the heteroatom is situated in the five-membered moiety. For benzoxazolyl, benzothiazolyl, benzoselenazolyl, and N-methylbenzimidazolyl functions the alkaline deuterodeprotonation of the 2-methyl group[271] gives values of σ_2^- of 1.7, 1.6, 1.8, and 1.5, respectively; the IR measurements[271] give σ_2^0 values of 1.4, 1.3, 1.2, and 0.7, respectively; and the PMR measurements[271] give σ_2^0 values of 1.2, 1.6, 1.9, and 0.5, respectively. Comparison of these values and those in Table XIII with those for the five-membered heterocycles alone (Table IX and Section III, A) reveals that the benzo substituent slightly reduces

---

[358] S. R. Sandler and K. C. Tsou, *J. Chem. Phys.* **39**, 1062 (1963).
[359] G. E. Philbrook and M. A. Maxwell, *Tetrahedron Lett.*, 1111 (1964).
[360] L. G. Tensmeyer and C. Ainsworth, *J. Org. Chem.* **31**, 1878 (1966).
[361] A. F. Cockerill, D. M. Rackham, and N. C. Franklin, *J. Chem. Soc., Perkin Trans. II*, 509 (1973).
[362] V. V. Zaitsev, S. P. Kozhevnikov, and S. J. Gaft, *Zh. Obshch. Khim.* **39**, 1835 (1969).
[363] C. Reichardt and W. Grahn, *Chem. Ber.* **103**, 1072 (1970).
[364] L. I. Baghal and M. S. Pevzner, *Khim. Geterotsikl. Soedin.* **7**, 272 (1971).
[365] D. S. Noyce and F. A. Forsyth, *J. Org. Chem.* **39**, 2828 (1974).
[366] S. Clementi, P. Linda, and C. D. Johnson, *J. Chem. Soc., Perkin Trans. II*, 1250 (1973).
[367] G. G. Smith and J. A. Kirby, *J. Heterocycl. Chem.* **8**, 1101 (1971).
[368] C. Eaborn and J. A. Sperry, *J. Chem. Soc.*, 4921 (1961).
[369] S. Clementi, P. Linda, and G. Marino, *J. Chem. Soc. B* 1153 (1970).
[370] S. Clementi, P. Linda, and G. Marino, *J. Chem. Soc. B*, 79 (1971).
[371] R. Baker, C. Eaborn, and R. Taylor, *J. Chem. Soc., Perkin Trans. II*, 97 (1972).

TABLE XIII
σ Constants for Five-Membered Heterocycles Fused to Benzo Substituents

Aryl group	σ	Type	Reaction[a]	Reference footnotes
Benzofuryl	0.8	σ_2^-	Deuterodeprotonation of endocyclic CH_3 in EtOK/EtOH	271
	0.7	σ_2^0	PMR and IR measurements	271
	−0.49	σ_2^+	Solvolysis of 1-arylethyl acetates in 30% aq. EtOH (25°)	249
	−0.48	σ_3^+		
Benzothienyl	0.6	σ_2^-	Deuterodeprotonation	271
	0.5	σ_2^0	PMR or IR	271
	−0.46	σ_2^+	Solvolysis of 1-arylethyl acetates	249
	−0.54	σ_3^+		
	−0.49	σ_2^+		
	−0.56	σ_3^+		
	−0.25	σ_4^+	Solvolysis of 1-(2-benzothienyl) ethyl chlorides	365
	−0.34	σ_5^+		
	−0.42	σ_6^+		
	−0.11	σ_7^+		
	−0.53	σ_2^+	Pyrolysis of 1-arylethyl acetates, 327°	366
	−0.43	σ_3^+		367
	−0.35	σ_2^+	Protodesilylation in $HClO_4$, MeOH, H_2O (51°)	366
	−0.35	σ_3^+		368
	−0.38	σ_2^+	Bromination in HBr, dioxane, H_2O (25°)	366
	−0.69	σ_3^+		369, 370
	−0.61	σ_2^+	Protodetritiation in CF_3COOH (70°)	366
	−0.62	σ_3^+		371
	−0.49	σ_2^+	Acetylation of Ac_2O, $SnCl_4$, dichloroethylene (25°)	366, 369, 370
	−0.58	σ_3^+		
	−0.61	σ_2^+	Chlorination in Cl_2, HOAc	366, 369, 370
	−0.77	σ_3^+		
	−0.61	σ_2^+	Bromination in Br_2, HOAc	366, 369, 370
	−0.77	σ_3^+		
N-Methylindolyl	0.5	σ_2^-	Deuterodeprotonation	271
	0.2	σ_2^0	PMR	271
	−1.17	σ_2^+	Solvolysis of 1-arylethyl acetates	249
	−1.93	σ_2^+		
Dibenzofuryl	−0.24	σ_1^+	Protodetritiation	371
	−0.40	σ_2^+		
	−0.28	σ_3^+		
	−0.25	σ_4^+		
Dibenzothienyl	−0.28	σ_1^+	Protodetritiation	371
	−0.37	σ_2^+		
	−0.30	σ_3^+		
	−0.29	σ_4^+		

[a] IR, infrared; PMR, proton magnetic resonance.

but does not remove the ability of the five-membered heteroaromatic rings to accept electrons, unless strong electron demand is made on them—in this latter case the negative σ^+ values then reveal the strong electron donation by resonance, although this is not nearly so pronounced as in the absence of the benzo substituent. Clementi et al. have commented on the reactivities of the α- and β-positions of benzofuran and benzothiophene toward electrophiles.[370] Their results are summarized in the σ^+ values of Table XIII; these demonstrate that the effect of the [b] annelation is to reduce the reactivity of the furan and thiophene rings in electrophilic reactions by very similar amounts. The difference in reactivity between the α- and β-positions is also considerably reduced, the β-position indeed becoming generally more reactive than α in benzothiophene. Another complicating point here is that the benzo substituent appears to confer on benzothiophene (there are not sufficient data to test this for the other aryl groups in Table XIII) a variable polarizability so that the Extended Selectivity Relationship applied to both α- and β-positions produces curved plots.[366] This parallels the case of electrophilic substitution in naphthalene and biphenyl and is in contrast to the essentially unique σ^+ values established for α- and β-substitution in both furan and thiophene, as well as for p-substitution in halogenobenzenes, phenol, phenoxybenzene, anisole, thioanisole, and aniline derivatives.[366,372]

The passage of substituent effects through these systems has also received much attention. Correlation of the rates of solvolysis of 4-, 5-, 6-, and 7-substituted 1-(2-benzothienyl) ethyl p-nitrobenzoates,[270] and 5- and 6-substituted 1-(2-benzofuryl)ethyl[373] and 1-(3-benzofuryl)ethyl p-nitrobenzoates[374] in 80% aq. ethanol at 75° with σ^+ values calculated using the Dewar–Grisdale approach [Eq. (2)] or modified methods[270] is very good. The ρ value for the first two sets of data is in good agreement with that for the equivalent benzenoid reaction (-5.8); for the latter data, however, interaction between the 3 side chain and the 4 peri-position reduces the sensitivity to substituent effects giving the ρ value of -3.8.

Effects of substituents on pK_a values and tautomerism have been considered for benzimidazoles[116,312,375,376] and 2-arylbenzimidazoles[312] (53, Y = NH) as well as the corresponding oxazoles (53, Y = O) and thiazoles (53, X = S).[312] Curiously, the latter two correlate with σ^+ for the X-substituent, and the former with σ; presumably this results from

[372] S. Clementi and P. Linda, J. Chem. Soc., Perkin Trans. II, 1887 (1973).
[373] D. S. Noyce and R. W. Nichols, J. Org. Chem. 37, 4306 (1972).
[374] D. S. Noyce and R. W. Nichols, J. Org. Chem. 37, 4311 (1972).
[375] H. Walba, D. L. Stiggal, and S. M. Coutts, J. Org. Chem. 32, 1954 (1967).
[376] H. Walba and R. Ruiz-Velazco, J. Org. Chem. 34, 3315 (1969).

(53)

the high resonance donation from Y = NH to the protonated unsaturated nitrogen which thus does not draw out the full resonance potential of donor X, a potential which is realized in the face of the smaller resonance donation from Y = O,S. The various types of correlation analysis have been classified as for the pK_a results in Table I.

Other pK_a studies include those on 6-substituted 7-azaindoles and 7-azaindolines,[377] benzoxazolines and benzoxazolthiones substituted in the 5-, 6-, and 7-positions,[378] purines, for which the basic centre appears to be located in the pyrimidine nucleus,[379] 4,4'-, 5,5'-, 6,6'-, and 7,7'-disubstituted leucothioindigos,[380] and dyes (54).[381]

(54) R = Me, Et; Y = C(Me)$_2$, S, CH=CH, NEt

The acid-base equilibria of 5- and 6-substituted indole carboxylic acids and coumarylic acids can be correlated with σ_m for groups in the 5-position and σ_p for groups in the 6-position,[382] confirming the insulating effect of the heteroatom to resonance effects.

For systems containing two heteroatoms in the five-membered component the use of "mixed" σ-constants must be considered. These have been used for the correlation of pK_a's of benzimidazoles for which $\frac{1}{2}$ (σ_m + σ_p) is required for the 5- and 6-substituted compounds and $\frac{1}{2}$ (σ_m + σ_o) for the 4 and 7.[383] This approach has been extended to the study of nucleophilic substitution in 2-chloro-N-methylbenzimidazoles,[384,385] 2-

---

[377] L. N. Yakhontov, M. A. Portnov, M. Y. Uritskaya, D. M. Krasnokutskaya, M. S. Sokolova, and M. V. Rubtsov, Zh. Org. Khim. 3, 580 (1967).
[378] N. A. Vorontsova, N. L. Poznanskaya, O. N. Vlasov, and N. I. Shvetsov-Shilovskii, Reakts. Sposobnost Org. Soedin. 5, 665 (1968).
[379] P. Tomasik, R. Zalewski, and J. Chodziński, Chem. Zvesti. (in press).
[380] V. V. Karpov and V. G. Abozin, Zh. Prikl. Khim. 37, 1165 (1964).
[381] E. B. Lifschits, L. M. Yagupolskii, D. Y. Naroditskaya, and E. S. Kozlova, Reakts. Sposobnost Org. Soedin. 6, 317 (1969).
[382] M. S. Metzer, J. Org. Chem. 27, 496 (1972).
[383] A. Ricci and P. Vivarelli, Boll. Sci. Fac. Chim. Ind. Bologna 24, 249 (1966).
[384] A. Ricci and P. Vivarelli, Gazz. Chim. Ital. 97, 758 (1967).
[385] A. Ricci, G. Seconi, and P. Vivarelli, Gazz. Chim. Ital. 99, 542 (1969).

chloro-*N*-methyl- and 2-chloro-*N*,*N'*-dimethylbenzimidazolium perchlorates,[386] and 2-halogenobenzthiazoles.[387-393] For the displacement of chloride from 5- and 6-substituted 2-chloro-*N*-methylbenzimidazoles with piperidine in 2-propanol and methoxide in methanol a combination of σ_m and σ_p values works well, but the influence of 4- and 7-substituents is complicated by the presence of steric effects.[384] Rates of reaction of thiophenol with 5(6)-substituted 2-chloro-*N*,*N'*-dimethylbenzimidazolium perchlorates[386] are governed by $\frac{1}{2}(\sigma_m + \sigma_p)$, while for the 5- and 6-substituted 2-chloro-*N*-methylbenzimidazolium perchlorates plots of (log k_5 + log k_6) and (log k_5 − log k_7) against $(\sigma_m + \sigma_p)$ and $(\sigma_m - \sigma_p)$, respectively, give good straight lines. This indicates that Eq. (24) best correlates the results, where ρ_1 and σ_1 refer to transmission through the protonated nitrogen and ρ_2 and σ_2 to transmission through the methylated nitrogen, which leads to $\rho_1 = 0.75$ and $\rho_2 = 0.48$.

$$\log k/k_0 = \rho_1 \sigma_1 + \rho_2 \sigma_2 \qquad (24)$$

σ_m and σ_p constants have been applied also to 5- and 6-substituents in benzothiazole-1,1-dioxides in accounting for their influence on the rate of decomposition to benzynes, nitrogen, and sulfur dioxide.[394] Substituent studies have been made on the dehydrogenation of 7-azaindolines[395] and the kinetics of reversible photochromic reactions of

(55) (56)

[386] G. Seconi, P. Vivarelli, and A. Ricci, *J. Chem. Soc. B*, 254 (1970).
[387] P. E. Todesco and P. Vivarelli, *Gazz. Chim. Ital.* **92**, 1221 (1962).
[388] P. E. Todesco, P. Vivarelli, and A. Ricci, *Tetrahedron Lett.*, 3703 (1964).
[389] P. E. Todesco and P. Vivarelli, *Gazz. Chim. Ital.* **94**, 372 (1964).
[390] A. Ricci, P. E. Todesco, and P. Vivarelli, *Gazz. Chim. Ital.* **95**, 478 (1965).
[391] P. E. Todesco and P. Vivarelli, *Boll. Sci. Fac. Chim. Ind. Bologna* **20**, 129 (1962).
[392] A. Ricci, M. Foa, P. E. Todesco, and P. Vivarelli, *Gazz. Chim. Ital.* **95**, 465 (1965).
[393] G. Di Modica, E. Barni, and B. M. Magrassi, *Ann. Chim.* **53**, 733 (1966).
[394] R. W. Hoffmann, W. Sieber, and G. Guhn, *Chem. Ber.* **98**, 3470 (1965).
[395] Y. J. Vainstein, J. G. Palant, L. N. Yakhontov, D. M. Krasnokutskaya, and M. B. Rubstov, *Khim. Geterotsikl. Soedin.*, 1106 (1969).

1,5-disubstituted 3,3-dimethyl-6'-nitro-8'-bromospiro [(2'H-1'-benzopyran 2,2'-indolines], **55** ⇌ **56**.[396] Information is also available on substituent effects in the decomposition of molecular ions of anilides of indole 2- and 3-carboxylic acids.[397]

In the region of spectral studies, Perjéssy has made extensive studies on indandione derivatives (**57**), demonstrating that the carbonyl

(57) Y = CH_2, O, SNH; Z = O, S

stretching frequency is correlated by σ.[398–400] NH stretching vibrations of benzimidazoles correlate with combinations of σ_m and σ_p values depending on substituent position,[401] while curved plots result for $\nu_{c=0}$ vs. σ for oxindoles and N-methyloxindoles which the authors suggest denotes the varying importance of the passage of inductive effects through the 3-methylene group.[402]

PMR studies on 1,3,4-triazaindolizines substituted in 2-, 5-, 6-, or 7-positions show a connection between σ_p and the proton shift relative to the 2- or 6-proton resonance in the unsubstituted compounds.[403] The PMR chemical shift in the N-Me group of 5-substituted 2-chloro-1-methylbenzimidazoles correlates with σ_p^- in $CDCl_3$ but with σ_p in MeCOOH demonstrating the reduction in through resonance on protonating the five-membered ring;[404] there is, however, no correspondence of chemical shift produced by 6-substitution with σ_m,[404] nor is there any connection between 4-, 5-, 6-, or 7-positional substitution of

---

[396] M. A. Halbershtam and M. B. Gorodin, *Photochem. Photobiol.* **17**, 103 (1973).
[397] Y. S. Nekrasov, P. A. Shartbatyan, R. S. Sagitullin, V. A. Puchkov, A. N. Kost, and N. S. Vulfson, *Izv. Akad. Nauk. SSSR, Ser. Khim.*, 2181 (1969).
[398] A. Perjéssy and P. Hrnčiar, *Collect. Czech. Chem. Commun.* **35**, 1120 (1970).
[399] P. Perjéssy, M. Laćova, and P. Hrnčiar, *Collect. Czech. Chem. Commun.* **36**, 275 (1971).
[400] A. Perjéssy and M. Laćová, *Collect Czech. Chem. Commun.* **36**, 2944 (1971).
[401] P. Dembech, S. Giorganni, G. Seconi, P. Vivarelli, and S. Ghersetti, *Boll. Sci. Fac. Chim. Ind. Bologna* **27**, 429 (1969).
[402] A. H. Beckett, R. W. Daisley, and J. Walker, *Tetrahedron* **24**, 6093 (1968).
[403] K. Kleinpeter, R. Borsdorf, G. Fischer, and M. J. Hofmann, *J. Prakt. Chem.* **314**, 515 (1972).
[404] P. Dembech, G. Seconi, P. Vivarelli, L. Schenetti, and F. Taddei, *J. Chem. Soc. B*, 1670 (1971).

benzothiazole and the PMR shift for the 2-position.[405] Linear relationships do exist, however, with varying degrees of accuracy for the PMR shift of methyl in 5- and 6-substituted 2-methyl benzazoles, benzothiazoles, benzoselenazoles, and benzoxazoles,[401,406,407] and correlations of very high quality between σ_p^+ or $\Sigma\sigma_p^+$ and the PMR shift of the 2-proton in 6-substituted purines[408] or the 8-proton in 2,6-disubstituted purines,[409] respectively, in dimethyl sulfoxide solution. Fluorine NMR chemical shifts of substituted benzenes have been used to describe σ parameters for entire aryl groupings of the type under discussion (see Table XIV), which confirm the electron-accepting properties of these rings.[410,411]

TABLE XIV

σ Constants for Fused Five- and Six-Membered Rings as Substituents

Substituent	σ_m	σ_p	σ_I	σ_R	Reference footnotes
2-Benzoxazolyl	0.26	0.34	0.27	0.06	410
2-Benzothiazolyl	0.30	0.32	0.28	0.03	410
2-(1-Phenylimidazolyl)	0.18	0.23	0.16	0.08	410
2-(1-Methylimidazolyl)	—	—	0.07	0.04	410
2-N-3,4-Benzotriazolyl	0.52	0.50	—	—	411
3,4-(2-Phenyltriazolyl)-4,5-ylenyl	—	0.46	—	—	411

Peak wavelengths in the UV spectra of 2-aryl-5-oxo-7-ethoxycarbonyl-1,3,4-oxadiazolo[3,2-a]pyrimidines (**58**) are influenced

(58)

[405] F. Taddei, P. E. Todesco, and P. Vivarelli, *Gazz. Chim. Ital.* **95**, 499 (1966).
[406] G. Di Modica, E. Barni, and A. Gasco, *J. Heterocycl. Chem.* **2**, 457 (1965).
[407] E. Barni, A. Gasco, and G. Di Modica, *Atti Accad. Sci. Torino* **101**, 1 (1966).
[408] W. C. Coburn, M. C. Thorpe, J. A. Montgomery, and K. Hewson, *J. Org. Chem.* **30**, 1114 (1965).
[409] W. C. Coburn, M. C. Thorpe, J. A. Montgomery, and K. Hewson, *J. Org. Chem.* **30**, 1110 (1965).
[410] V. E. Bystrova, Z. N. Belaya, B. E. Gruz, G. P. Syrova, A. I. Tolmatshev, L. M. Shylezhko, and L. M. Yagupolskii, *Zh. Obshch. Khim.* **38**, 1001 (1968).
[411] I. Cepciansky and J. Mayer, *Collect. Czech., Chem. Commun.* **34**, 72 (1969).

TABLE XV
σ VALUES FOR THREE-MEMBERED RINGS AND RELATED STRUCTURES

Substituent	σ_I	σ_R	Substituent	σ_I	σ_R
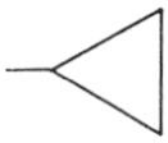 ▷— (cyclopropyl)	−0.08	−0.13	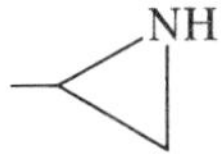 ▷—NH (aziridinyl)	−0.02	−0.09
Me	−0.08	−0.15	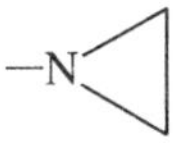 −N▷	0.07	−0.29
−CH=CH$_2$	0.01	−0.03	−CH$_2$NH$_2$	0.00	−0.15
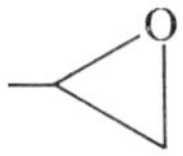 ▷—O (oxiranyl)	0.07	−0.04	−NH$_2$	0.01	−0.48
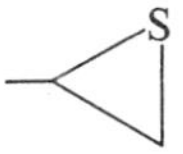 ▷—S (thiiranyl)	0.07	−0.06	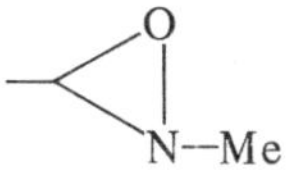 ▷(O)N−Me	0.07	0.05
			−CONH$_2$	0.11	0.11

by the resonance effect of X, giving a linear relationship with σ_R,[412] while λ_{max} in the visible spectra of the ethyl iodides of 5- or 6-substituted styrylbenzothiazoles has been related to σ.[413] σ_p^+ values express the effect of 3-substituents on the first maximum of the charge transfer spectra of thianaphthenes, although the integrated intensities of the 1264 cm^{-1} band in the IR spectra of these compounds, ascribed to a thiophene skeletal vibration, gives a linear correspondence with σ.[414]

V. Three-Membered Rings

Interest in the general area of three-membered rings has been prompted by their analogy with cyclopropane and the cyclopropyl group, the electronic effects of which have excited a good deal of attention and investigation. In the region, NMR has been much employed; F^{19}NMR of m- and p-fluorophenyl cyclopropanes, oxiranes, thiiranes, ethylenimines, and oxaziridines[415] have led, by standard equations due to Taft,[416] to the σ_I and σ_R values given in Table XV.

---

[412] P. Henhkin and G. Westphal, Z. Chem. 14, 19 (1974).
[413] G. Di Modica and E. Barni, Atti Accad. Sci. Torino 101, 1 (1966).
[414] N. Kucharczyk, B. Kakač, and V. Horak, Collect. Czech. Chem. Commun. 34, 2959 (1969).
[415] R. G. Pews, J. Amer. Chem. Soc. 89, 5605 (1967).
[416] R. W. Taft, E. Price, I. R. Fox, I. C. Lewis, K. K. Andersen, and G. T. Davis, J. Amer. Chem. Soc. 89, 3146 (1963).

Passage of effects through such rings has also received attention. Ionization constants of *m*- and *p*-substituted *trans*-2,3-epoxy-3-phenyl propionic acids yields a ρ value of 0.51[417]; cf. 0.47 for the corresponding phenylcyclopropane carboxylic acids. The correspondence with σ is excellent, showing that the interposition of the oxirane ring between phenyl and carboxylic acid group does not necessitate the use of Eq. (3). The NMR spectra of *trans*-1-(substituted phenyl)-2-benzoyl-3,3-dideuteriocyclopropanes (**59**), *trans*-2-(substituted phenyl)-3-benzoyloxiranes (**60**), and *trans*- and *cis*-1-cyclohexyl-2-(substituted phenyl)-3-benzoylaziridines (**61**) have been measured in CDCl$_3$ and benzene,[418] the benzene appearing to form a complex with the rings. The chemical shift differences $\Delta(= \delta_{CDCl_3} - \delta_{C_6H_6})$ for H$_a$ or H$_b$ with variation in X show good correspondence with σ, or $\mathscr{F}$ and $\mathscr{R}$ values using the Swain–Lupton approach [Eq. (1)] demonstrating the passage of resonance effects through these rings. The resultant ρ values indicate the "efficiency of transmission" of effects is cyclopropane $\simeq$ oxirane > aziridine.

XC$_6$H$_4$ Ha
Hb   COPh
  CD$_2$
  (59)

XC$_6$H$_4$ Ha
Hb   COPh
  O
  (60)

XC$_6$H$_4$ COPh
Hb   Ha
  NH
  (61)

General correlations of proton H$_a$ and H$_b$ and methyl PMR chemical shifts for 1-aryl-2-methyl oxiranes (**62**) together with the corresponding vinyl derivatives with σ for solvents CCl$_4$, CD$_3$COCD$_3$, and CD$_3$SOCD$_3$ have also been reported,[419] the most pronounced effect being on H$_b$, but other NMR experiments on aryloxiranes and oxetanes indicate that Hammett constants do not express all the effects on such protons.[420]

[417] L. Standoli, *J. Chem. Soc., Perkin Trans. II*, 371 (1975).
[418] A. B. Turner, R. E. Lutz, N. S. McFarlane, and D. W. Boykin, *J. Org. Chem.* **36**, 1107 (1971).
[419] R. Benassi, P. Lazzeretti, I. Moretti, F. Taddei, and G. Torre, *Org. Magn. Reson.* **5**, 391 (1973).
[420] C. Schaal, G. Aranda, and H. De Luze, *J. Chim. Phys. Physicochim. Biol.* **66**, 1597 (1969) [*CA* **72**, 66053n (1970)].

XC₆H₄\ /Ha
Hb/ \Me
    O
  (62)

IR measurements on N-substituted aziridines as pure liquids or in CCl$_4$ show a relation between asymmetrical CH$_2$ stretching frequencies and σ_I for the N substituent.[421]

The barriers to inversion of 1-aryl-2,2-dimethylaziridines have been investigated[422] by means of low-temperature NMR. Calculation of the $\Delta G^{\ddagger}$ values involves observation of the coalescence temperature for the protons of the methyl groups. The process is accelerated by π conjugation of the nitrogen lone pair in the transition state, and accordingly the

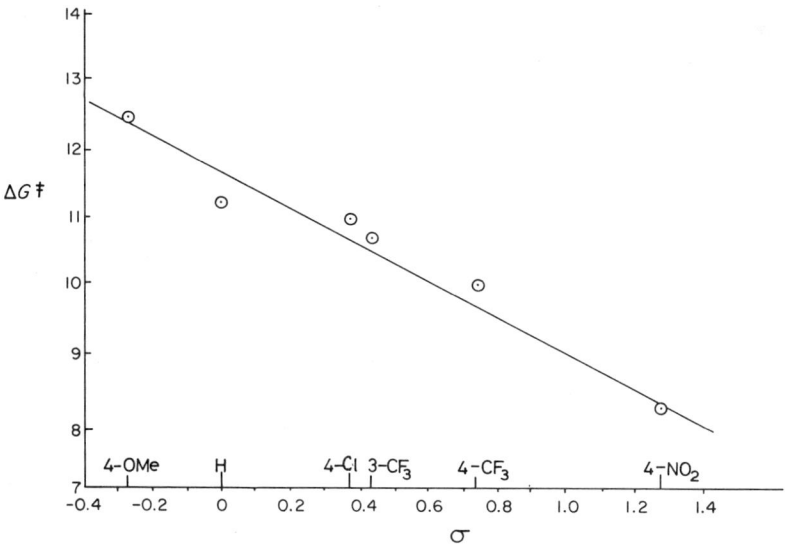

Fig. 9. Correlation of free energy of inversion with σ^- for 1-XC₆H₄-2,2-dimethylaziridines.

$\Delta G^{\ddagger}$ values plot approximately linearly with the σ^- values for the aryl substituents. The relevant plot is shown in Fig. 9. However, there is little substituent effect on such inversion values for sulfenylaziridines (**63**, R = —C₆H₄-*p*-X),[423] while the steric effect in such compounds (**63**,

---

[421] H. L. Spell, *Anal. Chem.* **39**, 185 (1967).
[422] J. D. Andose, J. M. Lehn, K. Mislow, and J. Wagner, *J. Amer. Chem. Soc.* **92**, 4050 (1970).
[423] D. Kost, W. A. Stacer, and M. Raban, *J. Amer. Chem. Soc.* **94**, 3233 (1972).

R = Me, Ph, *t*-Bu, CF$_3$, and CCl$_3$) does not correlate with the E$_s$ parameters of the groups.[424] This is probably because CF$_3$ and CCl$_3$ have lower activation energies than predicted on the basis of E$_s$ due to negative hyperconjugation stabilizing the inversion transition state.

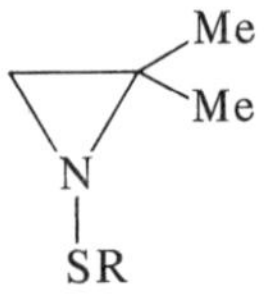

$$\underset{\underset{SR}{|}}{\underset{N}{\triangle}}\!\!\!\begin{array}{c}Me\\Me\end{array}$$

(63)

Substituent effects on aziridine formation from *N*-(2-halogenoethyl) arylsulphonamides by cyclization in basic media have been studied,[425] and Hammett equation studies have also been of utility in determination of mechanisms of thermal decomposition of 3-chloro-3-aryldiazirines,[426,427] the acid-catalyzed methanolysis of arylepoxides,[428] and the ring expansion reactions of 2-aryl-1,1-dimethylaziridinium salts with benzaldehyde to form 5-aryl-3,3-dimethyl-2-phenyloxazolidinium salts.[429]

---

[424] M. Raban and D. Kost, *J. Amer. Chem. Soc.* **94**, 3234 (1972).
[425] J. H. Coy, A. F. Hegarty, E. J. Flynn, and F. L. Scott, *J. Chem. Soc., Perkin Trans. II*, 53 (1974).
[426] M. T. H. Liu and K. Toriyama, *Can. J. Chem.* **50**, 3009 (1972).
[427] M. T. H. Liu and D. H. T. Chien, *J. Chem. Soc., Perkin Trans. II*, 937 (1974).
[428] J. Biggs, N. B. Chapman, and V. Wray, *J. Chem. Soc. B*, 55 (1971).
[429] T. R. Keenan and N. J. Leonard, *J. Amer. Chem. Soc.* **93**, 6567 (1971).

1,2,4-Oxadiazoles

LEALLYN B. CLAPP

Brown University,
Providence, Rhode Island

I. Synthesis: General Methods	66
II. Synthesis: Elaboration	67
A. From *O*-Acylamidoximes	67
B. From *N*-Acyliminoethers and Related Compounds	71
C. From *C*-Nitroso Derivatives	73
D. 1,3-Dipolar Cycloadditions	75
E. Oxidation of Dihydrooxadiazoles	76
F. Oxidation of Amidoximes and Oximes	77
G. Rearrangements	78
III. Synthesis of Oxadiazolines and Other Reduced Rings	79
A. Oxadiazolines and Oxadiazolidines	79
1. Δ^3-Oxadiazolines	79
2. Δ^3-Oxadiazolines	80
3. Δ^4-Oxadiazolines	81
4. Oxadiazolidines	81
B. Oxadiazolinones and Related Compounds	82
1. Oxadiazolinones	82
2. Oxadiazolidinediones	84
IV. Physical and Spectral Properties	85
A. Electron Density and Molecular Geometry	85
B. Thermodynamic Functions	86
C. Dipole Moments	87
D. Infrared Spectra	88
E. Ultraviolet Spectra	89
F. Nuclear Magnetic Resonance Spectra and Deuterium Exchange	90
G. Mass Spectra	92
V. Chemical Properties	93
A. Thermal Stability	93
B. Photochemical Effects	94
C. Stability in Solution: Hydrolysis	95
D. Ring Opening by Reduction	95
E. Functional Group Chemistry	97
F. Activation in the Side Chain	99
G. Electrophilic Substitutions	101
H. Nucleophilic Substitutions	102
I. Complexes with Inorganic Compounds	103
J. Rearrangements	104
K. Tautomerism and Hydrogen Bonding	109
VI. Uses	112
A. Chemotherapy	112
B. Agricultural and Other Uses	115

The chemistry of the four isomeric oxadiazoles (Scheme 1) was reviewed by Boyer in 1962[1] and that of 1,2,4-oxadiazoles by Behr[2] in

SCHEME 1. The Oxadiazoles

1962 and by Eloy[3] in 1965. Other references are given in the bibliography published in this series by Katritzky and Weeds.[4]

Throughout the present review, limited to the chemistry of 1,2,4-oxadiazoles and the corresponding reduced derivatives, this 1,2,4-isomer will be assumed when the complete numbering system is omitted. In the

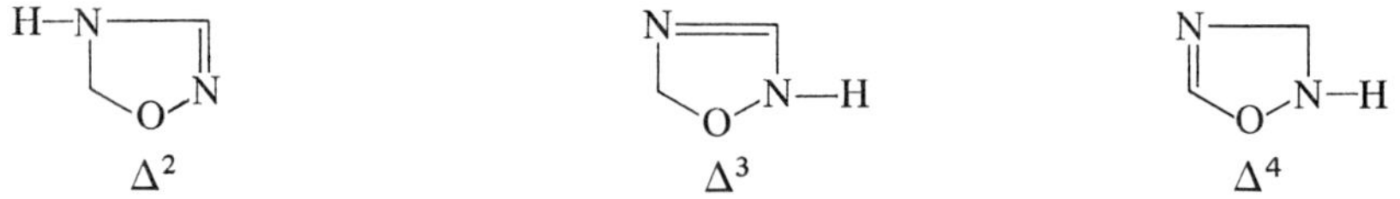

SCHEME 2

oxadiazoles and the reduced rings, substituents may be referred to as occupying positions N-2, C-3, N-4, or C-5. The double bonds in the partially reduced rings are designated Δ^2, Δ^3, and Δ^4, respectively (Scheme 2).

I. Synthesis: General Methods

Two widely used general methods of synthesizing 1,2,4-oxadiazoles embrace 95% of the practical preparations of these compounds: the conversion of amidoximes[5] by means of carboxylic acid derivatives to the cyclic structure [shown schematically as the combination of two skeletons (I)] and the cycloaddition of nitrile oxides[6] to nitriles (II). The

SCHEME 3

[1] J. H. Boyer, in "Heterocyclic Compounds" (R. C. Elderfield, ed.), Vol. 7, pp. 462–540. Wiley, New York, 1962.
[2] L. C. Behr, in "The Chemistry of Heterocyclic Compounds" (A. Weissberger, ed.), Vol. 17, pp. 245–262. Wiley (Interscience), New York, 1962.
[3] F. Eloy, Fortschr. Chem. Forsch. 4, 807–876 (1965).
[4] A. R. Katritzky and S. M. Weeds, this Series, 7, 225–299 (1966).
[5] F. Eloy and R. Lenaers, Chem. Rev. 62, 155 (1962).
[6] G. Leandri, Boll. Sci. Fac. Chim. Ind. Bologna 14, 80 (1956) [CA 51, 5771 (1957)].

Sec. II.A] 1,2,4-OXADIAZOLES 67

methods of building up the 4-atom fragment in I give the basis for subdivision of the synthetic methods in Sections II, A, B, and C. A few synthetic methods do not fit into the general pattern of I or II (Sections II, E, F, G).

II. Synthesis: Elaboration

A. From O-Acylamidoximes

The most widely used method of synthesizing oxadiazoles is the cyclization of O-acylamidoximes. Aldoximes (1) may be used as starting compounds for amidoximes (3) by the pathway shown in Eq. (1). The amidoximes may then be converted into O-acyl derivatives (4), long[7,8]

$$R-CH=NOH \xrightarrow{Cl_2} R-C\begin{pmatrix}NOH\\Cl\end{pmatrix} \xrightarrow{NH_3} R-C\begin{pmatrix}NOH\\NH_2\end{pmatrix} \xrightarrow{(R'CO)_2O}$$
(1) (2) (3)

$$R-C\begin{pmatrix}N-O-C-R'\\\|\\NH_2\;\;O\end{pmatrix} \longrightarrow \underset{R'}{\overset{R}{\text{oxadiazole}}} + H_2O \quad (1)$$
(4) (5)

assumed to be intermediates in the pathway to oxadiazoles (5) but only recently shown[9] to be in actuality. The O-acyl derivatives are not often isolated, and indeed cannot generally be isolated if the acylation is carried out above 100°. The mechanism of dehydration in the last step of Eq. (1) has not been determined.

Many variations on the pathway represented by Eq. (1) are successful. For example, the amidoxime can be made by heating a nitrile with hydroxylamine hydrochloride[10–17] and in other ways.

---

[7] F. Tiemann and P. Krüger, *Ber.* **17**, 1685 (1884).
[8] F. Tiemann, *Ber.* **17**, 1689 (1884).
[9] R. Buyle, R. Lenaers, and F. Eloy, *Helv. Chim. Acta* **46**, 1073 (1963).
[10] H. Simon, H. Lettau, H. Schubert, and A. Jumar, *Z. Chem.* **9**, 58 (1969).
[11] G. Palazzo and B. Silvestrini, *Boll. Chim. Farm.* **101**, 251 (1962) [*CA* **58**, 4548 (1963)].
[12] H. Goncalves and A. Secches, *Bull. Soc. Chim. Fr.* **7**, 2589 (1970).
[13] L. V. Phillips and A. J. Latham, U.S. 3,501,491 (1970), [*CA* **72**, 121542 (1970)].
[14] J. Barrans, *Ann. Fac. Sci. Univ. Toulouse Sci. Math. Sci. Phys.* **25**, 7 (1961) [*CA* **60**, 12004 (1964)].
[15] O. Treuner, Ger. Offen. 2,248,940 (May, 1973) [*CA* **79**, 18719 (1973)].
[16] G. Asato and G. Berkelhammer, U.S. 3,795,735 (May, 1974) [*CA* **80**, 120959 (1974)].
[17] E. W. Berndt, H. A. Fratzke, and B. G. Held, *J. Heterocycl. Chem.* **9**, 137 (1972).

$$R-C{\equiv}N + NH_2OH \cdot HCl \xrightarrow{\Delta} R-C\begin{smallmatrix}\nearrow NOH \\ \searrow NH_2\end{smallmatrix} + HCl \qquad (2)$$

In addition to the acid anhydride[12, 14, 17–19] shown as the source of the carbon at C-5 in the final product [Eq. (1)], ketene,[20] reactive acids such as formic[21] and acrylic,[22] acid chlorides,[10, 23–26] esters,[27] ortho esters,[28–34] ethyl oxalate,[35] amides,[35] and iminoethers[36] [Eqs. (12–14)] may also furnish the C-5 carbon.

The reaction of amides[35] with amidoxime salts is especially felicitous because no solvent is needed and recovery is simple. The two components are melted together at 160–180° for 10 minutes, and water is lost at the elevated temperature. Both aliphatic and aromatic (di- and monosubstituted) amidoximes give yields in the range 60–90%. Whether an O-acyl (or an N-acyl) intermediate (**4**) is formed has not been determined.

With very reactive acid chlorides, such as chloroacetyl chloride,[11] dichloroacetyl chloride,[13] α-chloropropionyl chloride,[37] and perfluoro-

---

[18] G. Palazzo and G. Corsi, *Gazz. Chim. Ital.* **93**, 1196 (1963).
[19] H. Hamano, K. Shimada, and S. Kuriyama, Japan. Patent 72 03,823 (1972) [*CA* **76**, 153752 (1972)].
[20] F. Eloy and R. Lenaers, *Bull. Soc. Chim. Belges* **72**, 91 (1963).
[21] R. Lenaers, C. Moussebois, and F. Eloy, *Helv. Chim. Acta* **45**, 441 (1962).
[22] K. Harsanyi, P. Kiss, D. Korbonits, I. R. Malyata, and K. Takacs, Austrian 261,608 (1968) [*CA* **69**, 106713 (1968)].
[23] G. Palazzo and G. Strani, *Gazz. Chim. Ital.* **90**, 1290 (1960).
[24] G. Corsi, B. Catanese, and B. Silvestrini, *Boll. Chim. Farm.* **103**, 115 (1964) [*CA* **61**, 3095 (1964)].
[25] J. A. Miller, *Ber.* **22**, 2790 (1889).
[26] L. H. Schubart, *Ber.* **19**, 1487 (1886).
[27] K. Harsanyi, P. Kiss, D. Korbonits, and I. R. Malyata, *Arzneim.-Forsch.* **16**, 615 (1966) [*CA* **70**, 37724 (1969)].
[28] G. I. Gregory, P. W. Seale, W. K. Warburton, and M. J. Wilson, *J. Chem. Soc., Perkin Trans. I,* 47 (1973).
[29] M. Arbasino and P. Grünanger, *Atti Accad. Naz. Lincei, Rend., Cl. Sci. Fis., Mat. Nat.* **34**, 532 (1963) [*CA* **60**, 4132 (1964)].
[30] M. Arbasino and P. Grünanger, *Chim. Ind. (Milan]* **45**, 1238 (1963) [*CA* **60**, 10692 (1964)].
[31] C. Ainsworth, W. E. Buting, J. Davenport, M. E. Callender, and M. C. McCowen, *J. Med. Chem.* **10**, 208 (1967).
[32] R. G. Micetich, *Can. J. Chem.* **48**, 2006 (1970).
[33] J. A. Claisse, M. W. Foxton, G. I. Gregory, A. H. Sheppard, E. P. Tiley, W. K. Warburton, and M. J. Wilson, *J. Chem. Soc., Perkin Trans. I,* 2241 (1973).
[34] A. D. Borthwick, M. W. Foxton, B. V. Gray, G. I. Gregory, P. W. Seale, and W. K. Warburton, *J. Chem. Soc., Perkin Trans I,* 2769 (1973).
[35] C. Moussebois, *Bull. Soc. Chim. Belges* **76**, 92 (1967).
[36] H. Weidinger and J. Kranz, *Chem. Ber.* **96**, 1049 (1963).
[37] M. Tavella and G. Strani, *Ann. Chim. (Rome),* **52**, 192 (1962).

acid chlorides,[38] the conversion of **2** to **4** can often be made at room temperature without isolating **3**.

Boron trifluoride[21,34] was found to catalyze the reaction of formic acid with oxamidoxime (**6**) to give the oxadiazolyloxadiazole (**7**) shown

$$\underset{(6)}{\underset{HON}{\overset{H_2N}{>}}\!\!C\!-\!C\!\overset{NOH}{\underset{NH_2}{<}}} + HCOOH \xrightarrow{BF_3} \underset{(7)}{[\text{oxadiazolyloxadiazole}]} + 2H_2O \quad (3)$$

in Eq. (3) and was useful in promoting the reactivity of ethyl orthoformate toward an amidoxime in the synthesis of 3-aryloxadiazoles.

Formation of the *O*-acylamidoxime (**4**) has also been accelerated by the well known dehydrating agent, dicyclohexylcarbodiimide.[29]

The unstable parent compound, 1,2,4-oxadiazole, was not reported until 1962.[39] Moussebois and co-workers treated formamidoxime with the mixed anhydride of formic and acetic acids to get the *O*-formylformamidoxime, which gave the oxadiazole upon heating. The product, very soluble in water and organic solvents, was isolated as the cadmium chloride complex, from which oxadiazole was obtained by vacuum distillation at 30°.

Carbonic acid derivatives were used in early work to put a hydroxy group on C-5.[40,41] Ethyl chloroformate reacts with amidoximes to give a carbonate ester that loses a molecule of ethanol in the cyclization. The expected oxadiazolone (**8**) is a tautomer of the 5-hydroxy form (**9**) (see

$$\underset{(3)}{R\!-\!C\!\overset{NOH}{\underset{NH_2}{<}}} + ClCO_2Et \longrightarrow R\!-\!C\!\overset{N\!-\!O\!-\!C\!-\!OEt}{\underset{NH_2\ \ \ \overset{\|}{O}}{<}} \longrightarrow$$

$$\underset{(8)}{[\text{oxadiazolone}]} \rightleftharpoons \underset{(9)}{[\text{5-hydroxy form}]}$$

Section V, K). More recently[42] an ArO group has been placed at C-3 in **9** by starting with ArO—C(=NOH)NH$_2$ in place of **3**. The aryloxy-

---

[38] J. P. Critchley, E. J. P. Fear, and J. S. Pippett, *Chem. Ind.* (*London*), 806 (1964).
[39] C. Moussebois, R. Lenaers, and F. Eloy, *Helv. Chim. Acta* **45**, 446 (1962).
[40] E. Falck, *Ber.* **18**, 2465 (1885); **19**, 1481 (1886).
[41] F. Tiemann, *Ber.* **18**, 2456 (1885); **19**, 1475 (1886).
[42] E. Grigat, R. Pütter, and C. König, *Chem. Ber.* **98**, 144 (1965).

amidoxime was synthesized from ArOCN, unknown until 1964.[43] With

$$\text{ArOH} + \text{ClCN} \xrightarrow[0°]{\text{Et}_3\text{N}} \text{ArOCN} \xrightarrow[\text{2. Na}_2\text{CO}_3]{\text{1. NH}_2\text{OH} \cdot \text{HCl}} \text{ArO}-\text{C}\underset{\text{NH}_2}{\overset{\text{NOH}}{\diagup}} \quad (4)$$

phosgene[44-46] both halogens are displaced in the first step and one molecule of the amidoxime is regenerated in the cyclization.

$$2\text{R}-\text{C}\underset{\text{NH}_2}{\overset{\text{NOH}}{\diagup}} + \text{COCl}_2 \longrightarrow \left(\text{R}-\text{C}\underset{\text{NH}_2}{\overset{\text{N}-\text{O}}{\diagup}}\right)_2 \text{C}=\text{O} \longrightarrow$$
(3)

$$\underset{\text{HO}}{\overset{\text{N}}{\diagup}}\underset{\text{O}}{\overset{\text{R}}{\diagdown}}\text{N} + \text{R}-\text{C}\underset{\text{NH}_2}{\overset{\text{NOH}}{\diagup}} \quad (5)$$
(9)   (3)

β-Ketoesters[47-50] are sufficiently activated to react with amidoximes, particularly at the boiling point of toluene,[51,52] to cyclize with the loss of azeotropic water and ethanol [Eq. (6)]. Probably O-acylamidoximes are intermediates.

$$\text{Ph}-\text{C}\underset{\text{NH}_2}{\overset{\text{NOH}}{\diagup}} + \text{MeCOCH}_2\text{COOEt} \xrightarrow{\text{toluene}}$$

$$\text{Me}-\text{CO}-\text{CH}_2\underset{\text{O}}{\overset{\text{N}}{\diagup}}\underset{\text{N}}{\overset{\text{Ph}}{\diagdown}} + \text{H}_2\text{O} + \text{EtOH} \quad (6)$$

It should not be taken for granted, however, that by using an amidoxime as starting compound the amidoxime carbon always becomes the C-3 carbon in the oxadiazole. Warburton[53] demonstrated the fallacy by showing that benzoyldicyandiamide (10) gave a mixture of two ureido derivatives as shown in Eq. (7), the first predominating.

[43] E. Grigat and R. Pütter, *Chem. Ber.* **97**, 3012 (1964).
[44] F. Tiemann and P. Krüger, *Ber.* **18**, 727 (1885).
[45] H. Müller, *Ber.* **22**, 2401 (1889).
[46] T. Bacchetti and A. Alemagna, *Rend. Accad. Naz. Lincei, Rend., Cl. Sci. Fis., Mat. Nat.* [8] **22**, 637 (1957) [*CA* **52**, 15511 (1958)].
[47] E. Richter, *Ber.* **22**, 2449 (1889).
[48] F. Tiemann, *Ber.* **22**, 2412 (1889).
[49] J. Weise, *Ber.* **22**, 2418 (1889).
[50] L. H. Schubart, *Ber.* **22**, 2433 (1889).
[51] R. Merckx, *Bull. Soc. Chim. Belges*, **56**. 339 (1947).
[52] R. Merckx, *Bull. Soc. Chim. Belges* **58**, 58 (1949).
[53] W. K. Warburton, *J. Chem. Soc. C*, 1522 (1966).

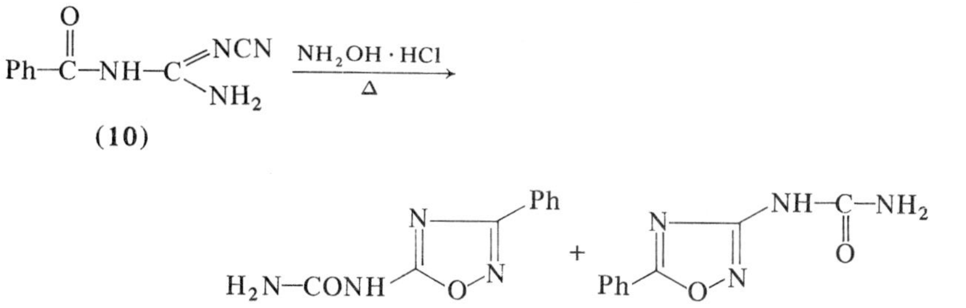

$$\text{Ph—C(=O)—NH—C(=NCN)(NH}_2\text{)} \xrightarrow[\Delta]{\text{NH}_2\text{OH·HCl}}$$

(10)

[structures: H$_2$N—CONH-substituted 3-Ph-1,2,4-oxadiazole + 3-Ph-5-(NH—C(=O)—NH$_2$)-1,2,4-oxadiazole] (7)

Warburton and co-workers[54] destroyed another unstated tenet of oxadiazole chemistry, namely, that only monoacylated amidoximes cyclize, when they found that **11** slowly cyclized to **12** merely on standing for 27 days.

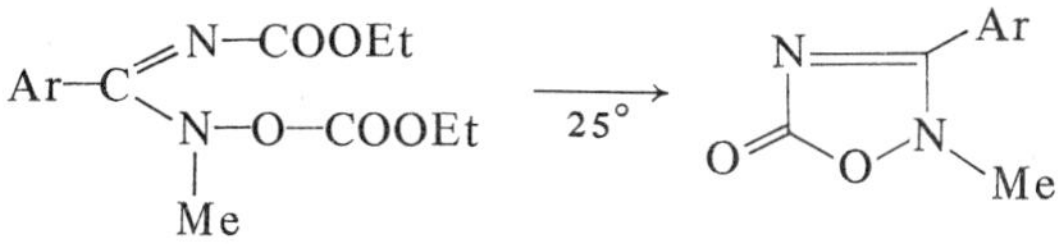

$$\text{Ar—C}\begin{pmatrix}=\text{N—COOEt}\\ \text{N(Me)—O—COOEt}\end{pmatrix} \xrightarrow{25°} \text{(12)}$$

(11)   (12)

B. FROM N-ACYLIMINOETHERS AND RELATED COMPOUNDS

N-Acyliminoethers may also be used as a source for the sp^2 carbon at position 3, bearing the two bonds to nitrogen in the oxadiazole. The iminoethers react with hydroxylamine and then cyclize. The intermediate shown in Eq. (8) has not been isolated.[55,56] Earlier Beckmann

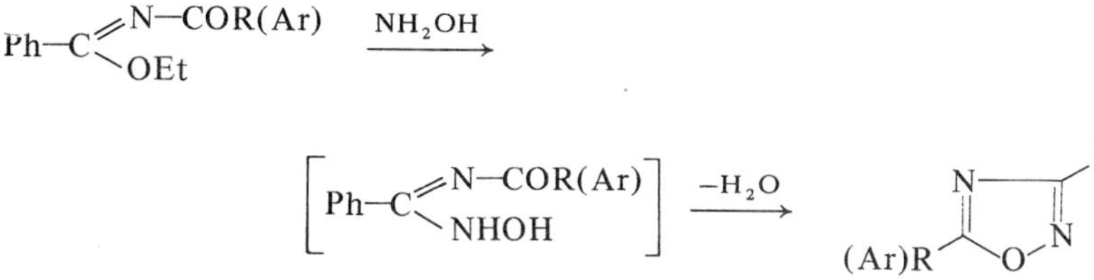

$$\text{Ph—C}\begin{pmatrix}=\text{N—COR(Ar)}\\ \text{OEt}\end{pmatrix} \xrightarrow{\text{NH}_2\text{OH}} \left[\text{Ph—C}\begin{pmatrix}=\text{N—COR(Ar)}\\ \text{NHOH}\end{pmatrix}\right] \xrightarrow{-\text{H}_2\text{O}} \text{[3-(Ar)R-5-Ph-1,2,4-oxadiazole]} \quad (8)$$

and Sandel[57] had used *N*-acyliminochlorides and *N*-acylbenzamidine in the same way with hydroxylamine, but these starting compounds are less stable and not so easily obtained as the *N*-acyliminoethers.

[54] J. A. Maddison, P. W. Seale, E. P. Tiley, and W. K. Warburton, *J. Chem. Soc., Perkin Trans I*, 81 (1974); P. W. Seale and W. K. Warburton, *ibid.*, 85 (1974).
[55] F. Eloy, R. Lenaers, and R. Buyle, *Bull. Soc. Chim. Belges* **73**, 518 (1964).
[56] B. G. Baccar and F. Mathis, *C.R. Acad. Sci.* **261**, 174 (1965).
[57] E. Beckmann and K. Sandel, *Ann.* **296**, 279 (1897).

$$\text{R-C} \begin{smallmatrix} \nearrow \text{N-CN} \\ \searrow \text{OR}' \end{smallmatrix} + \text{NH}_2\text{OH} \longrightarrow$$

$$\left[\text{R-C} \begin{smallmatrix} \nearrow \text{N-CN} \\ \searrow \text{NHOH} \end{smallmatrix} \right] \longrightarrow \underset{\text{H}_2\text{N}}{\overset{\text{N}\text{---}\text{R}}{\bigcirc_{\text{O}}\text{N}}} \qquad (9)$$

The amino function can be introduced at C-3 or C-5 by variations of this method as shown by Huffman and Schaeffer[58] [Eq. (9)] and by Eloy[55] [Eq. (10)]. Compare Eqs. (7) and (10).

$$\text{PhCONH-C} \begin{smallmatrix} \nearrow \text{NH} \\ \searrow \text{OEt} \end{smallmatrix} + \text{NH}_2\text{OH} \longrightarrow$$

$$\left[\text{PhCONH-C} \begin{smallmatrix} \nearrow \text{NOH} \\ \searrow \text{NH}_2 \end{smallmatrix} \right] \longrightarrow \underset{\text{Ph}}{\overset{\text{N}\text{---}\text{NH}_2}{\bigcirc_{\text{O}}\text{N}}} \qquad (10)$$

Using an iminothiolcarbonate,[59,60] a 3-methoxyoxadiazole has been synthesized [Eq. (11)]. Wittenbrook[60a] synthesized the unstable 5-amino-3-methylthio analog by the route shown in Eq. (11a).

$$\text{Ph-C} \begin{smallmatrix} \nearrow \text{O} \\ \searrow \text{Cl} \end{smallmatrix} \xrightarrow{\text{KSCN}} \text{Ph-C} \begin{smallmatrix} \nearrow \text{O} \\ \searrow \text{NCS} \end{smallmatrix} \xrightarrow{\text{MeOH}}$$

$$\text{Ph-C} \begin{smallmatrix} \nearrow \text{O} \\ \searrow \text{NH-C} \end{smallmatrix} \begin{smallmatrix} \nearrow \text{S} \\ \searrow \text{OMe} \end{smallmatrix} \xrightarrow{\text{MeI}} \text{Ph-C} \begin{smallmatrix} \nearrow \text{O} \\ \searrow \text{N=C} \end{smallmatrix} \begin{smallmatrix} \nearrow \text{SMe} \\ \searrow \text{OMe} \end{smallmatrix} \xrightarrow{\text{NH}_2\text{OH}}$$

$$\left[\text{Ph-C} \begin{smallmatrix} \nearrow \text{O} \\ \searrow \text{N=C} \end{smallmatrix} \begin{smallmatrix} \nearrow \text{OMe} \\ \searrow \text{NHOH} \end{smallmatrix} \right] \longrightarrow \underset{\text{Ph}}{\overset{\text{N}\text{---}\text{OMe}}{\bigcirc_{\text{O}}\text{N}}} \qquad (11)$$

$$\text{NCN=C(SK)}_2 \xrightarrow{\text{MeI}} \text{NCN=C(SMe)}_2 \xrightarrow{\text{NH}_2\text{OH}} \underset{\text{H}_2\text{N}}{\overset{\text{N}\text{---}\text{SMe}}{\bigcirc_{\text{O}}\text{N}}}$$
$$(11a)$$

An iminoether may also be used as the source of the C-5 carbon in the final product when the C-3 carbon comes from an amidoxime. Weidinger and Kranz[36] prepared 3-phenyl-5-methyloxadiazole in 87% yield and 5-phenyl-3-methyloxadiazole in 68% yield by this variation

---

[58] K. R. Huffman and F. C. Schaeffer, *J. Org. Chem.* **28**, 1812, 1816 (1963).
[59] S. T. Yang and T. B. Johnson, *J. Amer. Chem. Soc.* **54**, 2066 (1932).
[60] A. R. Katritzky, B. Wallis, R. T. C. Brownlee, and R. D. Topsom, *Tetrahedron* **21**, 1681 (1965).
[60a] L. S. Wittenbrook, *J. Heterocycl. Chem.* **12**, 37 (1975).

[Eqs. (12 and 13)]. The hydroxamic chloride has also been treated with the iminoether with a similar result[61,62] [Eq. (14)].

$$Me-C{\overset{+}{\underset{OEt}{\nwarrow}}}{NH_2Cl^-} + Ph-C{\underset{NH_2}{\nwarrow}}{NOH} \xrightarrow{\Delta} \underset{Me}{\overset{N\diagdown\diagup Ph}{\underset{O\diagdown N}{}}} \quad (12)$$

$$Ph-C{\overset{+}{\underset{OEt}{\nwarrow}}}{NH_2}\;Cl^- + Me-C{\underset{NH_2}{\nwarrow}}{NOH} \xrightarrow{\Delta} \underset{Ph}{\overset{N\diagdown\diagup Me}{\underset{O\diagdown N}{}}} \quad (13)$$

$$Ph-C{\underset{OEt}{\nwarrow}}{NH} + Ph-C{\underset{Cl}{\nwarrow}}{NOH} \longrightarrow \underset{Ph}{\overset{N\diagdown\diagup Ph}{\underset{O\diagdown N}{}}} \quad (14)$$

Benzamidine has been used recently in a novel way to synthesize oxadiazoles in good yields[63] [Eq. (15)]. A variation with the same starting compound may also find a limited use[64] [Eq. (16)].

$$Ph-C{\underset{NHCl}{\nwarrow}}{NH} + Me_2S \longrightarrow Ph-C{\underset{NH\overset{+}{S}Me_2}{\nwarrow}}{NH}\;Cl^- \xrightarrow{OH^-}$$

$$Ph-C{\underset{\overset{-}{N}\overset{+}{S}Me_2}{\nwarrow}}{NH} \xrightarrow{(RCO)_2O} Ph-C{\underset{\overset{-}{N}\overset{+}{S}Me_2}{\nwarrow}}{\overset{N-C-R}{\underset{}{\overset{\|}{O}}}} \xrightarrow{200°} \underset{R}{\overset{N\diagdown\diagup Ph}{\underset{O\diagdown N}{}}} \quad (15)$$

$$Ar-C{\underset{NH_2}{\nwarrow}}{NH} \xrightarrow{PhCOCl} Ar-C{\underset{NHCOPh}{\nwarrow}}{NH} \xrightarrow{Me_3COCl}$$

$$Ar-C{\underset{NHCOPh}{\nwarrow}}{N-Cl} \xrightarrow{NaOH} \underset{N\diagdown O\diagup}{\overset{Ar\diagdown\;\;\diagup N}{}}Ph \quad (16)$$

C. From C-Nitroso Derivatives

The nitrogen in C-nitroso derivatives has been utilized as a source of the N-2 atom in an oxadiazole. Nitrosation of an acylaminomalonic acid ester[65] gave N-acylamidoxime, which then cyclized to a carbethoxy oxa-

---

[61] F. Eloy and R. Lenaers, *Bull. Soc. Chim. Belges* **72**, 719 (1963).
[62] C. Musante, *Gazz. Chim. Ital.* **68**, 331 (1938).
[63] T. Fuchigami and K. Odo, *Chem. Lett.*, 247 (1974) [*CA* **80**, 120476 (1974)].
[64] T. Fuchigami and K. Odo, *Chem. Lett.*, 1139 (1974) [*CA* **82**, 16748 (1975)].
[65] H. Hellmann, H. Piechota, and W. Schwiersch, *Ber.* **94**, 757 (1961); Ger. Offen. 1,124,502 (Mar. 1962) [*CA* **57**, 3453 (1962)].

diazole, as shown in Eq. (17). The active hydrogen in an amidodi-

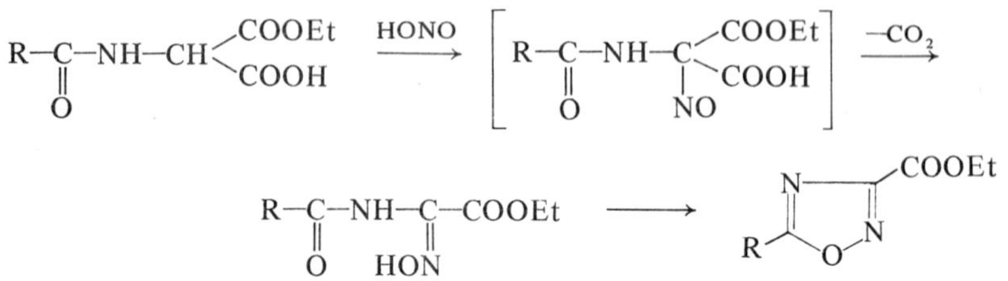

$$\text{R}-\underset{\underset{O}{\|}}{\text{C}}-\text{NH}-\text{CH}\underset{\text{COOH}}{\overset{\text{COOEt}}{<}} \xrightarrow{\text{HONO}} \left[\text{R}-\underset{\underset{O}{\|}}{\text{C}}-\text{NH}-\underset{\underset{\text{NO}}{|}}{\text{C}}\underset{\text{COOH}}{\overset{\text{COOEt}}{<}}\right] \xrightarrow{-\text{CO}_2}$$

$$\text{R}-\underset{\underset{O}{\|}}{\text{C}}-\text{NH}-\underset{\underset{\text{HON}}{\|}}{\text{C}}-\text{COOEt} \longrightarrow \underset{\text{R}}{\overset{\text{N}}{\underset{\text{O}}{\diagdown}}}\overset{\text{COOEt}}{\underset{\text{N}}{\diagup}} \qquad (17)$$

hydroresorcinol has been nitrosated in an analogous manner to give an oxadiazole (13) with the opened resorcinol ring now as side chain at C-3.[66]

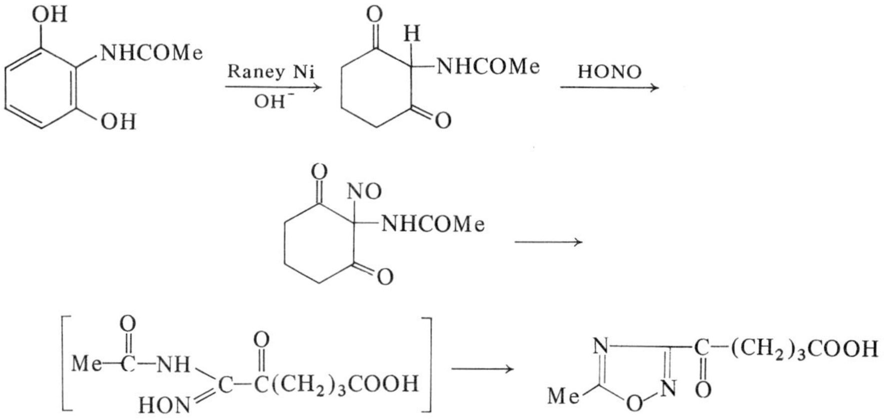

[diagram: 2-acetamidoresorcinol → (Raney Ni, OH⁻) → 2-acetamidocyclohexane-1,3-dione → (HONO) → 2-nitroso-2-acetamidocyclohexane-1,3-dione →]

$$\left[\underset{\text{HON}}{\overset{\text{Me}-\overset{O}{\overset{\|}{\text{C}}}-\text{NH}}{\diagdown}}\text{C}=\overset{\overset{O}{\|}}{\text{C}}-\text{C}(\text{CH}_2)_3\text{COOH}\right] \longrightarrow \underset{\text{Me}}{\overset{\text{N}}{\underset{\text{O}}{\diagdown}}}\overset{\text{C}-(\text{CH}_2)_3\text{COOH}}{\underset{\underset{\text{O}}{\|}}{\diagup}}$$

(13)

Nitrosoimidazoles are also hydrolyzed and recycled to oxadiazoles, apparently by the pathway given in Eq. (18) although the N-acylamidoxime was not isolated.[67] An aminooxadiazole has also been syn-

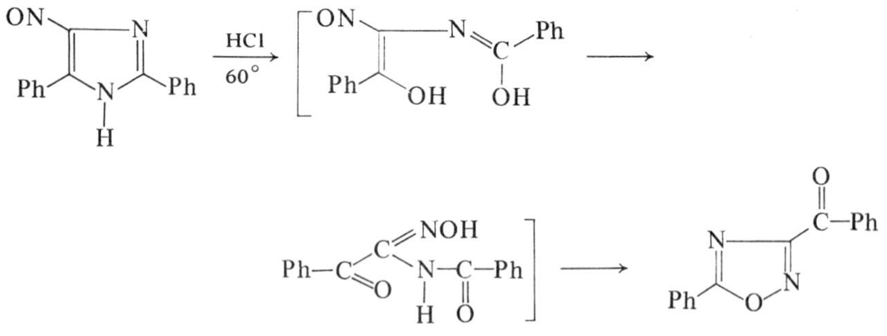

[diagram of nitrosoimidazole → HCl, 60° → intermediate →]

$$\left[\text{Ph}-\underset{\underset{O}{\|}}{\text{C}}\underset{\underset{H}{|}}{\overset{\diagup \text{NOH}}{\diagdown}}\underset{\underset{O}{\|}}{\overset{\text{N}-\text{C}-\text{Ph}}{}} \right] \longrightarrow \underset{\text{Ph}}{\overset{\text{N}}{\underset{\text{O}}{\diagdown}}}\overset{\overset{O}{\|}}{\underset{\text{N}}{\diagup}}\text{C}-\text{Ph} \qquad (18)$$

[66] H. Stetter and K. Hoehne, Ber. 91, 1123 (1958).
[67] S. Cusmano and M. Ruccia, Gazz. Chim. Ital. 85, 1686 (1955); 88, 463 (1958).

thesized by the nitrosation of a 2-aminoimidazole, followed by a similar rearrangement [68,69] (Section V, J, rearrangement 6).

D. 1,3-DIPOLAR CYCLOADDITIONS

Dipolar cycloadditions of nitrile oxides to nitriles have been widely exploited[70-81] since Leandri[6] first suggested the idea [Eq. (19)].

$$(Ar)R-C\equiv\overset{+}{N}-O^- + R'-C\equiv N \longrightarrow \underset{R'}{\overset{N}{\underset{O}{\bigvee}}}\overset{R(Ar)}{\underset{N}{}} \qquad (19)$$

Aromatic nitrile oxides may be added as such, but more commonly they are generated from hydroxamic chlorides by adding a base [Eq. (20)] or

$$Ar-C\underset{Cl}{\overset{NOH}{\diagdown}} \xrightarrow{Et_3N} Ar-C\equiv\overset{+}{N}-\bar{O} + Et_3\overset{+}{N}HCl^- \qquad (20)$$

by heat alone.[78,82] Aliphatic nitrile oxides[83] may be generated *in situ* from a primary nitroalkane[77] with the dehydrating action of phenyliso-cyanate [Eq. (21)]. Yields are as high as 88% for aryl nitrile oxides with

$$RCH_2NO_2 \xrightarrow[Et_3N]{2PhNCO} R-C\equiv\overset{+}{N}-\bar{O} + (PhNH)_2CO + CO_2 \qquad (21)$$

aryl or negatively substituted alkyl cyanides.[70] Yields of 35–40% were reported[71] for four aliphatic nitriles with benzonitrile oxide, but boron

---

[68] B. Cavalleri, P. Bellani, and G. Lancini, *J. Heterocycl. Chem.* **10**, 357 (1973).
[69] B. Cavalleri and G. Lancini, Ger. Offen. 2,331,058 (Jan. 1974) [*CA* **80**, 95961 (1974)].
[70] K. Bast, M. Christl, R. Huisgen, and W. Mack, *Chem. Ber.*, **105**, 2825 (1972).
[71] S. Morrocchi, A. Ricca, and L. Velo, *Tetrahedron Lett.*, 331 (1967).
[72] S. Morrocchi and A. Ricca, *Chim. Ind.* (Milan) **49**, 629 (1967).
[73] G. Leandri and M. Palotti, *Ann. Chim.* (Rome) **47**, 376 (1957).
[74] S. Morrocchi, A. Ricca, and L. Velo, *Chim. Ind.* (Milan) **49**, 168 (1967).
[75] S. Morrocchi, A. Ricca, and A. Zannarotti, *Chim. Ind.* (Milan) **50**, 352 (1968).
[76] R. Huisgen, *Proc. Chem. Soc.*, 357 (1961); R. Huisgen, W. Mack, and E. Anneser, *Tetrahedron Lett.*, 587 (1961).
[77] F. Eloy, *Bull. Soc. Chim. Belges* **73**, 793 (1964).
[78] R. Lenaers and F. Eloy, *Helv. Chim. Acta* **46**, 1067 (1963).
[79] G. Rembarz, E. Fischer, and F. Tittelbach, *J. Prakt. Chem.* **313**, 1065 (1971).
[80] P. Grünanger and P. Fingi, *Atti Accad Naz. Lincei, Rend., Cl. Sci. Fis., Mat. Nat.*
[82] S-J. Hong, *Daehan Hwahak Hwoejee* **15**, 121 (1971) [*CA* **76**, 4196 (1972)].
[81] L. Fabrini and G. Speroni, *Chim. Ind.* (Milan) **43**, 807 (1961).
[82] S-J. Hong, *Daehan Hwahak Hwoejee* **15**, 121 (1971) [*CA* **76**, 4196 (1972)].
[83] C. Grundmann and P. Grünanger, "The Nitrile Oxides," (pp. 44ff.) Springer-Verlag, Berlin, 1971.

trifluoride etherate was needed as catalyst. Addition of the nitrile oxide to an aldoxime to give the dihydro intermediate (14) which then lost water [Eq. (22)] was generally less satisfactory.[72]

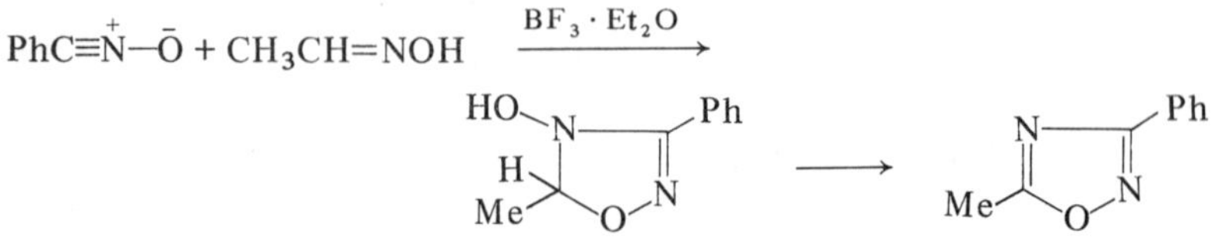

$$PhC≡\overset{+}{N}-\bar{O} + CH_3CH=NOH \xrightarrow{BF_3 \cdot Et_2O}$$

(14) → (22)

An interesting sandwich compound [Eq. (23)] bearing an oxadiazolyl side chain was recently synthesized.[84] By nitrosating the methylene

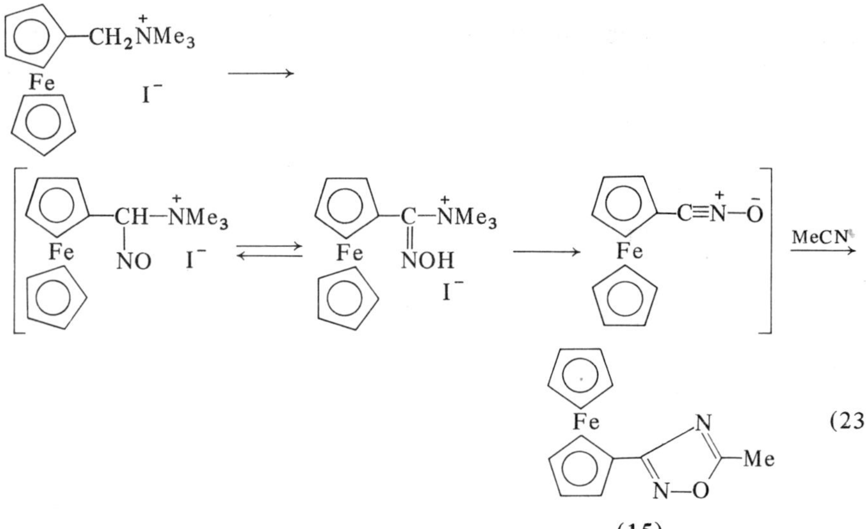

(15)

group adjacent to a quaternary ammonium salt, the nitrile oxide was generated in the presence of acetonitrile to give the 3-ferrocenyl-5-methyl-1,2,4-oxadiazole (15).

E. Oxidation of Dihydrooxadiazoles

Oxadiazolines are easily prepared by the condensation of an aldehyde and an amidoxime[14, 48–50, 85–88] [Eq. (24)]. This partially reduced ring

[84] T. Kondo, K. Yamamoto, H. Danda, and M. Kumada, *J. Organometal. Chem.* **61**, 361 (1973); T. Kondo, K. Yamamoto, and M. Kumada, *ibid.* **61**, 355 (1973).
[85] C. Malavaud, M-T. Boisden, and J. Barrans, *Bull. Soc. Chim. Fr.* **11**, Pt. 2, 2296 (1973).
[86] T. N'Gando M'Pondo, C. Malavaud, and J. Barrans, *C. R. Acad. Sci., Ser. C* **274**, 2026 (1972).
[87] D. Vorländer, *Ber.* **24**, 803 (1891).
[88] H. Zimmer, *Ber.* **22**, 3140 (1889).

PhCH$_2$CHO + Ph—C(=NOH)(NH$_2$) ⟶

[structure: HN—, H, O, N, Ph, PhCH$_2$] $\xrightarrow[\text{HOAc}]{\text{KMnO}_4}$ [structure: Ph—CH$_2$, O, N, N, Ph] (24)

may then be oxidized to the oxadiazole by such reagents as potassium permanganate,[48, 87, 88] nitrogen dioxide in ether,[86] sodium hypochlorite,[86] and N-chlorosuccinimide.[86]

F. Oxidation of Amidoximes and Oximes

The action of bromine, ferricyanide,[89] and other mild oxidizing agents[90] directly on benzamidoxime has been reported to give diphenyloxadiazole (16). While it is clear that one nitrogen has been oxidized, no intermediate was isolated.

Ph—C(=NOH)(NH$_2$) $\xrightarrow{[O]}$ [structure of 3,5-diphenyl-1,2,4-oxadiazole]

(16)

The reaction of nitrogen dioxide on benzaldoxime is used as a synthesis for phenyldinitromethane,[91] but milder conditions[92] and other reagents[93–98] give a compound known as "benzaldoxime peroxide."

Boyer[99] depicted this compound as benzaldoxime anhydride N-oxide (17a), but the alternative formula (17b)[93] cannot be excluded in view of the present known chemistry. In chloroform or benzene it can be converted into diphenyloxadiazole. Thermal decomposition also converts 17

[89] H. Krümmel, *Ber.* **28**, 2227 (1895).
[90] E. Beckmann, *Ber.* **22**, 1588 (1889).
[91] L. F. Fieser and W. E. Doering, *J. Amer. Chem. Soc.* **68**, 2252 (1946).
[92] C. Grundmann and G. F. Kite, *Synthesis* **3**, 156 (1973).
[93] L. Horner, L. Hockenberger, and W. Kirmse, *Chem. Ber.* **94**, 290 (1961).
[94] P. A. S. Smith and G. E. Hein, *J. Amer. Chem. Soc.* **82**, 5731 (1960).
[95] G. Ponzio and E. Durio, *Gazz. Chim. Ital.* **60**, 436 (1930); G. Ponzio, *ibid.* **62**, 860 (1932).
[96] J. P. Freeman, *J. Amer. Chem. Soc.* **82**, 3869 (1960).
[97] G. Just and K. Dahl, *Tetrahedron* **24**, 5251 (1968).
[98] J. Bougault and P. Robin, *C.R. Acad. Sci.* **169**, 341 (1919); P. Robin, *ibid.* **169**, 695 (19,19); **171**, 1150 (1920); *Ann. Chim. (Paris)* [9] **16**, 77 (1921).
[99] J. H. Boyer and H. Alul, *J. Amer. Chem. Soc.* **81**, 4237 (1959); H. Kropf and R. Lambeck, *Ann.* **700**, 18 (1966).

into diphenyloxadiazole and several other products, among which is the oxadiazole N-oxide (**18**).[92]

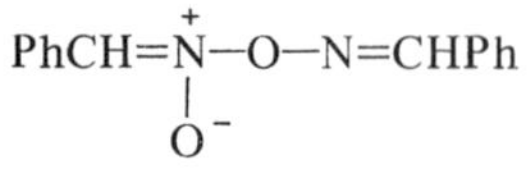

(17a) 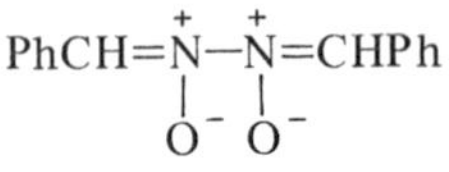　　　(17b) 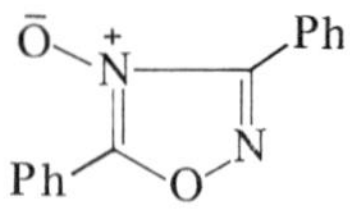　　　(18)

Van Meeteren and van der Plas[100] oxidized several imidazoles to oxadiazoles. They proposed the formylamidine intermediate shown in Eq. (25) since alkaline oxidation of imidazoles is known[101] to produce

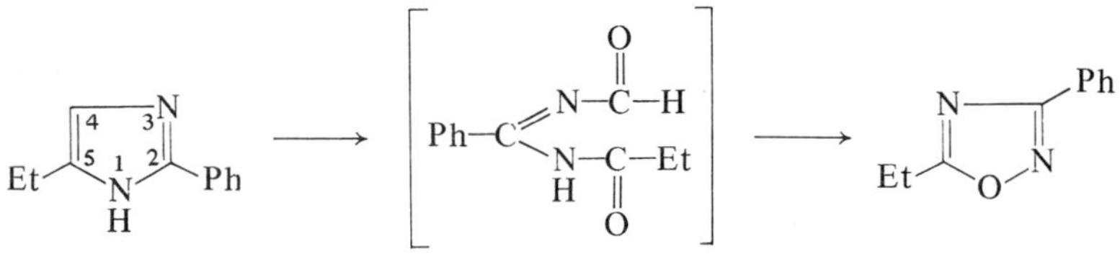

(25)

such derivatives. An additional fact supporting the intermediate was that imidazoles disubstituted at C-4, C-5 could not be oxidized to oxadiazoles.

None of the oxidation methods (Section II, E and F) are preferred synthetic routes to oxadiazoles.

G. Rearrangements

Beckman rearrangements[102,103] of α-dioximes sometimes lead to 1,2,4-oxadiazoles, although more often the dehydrating conditions produce 1,2,5-oxadiazoles (furazans). In fact, α-benzildioxime in polyphosphoric acid gives diphenyl-1,2,4-oxadiazole in 99% yield[104] [Eq. (26)].

Some 1,2,5-oxadiazoles can be transformed into 1,2,4-oxadiazoles but are themselves not easy to obtain.[95,102,105]

[100] H. W. van Meeteren and H. C. van der Plas, *Rec. Trav. Chim. Pays-Bas* **87**, 1089 (1968); **88**, 204 (1969).
[101] E. H. White and M. J. C. Harding, *J. Amer. Chem. Soc.* **86**, 5686 (1964); *Photochem. Photobiol.* **4**, 1129 (1965).
[102] A. H. Blatt, *Chem. Rev.* **12**, 215 (1933).
[103] F. D. Dodge, *Ann.* **264**, 178 (1891); E. Durio and A. Sburlati, *Gazz. Chim. Ital.* **62**, 1035 (1932).
[104] R. T. Conley and F. Mikulski, *J. Org. Chem.* **24**, 97 (1959).
[105] G. Ponzio, *Gazz. Chim. Ital.* **62**, 415 (1932).

$$Ph-\underset{HON}{\overset{\|}{C}}-\underset{NOH}{\overset{\|}{C}}-Ph \xrightarrow[120°]{H^+} \left[Ph-\underset{+N \overset{\|}{\swarrow} NOH}{\overset{\|}{C}}-\overset{\|}{C}-Ph \longrightarrow \underset{Ph}{\overset{N}{\|}}\overset{C-Ph}{\underset{C^+ NOH}{\|}} \right] \xrightarrow{-H^+}$$

$$\underset{Ph}{\overset{N}{\swarrow}}\overset{\text{---}}{\underset{O}{\text{---}}}\overset{Ph}{\underset{N}{\text{---}}} \quad (26)$$

A Curtius rearrangement has been suggested as a pathway to 5-hydroxyoxadiazoles[106] [Eq. (27)], but the method has not yet appealed to other workers.

$$PhCHO + MeCOCOOH \xrightarrow[MeOH]{NaOH} Ph-CH=CH-CO-COOH \xrightarrow[2.\ NH_2OH]{1.\ MeOH}$$

$$Ph-CH=CH-\underset{HO-N}{\overset{\|}{C}}-COOMe \xrightarrow{N_2H_4} Ph-CH=CH-\underset{HO-N}{\overset{\|}{C}}-CONHNH_2 \xrightarrow{HONO}$$

$$Ph-CH=CH-\underset{HON}{\overset{\|}{C}}-C\overset{O}{\underset{N_3}{\diagdown}} \xrightarrow[-N_2]{\Delta\ EtOH} \left[Ph-CH=CH-C\overset{NOH}{\underset{N=C=O}{\diagdown}} \right] \longrightarrow$$

$$\underset{HO}{\overset{N}{\swarrow}}\overset{\text{---}}{\underset{O}{\text{---}}}\overset{CH=CHPh}{\underset{N}{\text{---}}} \quad (27)$$

III. Synthesis of Oxadiazolines and Other Reduced Rings

A. Oxadiazolines and Oxadiazolidines

1. Δ^2-Oxadiazolines

The best method of synthesizing Δ^2-oxadiazolines is the cycloaddition of aliphatic[107-109] or aromatic[110] nitrile oxides to Schiff bases

$$ArCH=N-C_4H_9(t) + MeC\overset{+}{\equiv}\overset{-}{N}-\overset{-}{O} \longrightarrow \underset{Ar}{\overset{(t)C_4H_9-N}{\underset{H}{\diagup}}}\overset{Me}{\underset{O}{\diagdown}}\overset{}{\underset{N}{\diagup}} \quad (28)$$

[106] W. R. Vaughan and J. L. Spencer, *J. Org. Chem.* **25**, 1077 (1960).
[107] R. M. Srivastava and L. B. Clapp, *J. Heterocycl. Chem.* **5**, 61 (1968).
[108] G. Srimannarayana, R. M. Srivastava, and L. B. Clapp, *J. Heterocycl. Chem.*, **7**, 151 (1970).
[109] R. M. Srivastava and L. B. Clapp, *J. Heterocycl. Chem.* **5**, 735 (1968).
[110] C. Moussebois and J. F. M. Oth, *Helv. Chim. Acta* **47**, 942 (1964).

[Eq. (28)]. An unsubstituted imine[111] and an oxime[112] have also been used in place of the Schiff base as substrate, the latter to yield an *N*-hydroxy derivative.

Heterocyclic rings attached to the amidoxime function have been used to synthesize Δ^2-oxadiazolines[113] as exemplified by the synthesis shown below for compound **19**.

2. Δ^3-Oxadiazolines

Barrans and co-workers[14,85,86] used the ubiquitous benzamidoxime (Section II, A) as a starting point for the synthesis of Δ^2- and Δ^3-oxadiazolines [Eq. (29)]. By displacement of the hydrogen on N-4 in the Δ^2-oxadiazoline (**19**), the anion **20** was obtained. Alkylation of this anion with methyl iodide gave a quantitative yield of a 1 : 1 mixture of 3-phenyl-4,5-dimethyl-Δ^2-oxadiazoline (**21**) and the previously elusive Δ^3-isomer (**22**). With other starting compounds, different ratios of Δ^2- and Δ^3-isomers were obtained, but overall yields were quantitative.

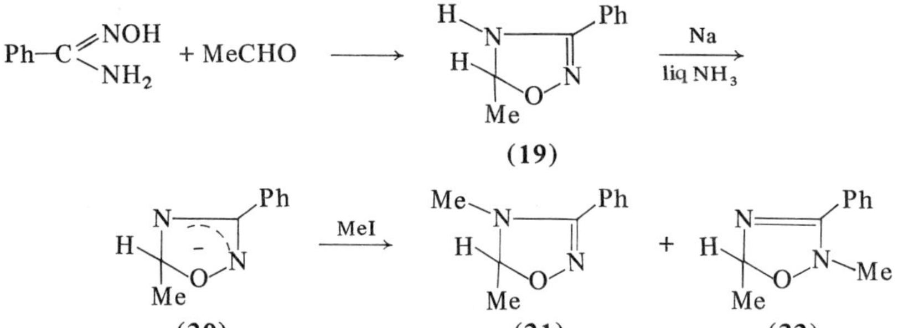

One other synthesis of a Δ^3-oxadiazoline was obtained unintentionally. Hull and Farrand[114] refluxed cyanamide and phenylhydroxylamine in acetone expecting to obtain the disubstituted hydroxylamine (**23**). Instead, the solvent acetone had condensed with the hypothetical intermediate to give 3-amino-5,5-dimethyl-2-phenyl-Δ^3-oxadiazoline (**24**).

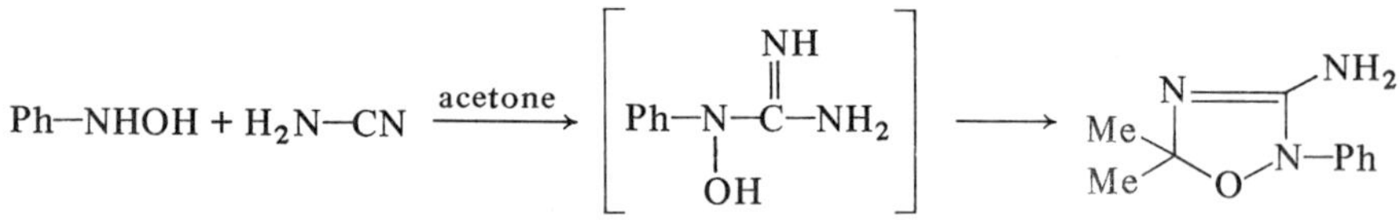

[111] M. Kamiya, *Bull. Chem. Soc. Jap.* **43**, 3344 (1970).
[112] J. L. Cotter, *J. Chem. Soc.*, 5491 (1964); *J. Chem. Soc. B*, 1271 (1967).
[113] E. A. Watts, Ger. Offen. 2,257,311 (June 1973) [*CA* **79**, 66366 (1973)].
[114] R. Hull and R. Farrand, *J. Chem. Soc.*, 6028 (1963).

3. Δ⁴-Oxadiazolines

Grigat and Pütter[115] succeeded in synthesizing a Δ⁴-oxadiazoline by cycloaddition of a nitrone to an arylcyanate[43] [Eq. (30)].

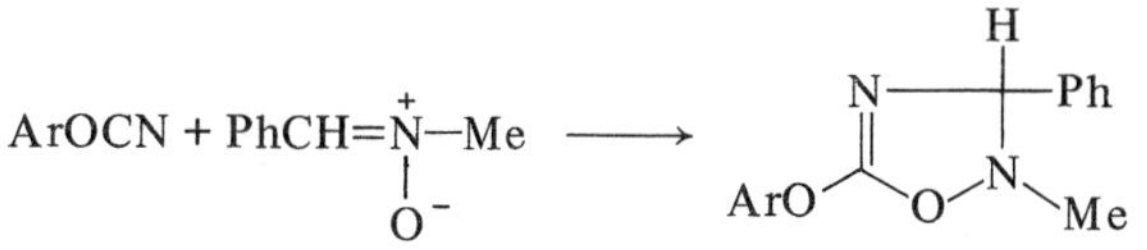

$$\text{ArOCN} + \text{PhCH}=\overset{+}{\underset{\underset{O^-}{|}}{N}}-\text{Me} \longrightarrow \underset{\text{ArO}}{\overset{\overset{H}{\underset{|}{N}-\overset{|}{\underset{N}{C}}-\text{Ph}}}{\diagdown}}\hspace{-1em}\diagdown\hspace{-0.5em}\text{O}\diagdown\text{Me} \quad (30)$$

4. Oxadiazolidines

The completely reduced oxadiazole ring (**26**) was recently reported[116] as an air oxidation product of a nitrimine. Presumably it was formed by a 1,3-dipolar addition of the intermediate nitrimine nitrone (**25**) to the nitrimine. The nitrimine results from the deoxygenation of a *gem*-nitrosonitro compound by triethyl phosphite [Eq. (31)].

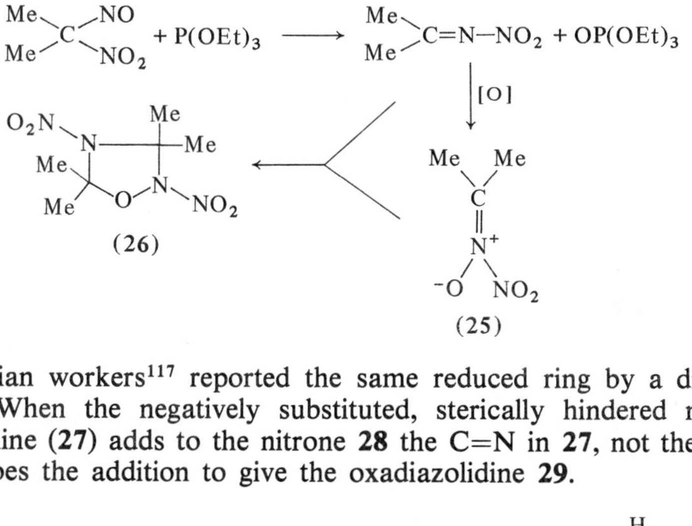

Russian workers[117] reported the same reduced ring by a different route. When the negatively substituted, sterically hindered mesitylketenimine (**27**) adds to the nitrone **28** the C=N in **27**, not the C=C, undergoes the addition to give the oxadiazolidine **29**.

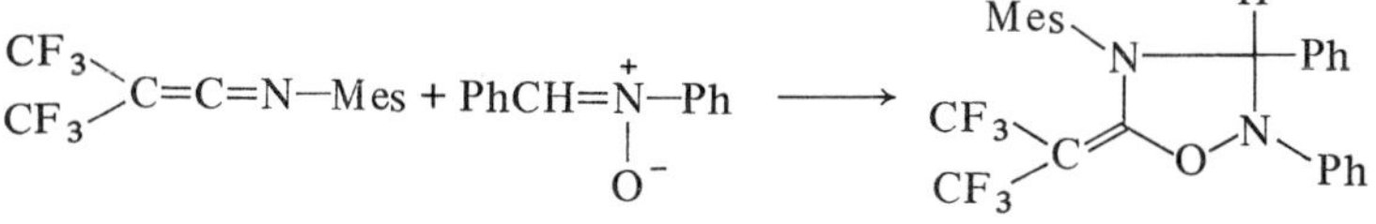

[115] E. Grigat, R. Pütter, and E. Muhlbauer, *Ber.* **98**, 3777 (1965).
[116] J. Burdon and A. Ramirez, *Tetrahedron* **29**, 4195 (1973).
[117] D. P. Del'tsova and N. P. Gambaryan, *Izv. Akad. Nauk SSSR, Ser. Khim.* **11**, 2566 (1973); *Bull. Acad. SSR, Chem. Sci.* **22**, 2500 (1973).

Incidentally, compound **22** was not reduced to the corresponding oxadiazolidine by lithium aluminum hydride.[85]

B. OXADIAZOLINONES AND RELATED COMPOUNDS

1. *Oxadiazolinones*

When the hydroxy proton in an amidoxime is replaced by a carbethoxy group (**30**), cyclization leads to an oxadiazolin-5-one (**31**) as shown in Eq. (32). An obvious extension is to use an *N*-alkyl-amidoxime[118] to give an *N*(4)-alkyloxadiazolin-5-one.

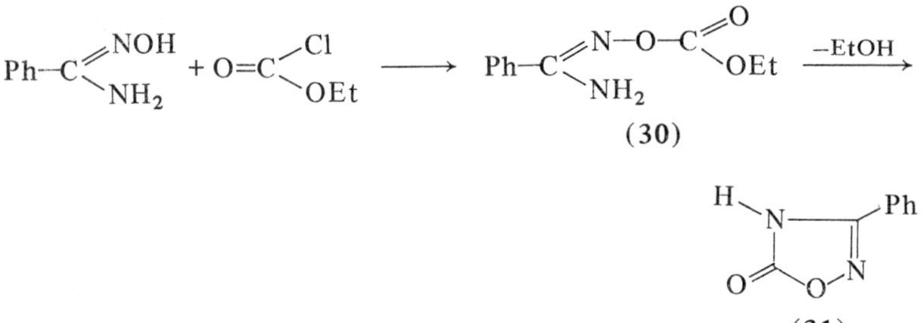

$$Ph-C(NOH)(NH_2) + O=C(Cl)(OEt) \longrightarrow Ph-C(N-O-C(=O)OEt)(NH_2) \xrightarrow{-EtOH}$$
(30)

$$\text{(31)} \quad \text{1,2,4-oxadiazolin-5-one} \tag{32}$$

A similar type of reaction was recently used to introduce an amine group at C-3 [Eq. (33)].[119]

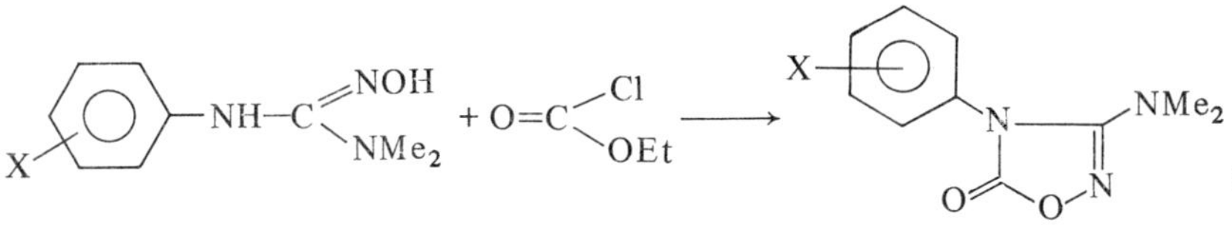

$$X-C_6H_4-NH-C(NOH)(NMe_2) + O=C(Cl)(OEt) \longrightarrow \tag{33}$$

A further stage of reduction is reached when an oxaziridine (**32**) opens in the presence of phenyl isocyanate. This can be accomplished [Eq. (34)] in a sealed tube at 85° in benzene to give a 94% yield of a

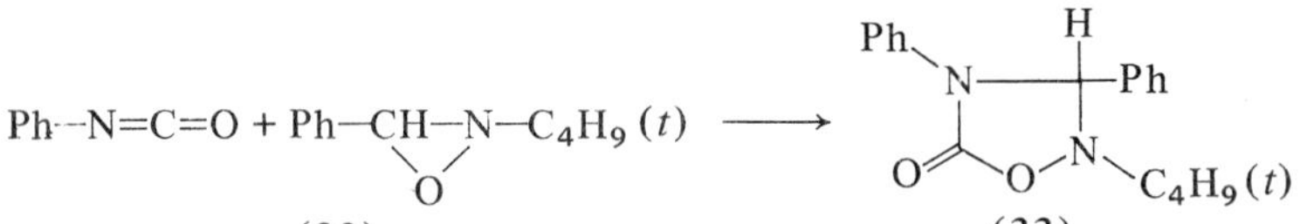

$$Ph-N=C=O + Ph-CH-N-C_4H_9\,(t) \longrightarrow \tag{34}$$
$$\quad\quad\quad\quad\quad\quad\quad\quad\,\backslash O /$$
(32)   (33)

1,2,4-oxadiazolidin-5-one (**33**).[120] With phenyl isothiocyanate the cor-

---

[118] G. D'Alo, M. Perghum, and P. Grünanger, *Ann. Chim. (Rome)* **53**, 1405 (1963).
[119] M. Gross, P. Held, and H. Schubert, *J. Prakt. Chem.* **316**, 434 (1974).
[120] M. Komatsu, Y. Ohshiro, H. Hotta, M. Sato, and T. Agawa, *J. Org. Chem.* **39**, 948 (1974).

responding 5-thione derivative was obtained in 68% yield.[121] With the cumulative double-bond system of diphenyl carbodiimide (34) and the nitrone (35) (an isomer of 32), the analogous product (36) was obtained

$$\text{Ph}-\text{N}=\text{C}=\text{N}-\text{Ph} + \text{Ph}-\text{CH}=\overset{+}{\underset{\underset{\text{O}^-}{|}}{\text{N}}}-\text{C}_4\text{H}_9(t) \longrightarrow \underset{(36)}{\text{PhN}\overset{\text{Ph}}{\underset{\text{O}}{\diagup}}\overset{\text{H}}{\underset{\text{N}}{|}}\text{Ph}\atop\text{C}_4\text{H}_9(t)} \quad (35)$$

(34) (35) (36)

in quantitative yield.[122] Other examples of the reactions shown in Eqs. (34) and (35), generally with lower yields, are given by the Japanese workers. Yet other cases are known.[123,124] Phenyl isocyanate also adds to nitrones[125] and hindered nitrile oxides[70] as other examples of 1,3-dipolar additions.

A nitrogen common to two rings was obtained by cycloaddition of phenyl isocyanate to 3-picoline N-oxide.[126] Isomeric products 37 and 38 were obtained, the first predominating. Phosgene, acting on an aminoquinoxaline N-oxide, gave a comparable fused ring system.[127] Hamana

and Kumadaki[128] recently reported a similar accomplishment with 2-aminoquinoline N-oxide and cyanogen bromide [Eq. (36)] though the same product had been obtained earlier by heating compound 39.[129]

---

[121] M. Komatsu, Y. Ohshiro, K. Yasuda, S. Ichijima, and T. Agawa, *J. Org. Chem.* **39**, 957 (1974).
[122] M. Komatsu, Y. Ohshiro, and T. Agawa, *J. Org. Chem.* **37**, 3192 (1972).
[123] R. B. Moffett, *J. Org. Chem.* **39**, 568 (1974).
[124] G. Zimmer and O. Hantelmann, *Arch. Pharm. (Weinheim)* **307**, 780 (1974) [*CA* **82**, 4182 (1975)].
[125] G. Zimmer, O. Hantelmann, and U. Dybowski, *Chem-Ztg.* **97**, 205 (1973) [*CA* **79**, 18640 (1973)].
[126] T. Hisano, S. Yoshikama, and K. Muraoka, *Chem. Pharm. Bull.* **22**, 1611 (1974).
[127] F. Seng and K. Ley, Ger. Offen. 2,232,468 (Jan. 1974) [*CA* **80**, 83067 (1974)].
[128] M. Hamana and S. Kumadaki, *Chem. Pharm. Bull.* **22**, 1506 (1974).
[129] H. Tanida, *Yakugaku Zasshi* **79**, 1063 (1959) [*CA* **54**, 4587 (1960)]; see also A. R. Katritzky, *J. Chem. Soc.*, 2063 (1956).

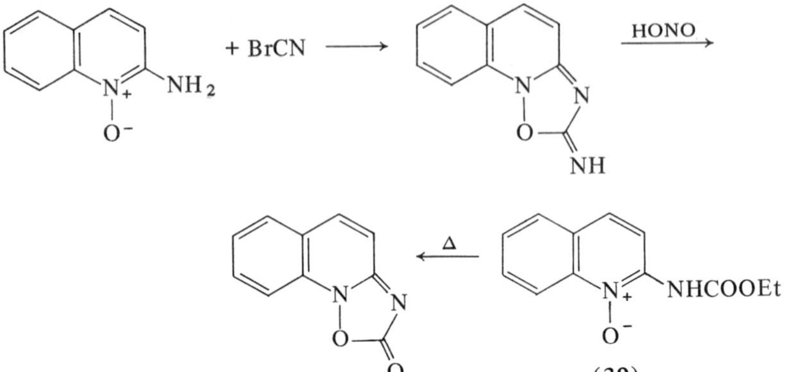

2. Oxadiazolidinediones

When the carbons in reactants that end up at C-3 and C-5 in the ring are both in oxidation state 4+, then oxadiazolidinediones may be formed. For example,[130] the very reactive isocyanate **40** reacts with hydroxylamine to give an intermediate (**41**) that cyclizes in alkaline solution to yield the parent compound of this series, oxadiazolidine-3,5-dione (**42**). Other N-substituted derivatives[131–133] and oxygen replacement compounds[132] have more recently been reported [Eq. (38)]. The sulfur-containing compound **43** can be converted into **45** by the use of mercury(II) oxide, and the oxygen analog of **43** can be hydrolyzed in acid solution to the N2,N4-disubstituted oxadiazolidine-3,5-dione (**44**).

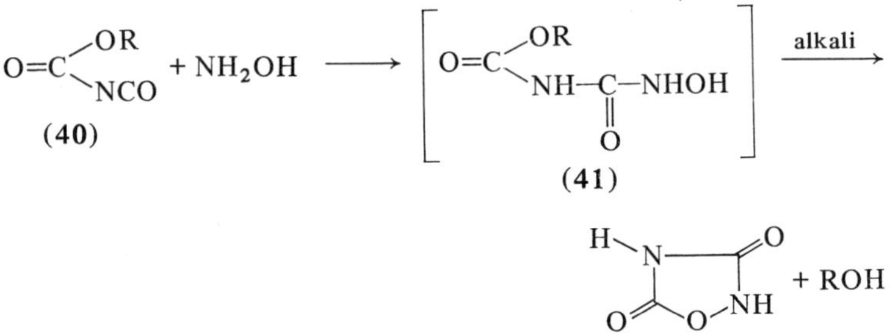

---

[130] G. Zimmer and R. Stoffel, *Arch. Pharm. (Weinheim)* **302**, 691 (1969) [*CA* **72**, 21631 (1970)].
[131] H. Moser and J. Rumpf, S. Afr. 67-06664 [*CA* **70**, 77976 (1969)].
[132] G. Voss, E. Fischer, and H. Werchau, *Z. Chem.* **13**, 102 (1973).
[133] G. W. Ivie, H. W. Dorough, and R. A. Cardona, *J. Agr. Food Chem.* **21**, 386 (1973).

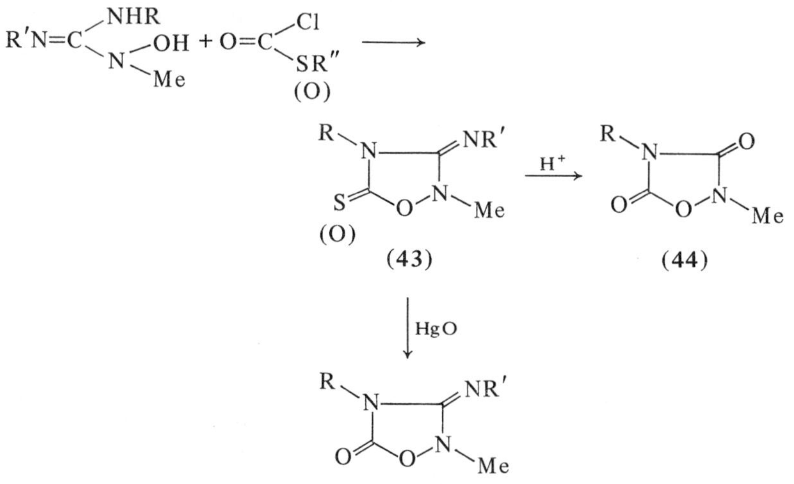

Sec. IV.A]   1,2,4-OXADIAZOLES   85

$$R'N=C\begin{smallmatrix}NHR\\N-OH\\Me\end{smallmatrix} + O=C\begin{smallmatrix}Cl\\SR''\end{smallmatrix} \longrightarrow$$
(O)

(43) → [H⁺] → (44)   (38)

(43): R-N, S=C(O), O-N(Me), =NR'
(44): R-N, O=C, O-N(Me), =O

↓ HgO

(45): R-N, O=C, O-N(Me), =NR'

IV. Physical and Spectral Properties

The parent compound in the oxadiazole series has a boiling point of 87°,[39] which falls within the normal range for two-carbon compounds of comparable molecular weights. The heats of vaporization of the first three homologs lie in the range 9.5–10 kcal/mole, comparable to ethanol.[3]

The aliphatic derivatives of oxadiazole have an ethereal odor while the aromatic ones are more like aromatic esters in this respect.[3]

Oxadiazole is soluble in water and organic solvents. Other homologs follow normal solubility rules.[134] However, oxadiazole codistills with organic solvents, even ether; this means that there is some van der Waals attraction with other molecules. It cannot be recovered in pure form from a solution by distillation (Section V, I).

The boiling points or melting points of more than 400 oxadiazoles are recorded by Eloy[3] with references.

A. Electron Density and Molecular Geometry

X-ray diffraction measurements have not been reported on oxadiazole so bond lengths, bond angles, and planar geometry have been

[134] R. L. Shriner, R. C. Fuson, and D. Y. Curtin, "Identification of Organic Compounds," 5th ed., Chapter 6, p. 67ff. Wiley, New York, 1964.

assumed for theoretical calculations.[111,135-138] The estimated bond lengths and bond angles given in Scheme 4 were used for a theoretical

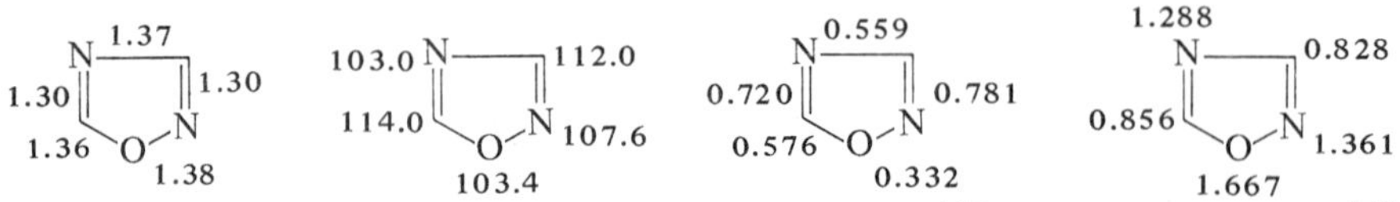

Estimated bond lengths[137]: N–N 1.37, C=N 1.30, C=N 1.30, C–O 1.36, C–O 1.38

Estimated bond angles[137]: 103.0, 112.0, 114.0, 107.6, 103.4

Bond orders[111]: 0.559, 0.720, 0.781, 0.576, 0.332

Electron densities[111]: 1.288, 0.828, 0.856, 1.361, 1.667

SCHEME 4

study of electric dipole moments (Section IV, C) by the VE SCF (variable electronegativity-self-consistent field) method.[137] Kamiya[111] calculated electron densities for the six π-electrons by the standard Pariser–Parr–Pople SCF method using two different assumptions. The results are shown in Scheme 4 for the first method, which included $\beta_{\mu\nu}$. Other calculations have been made by an improved linear combination of atomic orbitals (LCAO) method,[138] a complete neglect of differential overlap (CNDO)/2 method,[139] Hückel molecular orbital (MO) method,[136] and an extended Hückel method[135] to estimate dipole moments of oxadiazole (Section IV, C).

Paolini and Cignitti[140] and Kamiya[111] attempted to correlate the electronic structure of 1,2,4-oxadiazole with its chemical properties.

The crystal structure of a related sulfur compound, 2,4-dimethyl-1,2,4-thiadiazolidine-3,5-dione was recently determined.[141]

B. THERMODYNAMIC FUNCTIONS

Soptrajanov[142] calculated thermodynamic functions for oxadiazole over the range 298.16–1000°K, based on the infrared spectral assignments of Zecchina and co-workers,[143] with which he agreed ex-

[135] W. Adam and A. Grimison, *Theor. Chim. Acta* **7**, 342 (1967).
[136] B. Zurawski, *Bull. Acad. Pol. Sci., Ser. Sci. Chim.* **14**, 481 (1966) [*CA* **66**, 14842 (1967)].
[137] R. D. Brown, B. A. W. Coller, and J. E. Kent, *Theor. Chim. Acta* **10**, 435 (1968).
[138] M. Roche, F. D'Amato, and M. Benard, *J. Mol. Struct.* **9**, 183 (1971).
[139] D. W. Davies and W. C. Mackrodt, *Chem. Commun.* 345 and 1226 (1967).
[140] L. Paolini and M. Cignitti, *Tetrahedron* **24**, 485 (1968).
[141] C. L. Raston, A. H. White, A. C. Willis, and J. N. Varghese, *J. Chem. Soc., Perkin Trans. II*, 1096 (1974).
[142] B. Soptrajanov, *Croat. Chem. Acta* **41**, 223 (1969).
[143] A. Zecchina, G. E. Andreoletti, and P. Sampietro, *Spectrochim. Acta, Part A* **23**, 2647 (1967).

cept for one absorption (Section IV, D). The three functions and the heat capacity are compared with other heterocycles in Table I at standard temperature.

TABLE I

THERMODYNAMIC FUNCTIONS OF FIVE HETEROCYCLES (kcal/mole) AT 298.16°[a]

	Isoxazole	Oxazole	1,2,5-Oxadiazole	1,3,4-Oxadiazole	1,2,4-Oxadiazole	
$\Delta H°$	9.56	9.55	9.41	9.16	9.23^b	9.19^c
$\Delta G°$	55.17	55.11	53.68	53.56	54.96	54.95
$T\Delta S°$	64.73	64.66	63.09	62.72	64.19	64.14
C_p^0	14.39	14.37	13.81	12.91	13.11	12.99

[a] From Soptrajanov.[142]
[b] Calculated from Soptrajanov's infrared (IR) assignments.[142]
[c] Calculated from Zecchina's IR assignments.[143]

C. DIPOLE MOMENTS

Sheridan and co-workers[144] took the microwave spectra of oxadiazole after its synthesis in 1962.[39] From an analysis of Stark effects for a number of transitions, they concluded that the dipole moment should be 1.2 ± 0.3 D (Table II). Davies and Mackrodt[139] calculated the dipole moment to be 1.34 D, within the experimental error of Sheridan's value, by the CNDO/2 method of Pople and Segal. Other calculations[135,136,138] indicate that the dipole moment of 1,2,4-oxadiazole should be notably less than that of the 1,2,5- and 1,3,4-isomers. Direct measurements of dipole moments by Milone[145] had portended this much earlier. Besides the ones given in Table II, Milone had also found the same range of dipole moments for 3-methyl-5-phenyl, 5-methyl-3-phenyl, and other derivatives of the three sets of isomeric oxadiazoles.

With these wide differences (Table II) among the three sets of isomers, it is surprising that dipole moment measurements were not used in the 1930s for identification. Ponzio's work,[95, 101, 103, 146] in particular,

[144] J. H. Griffiths, A. Wardley, V. E. Williams, N. L. Owen, and J. Sheridan, *Nature (London)* **216**, 1301 (1967).
[145] M. Milone, *Gazz. Chim. Ital.* **65**, 152 (1935).
[146] J. Boeseken and D. P. Ross van Lennep, *Rec. Trav. Chim. Pays-Bas* **31**, 196 (1912).

TABLE II
Dipole Moments of Oxadiazoles

Oxadiazole	Dipole moment (Debye units)	Method
1,2,4-	1.2 ± 0.3	Microwave spectra, Stark effect[144]
	1.34	CNDO/2[139] (Pople and Segal) and quadrupole (^{14}N) coupling constants
3,5-Diphenyl-1,2,4-	1.56	Measured in dioxane[145]
3,4-Diphenyl-1,2,5-	4.74	Measured in dioxane
2,5-Diphenyl-1,3,4-	3.86	Measured in dioxane
5-Methyl-3-phenyl-1,2,4-	1.41	Measured in dioxane
3-Methyl-5-phenyl-1,2,4-	1.63	Measured in dioxane

on cyclizations of dioximes generally gave mixtures of 1,2,4- and 1,2,5-oxadiazoles. Unequivocal identification was therefore a problem. During the same decade, the pathway of the Beckmann rearrangement had been reinterpreted. Dipole moments might therefore have been used to decide which of two ways the dehydration of dioximes proceeded [Eq. (39)].

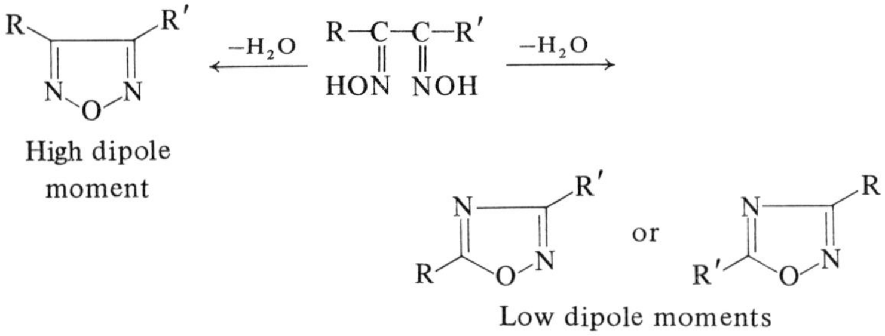

$$\underset{\text{High dipole moment}}{\underset{N\diagdown_{O}\diagup N}{R\diagup\diagdown R'}} \xleftarrow{-H_2O} \underset{HON\quad NOH}{R-\overset{\|}{C}-\overset{\|}{C}-R'} \xrightarrow{-H_2O} \underset{\text{Low dipole moments}}{\underset{R\diagdown_{O}\diagup N}{N\diagup\diagdown R'}\quad\text{or}\quad\underset{R'\diagdown_{O}\diagup N}{N\diagup\diagdown R}} \quad (39)$$

D. Infrared Spectra

Vibrational assignments from infrared spectra of oxadiazole, assuming planar geometry (symmetry C_s) and reasonable bond lengths, were made by Zecchina and co-workers.[143] These suggested assignments have been revised in only one respect (Table III) by one theoretical worker.[142]

Spectra are published for 3,5-diphenyl-, 3-methyl-5-phenyl-, and 5-methyl-3-phenyl-1,2,4-oxadiazoles.[147] Barrans[148] suggested 1560–1590 cm^{-1} as the range for C=N absorption and 885–915 cm^{-1}

[147] M. Milone and E. Borello, *Gazz. Chim. Ital.* **81**, 677 (1951).
[148] J. Barrans, *C.R. Acad. Sci.* **249**, 1096 (1959).

TABLE III

Infrared Spectral Assignments for 1,2,4-Oxadiazole

3147, 3076 cm^{-1}	C—H stretch[a]
1560, 1430, 1365	Ring mode
1289	Ring mode [in-plane mode]
1229	In-plane bend [ring mode]
1125	Ring-breathing
1093	In-plane bend (CH)
956, 858	Ring mode
941	Out-of-plane bend (CH) [886]
886	Out-of-plane bend (CH) [855]
649, 618	Out-of-plane bend (ring)

[a] Assignments of Zecchina et al.[143] except those in brackets, which are from Soptrajanov.[142]

for the N—O band in oxadiazoles. There is now general agreement[70, 149] on these assignments. Huisgen[70] found 1550–1565 cm^{-1} for C=N absorption in several Δ^2-oxadiazolines. Barrans and co-workers[86] reported NH bands at 3320 and 3220–3250 cm^{-1} in nine Δ^2-oxadiazolines with alkyl groups at C-3 and C-5.

Katritzky and co-workers[60] used infrared spectra to distinguish O-methyl from N-methyl derivatives of 3-hydroxy-5-phenyl-1,2,4-oxadiazole and suggested assignments for the ring and the phenyl group absorptions.

Najer and co-workers[150, 151] reported a higher frequency (1660 cm^{-1}) for the C=N band in 5-amino-3-phenyl-1,2,4-oxadiazole.

E. Ultraviolet Spectra

Moussebois and Oth[110] studied the ultraviolet spectra of a number of aryl-substituted oxadiazoles and other functions conjugated with the double bonds of the ring. They found that the wavelength for maximum absorption (λ_{max}) in 3-phenyl- (238 nm) and 5-phenyl- (250 nm) bracketed the λ_{max} for 3,5-diphenyl-1,2,4-oxadiazole (245 nm). From these and other results they concluded that the ring does not make a strong mesomeric contribution to aromaticity even though the oxadiazole ring nominally fits the $(4n + 2)$ rule. The ring is better described as a conjugated diene. In a series of 5-substituted amino oxadiazoles,

---

[149] H. Goncalves and F. Mathis, *C.R. Acad. Sci.* **259**, 1819 (1964).
[150] H. Najer, J. Menin, D. Caillaux, and G. Petry, *C.R. Acad. Sci., Ser. C*, 628 (1968).
[151] H. Najer, J. Menin, and G. Petry, *C.R. Acad. Sci., Ser. C.*, 1587 (1968).

Selim and Selim[152] found that the λ_{max} in eight different compounds varied only within the limits 225–235 nm. They concluded that none of these derivatives existed in the tautomeric imine form. The absorption coefficients (log ϵ) were all > 4. Five UV spectra of other derivatives are published.[153] Katritzky and co-workers[60] have published spectra of potentially tautomeric derivatives.

F. Nuclear Magnetic Resonance Spectra and Deuterium Exchange

The chemical shift of the protons in oxadiazole are downfield from those in benzene,[3] the C-5 proton being farther downfield than the C-3 proton (Table IV). When there is an aryl group at C-3 or C-5, the

TABLE IV

Chemical Shifts in Oxadiazoles

	δC_5-H	δC_3-H	δC_5-Me	δC_3-Me
1,2,4-Oxadiazole	8.7	8.2	—	—
3-Methyl-1,2,4-oxadiazole	8.6	—	—	2.4
5-Methyl-1,2,4-oxadiazole	—	8.0	2.7	—
3-Phenyl-1,2,4-oxadiazole	8.3	—	—	—
5-Phenyl-1,2,4-oxadiazole	—	8.1	—	—
3,5-Dimethyl-1,2,4-oxadiazole	—	—	2.6	2.3
5-Methyl-3-phenyl-1,2,4-oxadiazole	—	—	2.6	—

remaining hydrogen at C-5 or C-3 displays a discrete singlet, downfield from the aromatic ring protons and not coupled with them. This along with other evidence (Section IV, E) suggested to Moussebois and Oth[110] the absence of aromaticity in the oxadiazole ring.

Brown and Ghosh[154] found a satisfactory way of determining pK_a values of weak bases (oxazoles) by plotting chemical shifts against the pH of appropriate solutions. They were unable to apply the method to oxadiazoles because there was too little change in the spectra of these compounds in DCl-D$_2$O solutions. However, by extrapolating the plot of pH against log $t_{1/2}$ values for base-catalyzed deuteriations of oxazoles (of known pK_a's) and oxadiazoles, the relative ease of displacing hydrogen on the oxadiazole ring and hydrogen on a methyl side chain

[152] M. Selim and M. Selim, *Bull. Soc. Chim. Fr.*, 1219 (1967).
[153] M. Selim and M. Selim, *Bull. Soc. Chim. Fr.*, 823 (1969).
[154] D. J. Brown and P. B. Ghosh, *J. Chem. Soc. B*, 270 (1969).

were estimated (Scheme 5). They also found that 2,5-dimethyl-1,3,4-oxadiazole did not exchange deuterium for hydrogen at pH 11–14. The

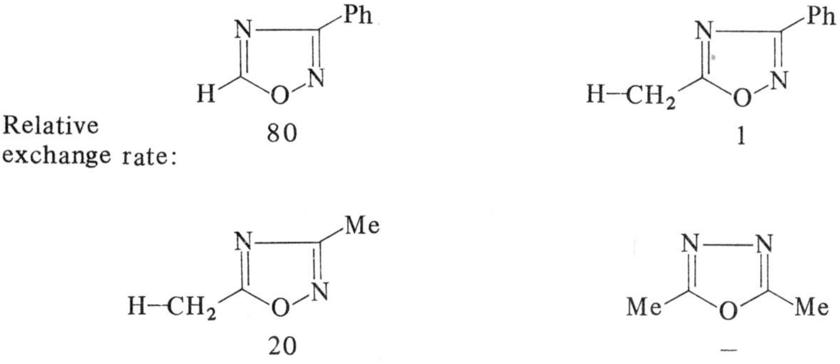

Relative exchange rate: 80, 1, 20, —

SCHEME 5

explanation of this difference in the lability of the hydrogen at C-5 and C-3 in 3,5-dimethyl-1,2,4-oxadiazole and the nonlability of hydrogen in the corresponding 1,3,4-isomer is apparent upon looking at the mesomerism possible in the anions (Scheme 6). The anion **46** can accommodate the negative charge on both nitrogens, but anion **47** cannot, and neither can the anion from 2,5-dimethyl-1,3,4-oxadiazole.

(46)

(47)

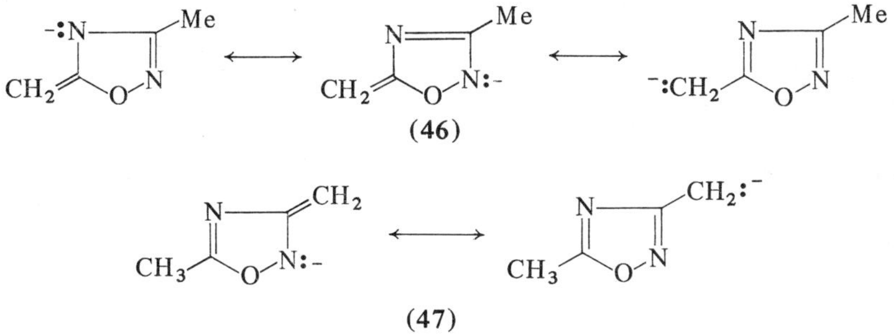

SCHEME 6

Selim and Selim[152] reported nuclear magnetic resonance (NMR) data on various aminooxadiazole derivatives. Komatsu and co-workers[120–122] recorded NMR data on oxadiazolidones.

Barrans[85, 86] used NMR spectra to identify the first Δ^3-oxadiazoline and to distinguish it from the Δ^2-isomer (Scheme 7). See Eq. (29).

(δ 4.97), (δ 6.0)

SCHEME 7

Burdon and Ramirez[116] postulated an intermediate (**25**) in a reaction to form an oxadiazolidine (**26**) [Eq. (31)] by using an NMR spectrometer to monitor the reaction.

G. Mass Spectra

Selva and co-workers[155] studied the mass spectral decomposition patterns of fifteen 5-aryl-3-phenyl- and four 3-aryl-5-phenyloxadiazoles and found the principal cleavage to be a retro 1,3-dipolar cycloaddition. In the latter group, the positive charge resided mainly on the 3-aryl fragment as shown by pathway 1a of Eq. (40) when X was varied from

$$
\begin{array}{l}
\text{a. } X-C_6H_4CNO^{+\cdot} \\
\quad + Y-C_6H_4-CN \\
\text{b. } Y-C_6H_4-CN^{+\cdot} \\
\text{c. } Y-C_6H_4-CO^+
\end{array}
\quad (40)
$$

electron-donating groups (NH_2, CH_3) through electron-withdrawing groups (NO_2, Cl). Fragmentation patterns (1b) and (2c) were minor processes for both Y = H and Y = D. In the 5-aryl substituted oxadiazoles there were minor substituent effects for Y (NO_2, Cl, and CH_3 groups) in both *m*- and *p*-positions. Pattern 1a still predominated. Only with stronger electron-donating groups (Y = OCH_3 or NH_2 in *m*- or *p*-positions) was there a sharp decrease in pattern 1a and an increase in both 1b and 2c.

Cotter and co-workers[112,156] had earlier found a close resemblance between electron impact and thermal decomposition of diphenyloxadiazole (Section V, A). They suggested a migration of the phenyl group as shown in the pattern of Eq. (41) to account for the fragments *m*/*e* 103

$$
\begin{array}{l}
PhNCO^+ + PhCN \\
m/e\ 119 \\
\\
PhNCO + PhCN^+ \\
m/e\ 103
\end{array}
\quad (41)
$$

---

[155] A. Selva, L. F. Zerilli, B. Cavalleri, and G. G. Gallo, *Org. Mass Spectrom.* **6**, 1347 (1972); **9**, 558 (1974).

[156] J. L. Cotter, G. J. Knight, and W. W. Wright, *J. Gas Chromatogr.* **5**, 86 (1967); G. J. Knight and J. L. Cotter, *U.S. Clearinghouse Fed. Sci. Inform. AD470669*, 12 pp. (1965) [*CA* **67**, 116387 (1967)]; J. L. Cotter and G. J. Knight, *Chem. Commun.*, 336 (1966).

and 119. This early suggestion had to be modified after Selva's definitive experiment with the deuteriated compound [Eq. (40), $Y = D$].[155]

Boulton and co-workers[156a] identified peaks in the mass spectrum of 3,5-bis(p-chlorophenyl)-1,2,4-oxadiazole 4-oxide corresponding to the fragments $ArCNO^+$, $ArCN^+$, and $ArCO^+$ in addition to an intense parent peak.

Oxadiazolines, on the other hand, give a more complex mass spectra with three initial scissions, two of which, (a) and (b), involve the weak N—O bond (53 kcal/mole) and hence were more probable than the third

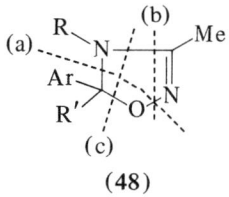

(48)

(c) in **48**. Srivastava and Clapp[107] discuss the cracking patterns of seven oxadiazolines.

V. Chemical Properties

A. Thermal Stability

Remarkable thermal stability was attributed to 3,5-diphenyloxadiazole by the first workers in the field in 1884. Tiemann and Krüger[7] observed that this compound could be distilled without decomposition at normal pressure (b.p. 296°). Ainsworth[157] modified that position when he found slow decomposition occurred at 250° with an open flame as heat source. In contrast, 2,4-diphenyl-1,3,4-oxadiazole decomposed only at 500°.[156] With substitution on aryl groups, Ainsworth[157] found that 3-

$$\text{Ph-oxadiazole} \xrightarrow{250°} PhCN + PhNCO \tag{42}$$

(p-chlorophenyl)-5-(p-methoxyphenyl)-1,2,4-oxadiazole gave p-chlorophenyl isocyanate and p-methoxybenzonitrile at 250°, but the experiment with C-3 and C-5 substituents reversed was not done. Conse-

---

[156a] A. J. Boulton, P. Hadjimihalakis, A. R. Katritzky, and A. Majid Hamid, *J. Chem. Soc. C*, 1901 (1969).
[157] C. Ainsworth, *J. Heterocycl. Chem.* **3**, 470 (1966).

quently, no generalization concerning the influence of substituents on the mode of thermal ring cleavage can be made.

B. Photochemical Effects

Newman[158] found that photochemical decomposition of 3,5-diphenyloxadiazole occurred in ether to give *N*-benzoylbenzamidine in 31% yield. In contrast to thermal decomposition [Eq. (42)] cleavage occurred at the weak N–O bond [Eq. (43)]. Ether furnished the extra hydrogens found in the product.

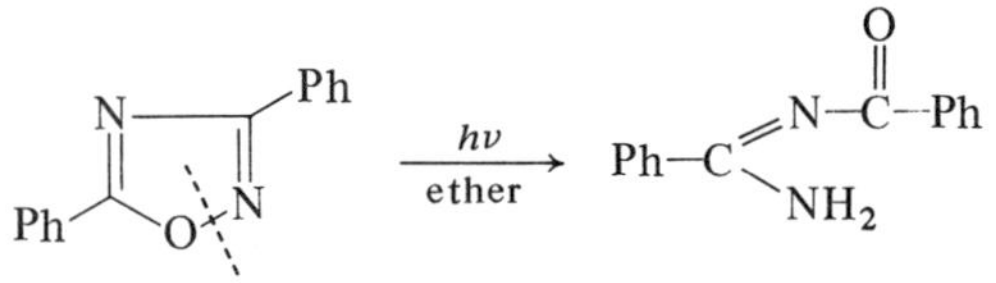

$$\text{[Ph-oxadiazole]} \xrightarrow[\text{ether}]{h\nu} \text{Ph–C}(=\text{N–C(=O)–Ph})(\text{NH}_2) \quad (43)$$

Ring opening by photolysis also occurs in the oxadiazoline ring. Srivastava and Clapp[109] noted that a simple rearrangement [Eq. (44)] of 4-β-ethoxyethyl-3-methyl-5-*p*-nitrophenyl-1,2,4-oxadiazoline to a substituted amidine (**50**) took place on a bench top in ordinary daylight or more rapidly with "black light" (3560 Å). More stable oxadiazolines[108] were found to decompose in analogous fashion with a higher energy source (2537 Å).

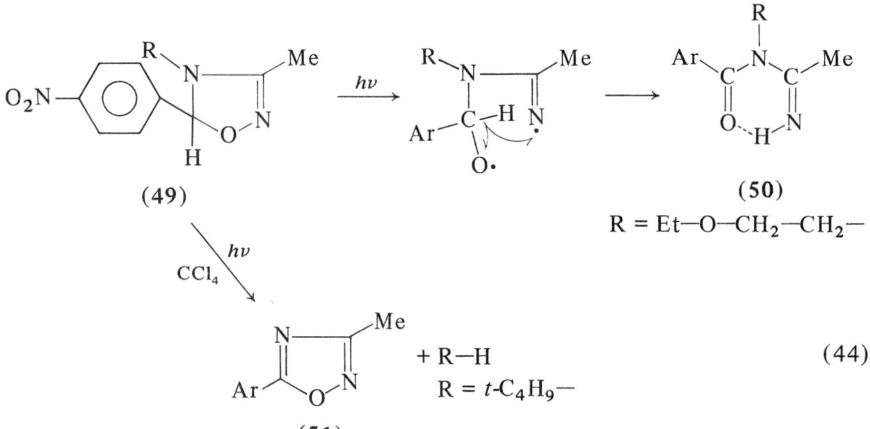

(49) → (50) R = Et–O–CH$_2$–CH$_2$–

(49) $\xrightarrow{h\nu, \text{CCl}_4}$ (51) + R–H, R = *t*-C$_4$H$_9$– (44)

Isobutane was lost from **49** (R = *t*-butyl) to give an oxadiazole (**51**) under the same conditions. The *t*-butyl free radical was shown to be a

---

[158] H. Newman, *Tetrahedron Lett.*, 2417, 2421 (1968).

probable link in the chain by the trapping of *t*-butyl chloride in the products when carbon tetrachloride was used as a solvent for the irradiation.[108]

Evans and co-workers[159] reported that 5-*p*-methoxyphenyl-3-phenyl-1,2,4-oxadiazole fluoresces in acetonitrile as solvent and was quenched by olefins.

C. STABILITY IN SOLUTION: HYDROLYSIS

Tiemann and Krüger[7] also based the contention of remarkable stability of oxadiazoles (Section V, A) on their ability to heat a 3,5-diphenyloxadiazole in concentrated sulfuric acid without decomposition. It could be recrystallized from fuming nitric acid. However, this inertness of oxadiazoles is lost when there is a hydrogen at C-3 or C-5.

On standing at room temperature, 3-methyloxadiazole decomposes into acetonitrile and cyanuric acid (Eq. 45) whereas in warm

$$Me-C\underset{NH_2}{\overset{NOH}{\diagup}} \xleftarrow{HCl \atop HOH} \underset{O-N}{\overset{N---Me}{\diagdown}} \begin{array}{c} \xrightarrow{25°} CH_3CN + (HOCN)_3 \\ \xrightarrow{OH^-} CH_3CN + NH_3 \end{array} \quad (45)$$

hydrochloric acid acetamidoxime is generated. The 5-methyl isomer gives acetic acid and formamidoxime under the same conditions. In basic solution the 5-methyl isomer yields acetylcyanamide and the 3-methyl isomer yields acetonitrile and ammonia.[160]

The 5-phenyl derivative slowly isomerizes to benzoylcyanamide (Eq. 46) on standing and more rapidly in alkaline solution.[21,160]

$$\underset{Ph}{\overset{N---}{\diagdown}}\underset{O-N}{} \xrightarrow[\text{or OH}^-]{\text{standing}} Ph-\underset{\underset{O}{\|}}{C}-NHCN \quad (46)$$

D. RING OPENING BY REDUCTION

Reductions of oxadiazoles to oxadiazolines and oxadiazolidines were conspicuous by their absence in the description of their syntheses (Section III). Reagents vigorous enough to reduce the double bonds in an oxadiazole ring also break the ring. It was early reported[8] that zinc and hydrochloric acid acted on 3,5-diphenyloxadiazole to give benzonitrile.

---

[159] T. R. Evans, R. W. Wake, and M. M. Sifain, *Mol. Photochem.* **3**, 275 (1971) [*CA* **76**, 29487 (1972)].

[160] F. Eloy, R. Lenaers, and C. Moussebois, *Chem. Ind. (London)*, 292 (1961).

Catalytic reduction[161, 162] with various catalysts (PtO$_2$, Pd/C, or Raney Ni) on other oxadiazoles all broke the N—O bond [Eqs. (47–49)]. The product in one case (52) was acetamidine benzoate, which suggested the reduction pathway shown in Eq. (47).

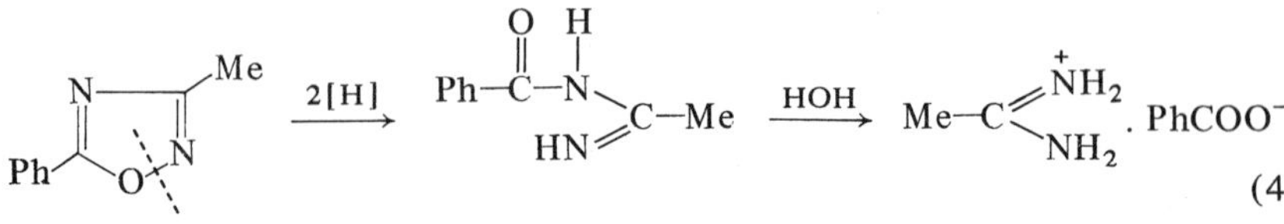

$$\underset{(52)}{\text{Ph-oxadiazole}} \xrightarrow{2[H]} \underset{HN=C-Me}{\overset{O\;H}{\underset{}{Ph-C-N}}}\! \xrightarrow{HOH} \text{Me}-C\!\!\begin{array}{c}\overset{+}{N}H_2\\NH_2\end{array}\!\cdot\text{PhCOO}^- \quad (47)$$

The 5-amino derivative (53) was also reduced catalytically breaking the N—O bond as shown in Eq. (48).[163]

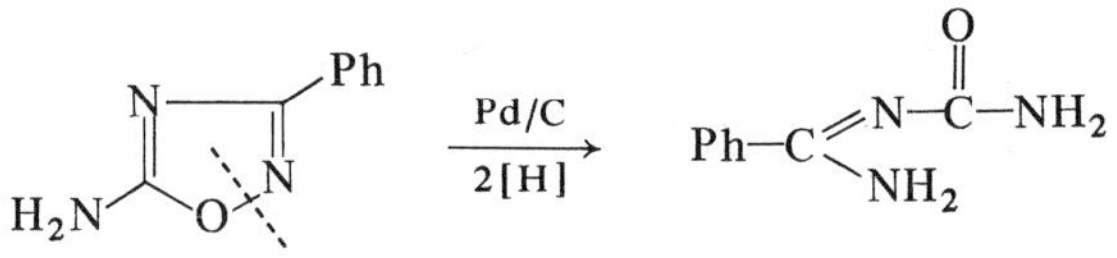

$$\underset{(53)}{H_2N\text{-oxadiazole-Ph}} \xrightarrow[2[H]]{Pd/C} \text{Ph}-C\!\!\begin{array}{c}N-C-NH_2\\NH_2\end{array} \quad (48)$$

In the reduction of the oxadiazoline (54), the products were an amidine (55) and benzyl alcohol [Eq. (49)].[108]

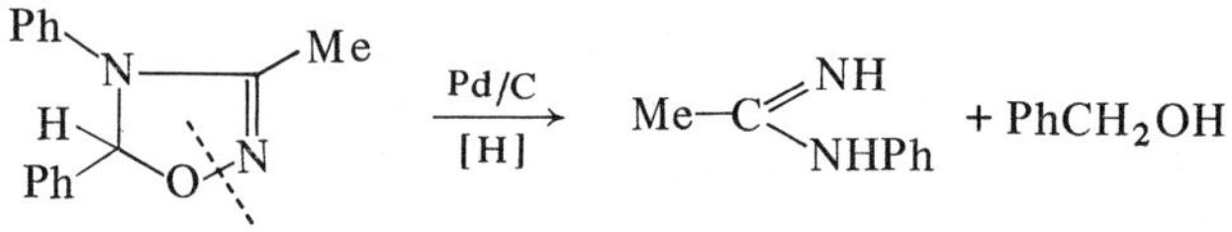

$$\underset{(54)}{\text{oxadiazoline}} \xrightarrow[[H]]{Pd/C} \underset{(55)}{\text{Me}-C\!\!\begin{array}{c}NH\\NHPh\end{array}} + \text{PhCH}_2\text{OH} \quad (49)$$

In contrast to catalytic reduction, lithium aluminum hydride appears always to break the C—O bond in the ring. For example, oxadiazoles[164] yield substituted amidoximes [Eq. (50)]. The same reagent with the

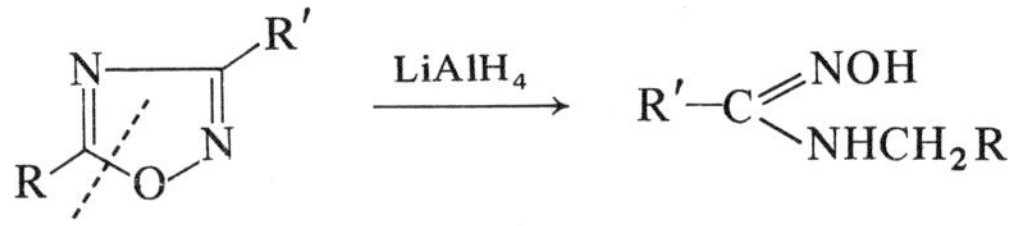

$$\xrightarrow{\text{LiAlH}_4} R'-C\!\!\begin{array}{c}NOH\\NHCH_2R\end{array} \quad (50)$$

isomeric oxadiazolones 56 and 57 gave the products indicated in Eqs. (51) and (52), respectively. Selim and co-workers[165] pointed out that the

---

[161] G. Palazzo, G. Strani, and M. Tavella, *Gazz. Chim. Ital.* **91**, 1085 (1961).
[162] G. Palazzo and G. Strani, *Ann. Chim. (Rome)* **51**, 130 (1961).
[163] G. Palazzo and G. Strani, *Gazz. Chim. Ital.* **91**, 216 (1961).
[164] M. Tavella and G. Strani, *Ann. Chim. (Rome)* **51**, 361 (1961).
[165] Y. Royer, M. Selim, and P. Rumpf, *Bull. Soc. Chim. Fr.*, 1060 (1973).

N-hydroxy compounds (**58**) are not so easily made by any other route.

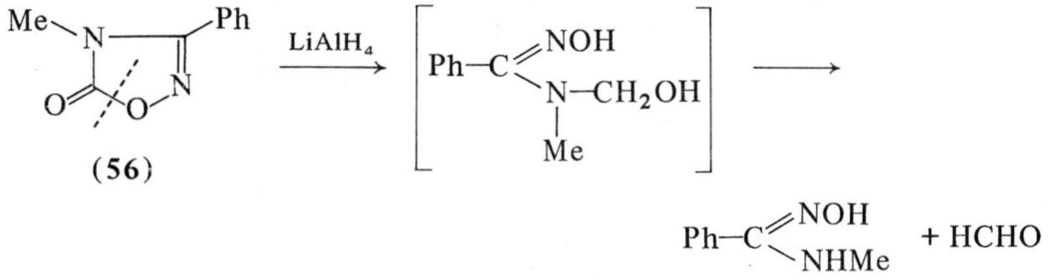

[Scheme showing compound (56) → LiAlH₄ → [Ph–C(=NOH)–N(Me)–CH₂OH] → Ph–C(=NOH)–NHMe + HCHO (51)]

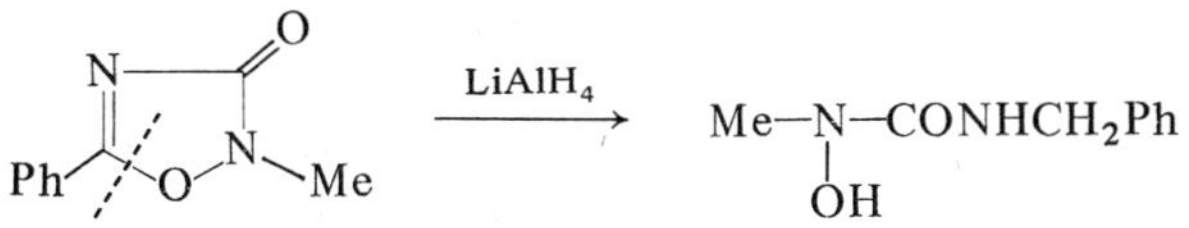

[Scheme showing compound (57) → LiAlH₄ → Me–N(OH)–CONHCH₂Ph (58) (52)]

E. Functional Group Chemistry

As a neighboring group the oxadiazole ring does not appreciably modify the chemistry of many functional groups, but there is frequently a difference in activity of the same group at C-3 and C-5 (Section V, F). The transformations shown in Eq. (53), for example,[23] are not different

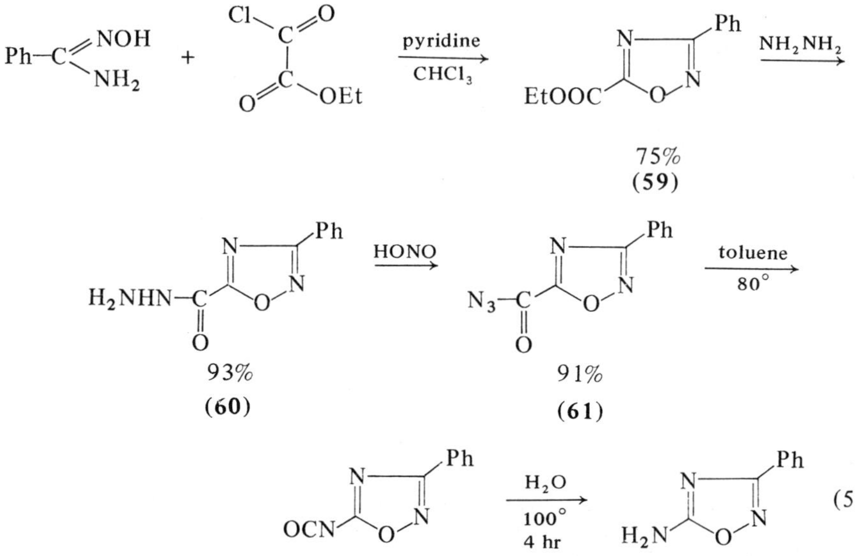

[Reaction scheme (53):

Ph–C(=NOH)–NH₂ + Cl–C(=O)–C(=O)–OEt → (pyridine, CHCl₃) → EtOOC-oxadiazole-Ph (75%) (**59**) → NH₂NH₂ → H₂NHN–C(=O)-oxadiazole-Ph (93%) (**60**) → HONO → N₃–C(=O)-oxadiazole-Ph (91%) (**61**) → toluene, 80° → OCN-oxadiazole-Ph (93%) (**62**) → H₂O, 100°, 4 hr → H₂N-oxadiazole-Ph (60%) (**63**)]

in character from the same well-known series in aliphatic or aromatic compounds. Yields are shown for each of the five products, **59–63**. The acyl azide (**61**) was also converted back to the ester (**59**) in refluxing ethanol. The carbamate was also obtained from the isocyanate (**62**) by normal ethanolysis. Other workers[28, 166] have used the same reactions in oxadiazole chemistry.

In addition to the series of reactions of Eq. (53) to introduce the amino group at C-5, the amine function has been introduced directly[167–169] [Eqs. (9) and (10)]. The hydroxy group as well can be introduced directly [Eq. (5)]. The amido function at C-3 may be obtained by ammonolysis of the carbethoxy group and subsequently dehydrated to the nitrile.[28] The carbethoxy group at C-3 may also be hydrolyzed to the carboxy group[166] in alkaline methanol without opening the oxadiazole ring. However, see Eq. (61) below.

Chlorine may be placed on the α-carbon of a side chain on the oxadiazole ring by allowing an amidoxime to react with an α-chloroacyl chloride.[11, 37] The chloromethyl group in **64** has reactivity of the order of

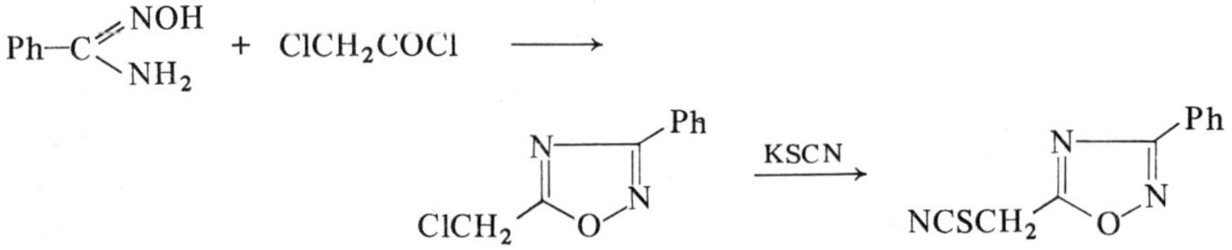

a benzyl chloride. Jaunin[170] reported displacement of the halogen by potassium thiocyanate to give **65**. However, in dimethyl sulfoxide

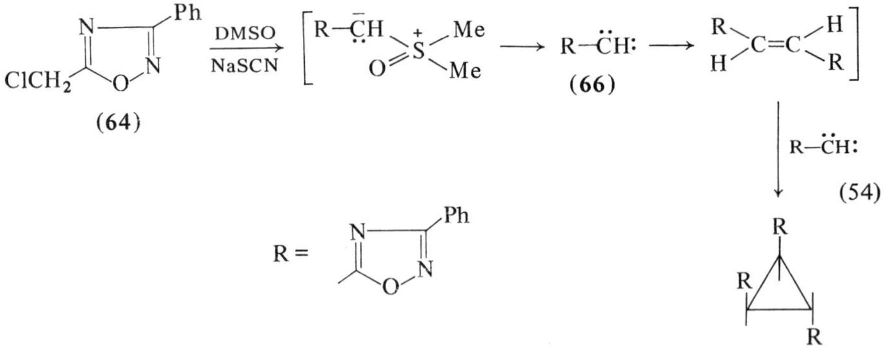

[166] G. Strani and A. M. Garau, *Gazz. Chim. Ital.* **93**, 482 (1963).
[167] F. Eloy and A. Deryckere, *Bull. Soc. Chim. Belges* **78**, 41 (1969).
[168] G. Westphal and R. Schmidt, *Z. Chem.* **14**, 94 (1974).
[169] R. Schmidt, G. Westphal, and B. Frölich, *Z. Chem.* **14**, 270 (1974).
[170] R. Jaunin, *Helv. Chim. Acta* **49**, 412 (1966).

(DMSO) as solvent with the more soluble sodium thiocyanate, Jaunin found a cyclopropane (67) as product, carrying three phenyloxadiazolyl groups, R. He suggested that a carbene (66) intermediate must have formed from 64 with the solvent as shown in the sequence of Eq. (54). An ionic mechanism can be written, however.

Crovetti and co-workers[171] were also able to displace the halogen in a 5-nitrofuryl derivative (68) of an oxadiazole in dimethylformamide (DMF) [Eq. (55)] although the displacement could not be accomplished with liquid ammonia or anhydrous dimethylamine.

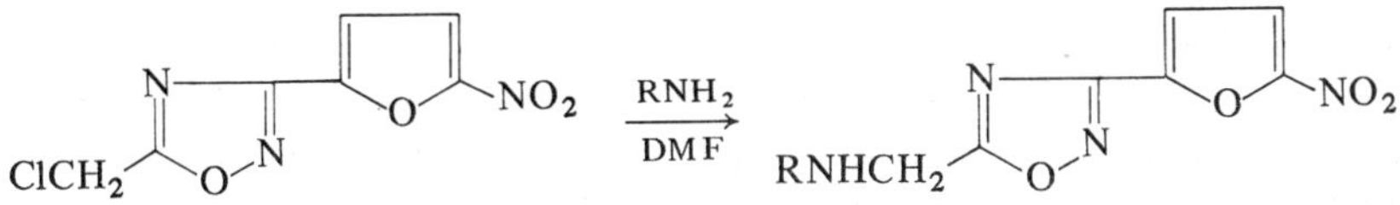

$$\text{(68)} \xrightarrow[\text{DMF}]{\text{RNH}_2} \text{(55)}$$

A halogen directly attached to the oxadiazole ring may be placed there by diazotization[172] of an amine in hydrochloric acid, for example

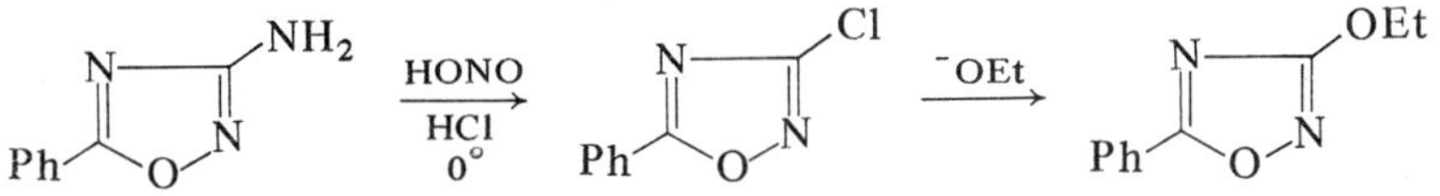

$$\text{(69)} \xrightarrow[\substack{\text{HCl} \\ 0°}]{\text{HONO}} \text{(70)} \xrightarrow{^-\text{OEt}} \text{(71)} \quad (56)$$

69. The halogen in 70 may in turn be displaced by the hydroxy or ethoxy group [Eq. (56)] (Section V, H).

F. ACTIVATION IN THE SIDE CHAIN

The methyl group as a side chain on the oxadiazole ring acts in a different way at C-5 than at C-3 because of the difference in the inductive and mesomeric effects at these positions (Section IV, F). On 5-methyl-3-phenyl-1,2,4-oxadiazole (72), treatment with lithium n-butyl at −70° gave what Micetich[32] called "lateral lithiation" (73). Subsequent

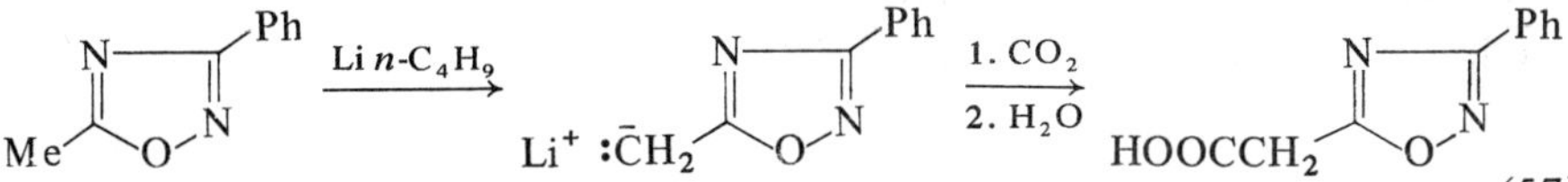

$$\text{(72)} \xrightarrow{\text{Li } n\text{-C}_4\text{H}_9} \text{(73)} \xrightarrow[\text{2. H}_2\text{O}]{\text{1. CO}_2} \text{(74)} \quad (57)$$

[171] A. J. Crovetti, A. M. Von Esch, and R. J. Thill, *J. Heterocycl. Chem.* **9**, 435 (1972).
[172] F. Eloy, A. Deryckere, and A. Van Overstraeten, *Bull. Soc. Chim. Belges* **78**, 47 (1969).

reaction with carbon dioxide (Eq. 57) gave the 5-carboxymethyl derivative. But with the isomer, 3-methyl-5-phenyl-1,2,4-oxadiazole (75), addition occurred at the Δ^4-double bond [Eq. (58)]. Then carbonation

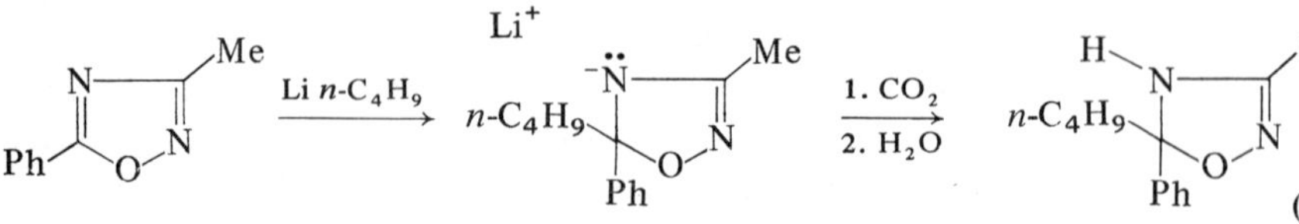

and aquation resulted only in the isolation of a Δ^2-oxadiazoline (77). Micetich invoked the inductive effect of oxygen and nitrogen at C-5 in 72 as opposed to two nitrogens at C-3 in 75 to account for the greater activation of the methyl group at C-5. It may also be noted, however, that in addition to the inductive effect the anion 76 is stabilized by smearing out the negative charge over both nitrogens (Scheme 8).

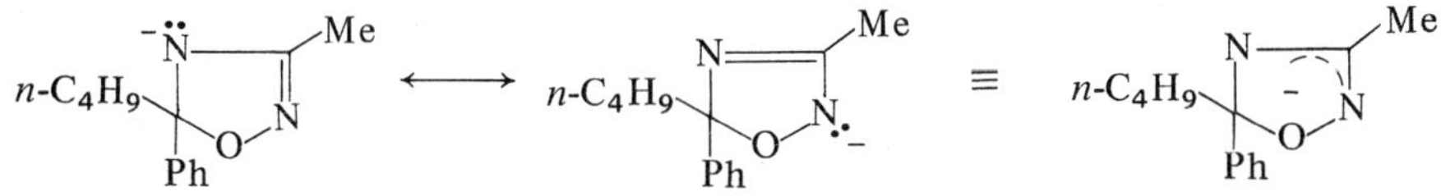

SCHEME 8

The methyl group in 5-methyl-3-phenyl-1,2,4-oxadiazole is likewise sufficiently activated to condense with benzaldehyde in the presence of zinc chloride to give a styryl derivative [Eq. (59)] in 65% yield.[173] This

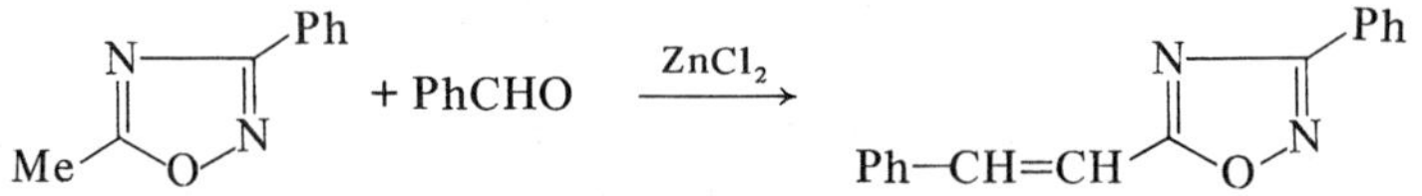

activity is reduced even by substituting ethyl for methyl at C-5 so that the yield in the same reaction is only 40%. The methylene group in a 5-cyanomethyloxadiazole gives a similar condensation with aldehydes.

Ethyl oxalate[51, 174] undergoes a Claisen condensation with a C-5 methyl group as shown in Eq. (60).

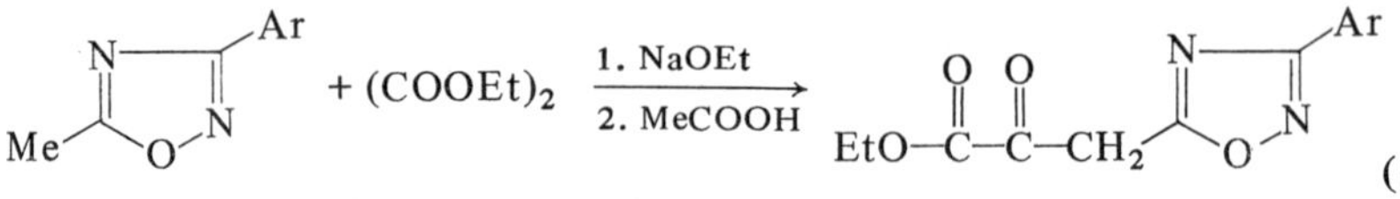

[173] C. Moussebois and F. Eloy, *Helv. Chim. Acta* **47**, 838 (1964).
[174] R. Merckx, *Bull. Soc. Chim. Belges* **58**, 460 (1949).

A methylene group between the oxadiazole ring and a carboxyl group acts like a malonic acid with respect to ease of loss of the carboxyl group. Likewise the acetyl group from an acetonyl side chain at C-5 is lost in basic solution, analogous to the ketone split in a β-keto ester or β-diketone. This was recognized in the early work on oxadiazoles by Tiemann,[48] Wiese,[49] Richter,[47] and Schubart.[50]

Although 3-carbethoxy groups and other derivatives (Eq. 53) are stable on the oxadiazole ring, carboxyl groups directly attached at C-3 or C-5 are easily lost. In fact, Brachwitz[175] was unable to prepare the 3-carboxy compound **81** by the pathway [Eq. (61)] that might have been

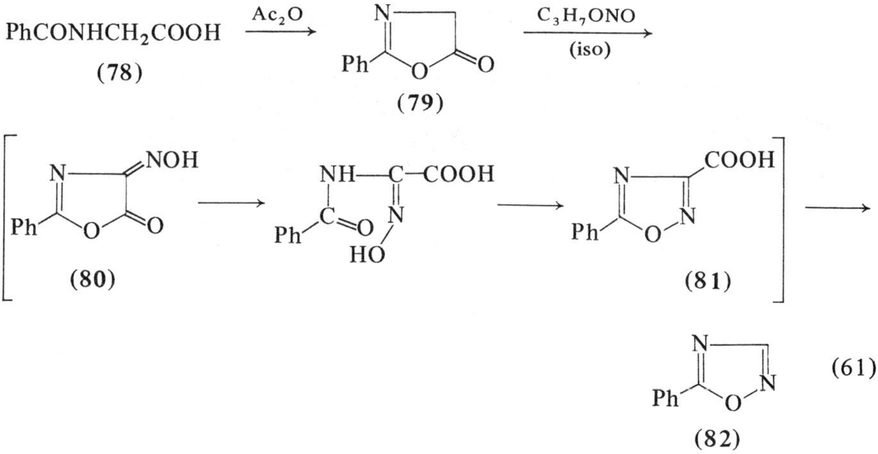

expected to give it. When the azlactone (**79**) was allowed to react with isopropyl nitrite, a precipitate formed from which only benzoic acid and benzonitrile were obtained. Alkaline treatment of the precipitate gave benzoylcyanamide, which is what would be expected of such treatment of **82** [see Eq. (46)]. Compound **82**, however, has been prepared from formamidoxime and benzoic anhydride.[160]

G. Electrophilic Substitutions

Electrophilic reagents do not fare well with the oxadiazole ring since it does not have much aromatic character (Section IV, E). Instead, the oxadiazole ring acts as an *m*-directing group for electrophilic substitution when there is an aryl group at either C-3 or C-5. Nitration of 3-phenyl-oxadiazole, for example, gives a mixture of *m*- and *p*-nitrophenyloxa-

[175] H. Brachwitz, *Z. Chem.* **12**, 130 (1972).

diazoles. With an additional methyl substitution at C-5, Corsi[18] obtained a 98% yield of 5-methyl-3-(*m*-nitrophenyl)-1,2,4-oxadiazole.

At present the only successful electrophilic agent for oxadiazoles is mercury(II) chloride[173] (see also Section V, I). Chlorine in turn will displace the —HgCl group [Eq. (62)].

$$\underset{O}{\overset{N}{\diagup}}\hspace{-0.3em}\underset{N}{\overset{Ph}{\diagdown}} + HgCl_2 \longrightarrow \underset{Cl-Hg}{\overset{N}{\diagup}}\hspace{-0.3em}\underset{O-N}{\overset{Ph}{\diagdown}} \xrightarrow{Cl_2} \underset{Cl}{\overset{N}{\diagup}}\hspace{-0.3em}\underset{O-N}{\overset{Ph}{\diagdown}} \quad (62)$$

H. Nucleophilic Substitutions

Nucleophilic displacements on the sp² carbons at C-3 or C-5, in contrast to electrophilic reactions, are plentiful. Chlorine, for example, may be displaced by amino, hydroxy, or alkoxy groups[173] (Eq. 63). The ether group in **84** may in turn be displaced by hydroxy or amino (as shown) groups.

$$\underset{Cl}{\overset{N}{\diagup}}\hspace{-0.3em}\underset{O-N}{\overset{Me}{\diagdown}} \xrightarrow{OH^-} \underset{HO}{\overset{N}{\diagup}}\hspace{-0.3em}\underset{O-N}{\overset{Me}{\diagdown}} \xrightarrow[\text{2. EtI}]{\text{1. Ag}^+} \underset{EtO}{\overset{N}{\diagup}}\hspace{-0.3em}\underset{O-N}{\overset{Me}{\diagdown}}$$

$$(83) \hspace{6em} (84)$$

$$\searrow RNH_2 \hspace{3em} RNH_2 \swarrow \hspace{3em} +\hspace{3em} (63)$$

$$\underset{RNH}{\overset{N}{\diagup}}\hspace{-0.3em}\underset{O-N}{\overset{Me}{\diagdown}} \hspace{4em} \underset{O}{\overset{Et-N}{\diagup}}\hspace{-0.3em}\underset{O-N}{\overset{Me}{\diagdown}}$$

$$(86) \hspace{6em} (85)$$

When ethyl iodide acts on the silver salt of the 5-hydroxy derivative (**83**), a 1 : 1 mixture of the *O*- and *N*-ethyl derivatives (**84** and **85**) are obtained. Evidently **85** is thermodynamically the more stable of the two isomers, since on standing for 3 days as much as 90% of 4-ethyl-3-methyl-1,2,4-oxadiazolin-5-one (**85**) is formed. Only **84** can be converted to the amino derivative **86**.

The trichloromethyl group is said to be easier to displace with nucleophiles than chlorine itself when it is at C-5. This electron-withdrawing group enables one to demonstrate a difference in ease of displacement at C-5 with respect to C-3. Guanidine[78, 176] will displace

---

[176] F. Eloy and R. Lenaers, *Helv. Chim. Acta* **49**, 1430 (1966).

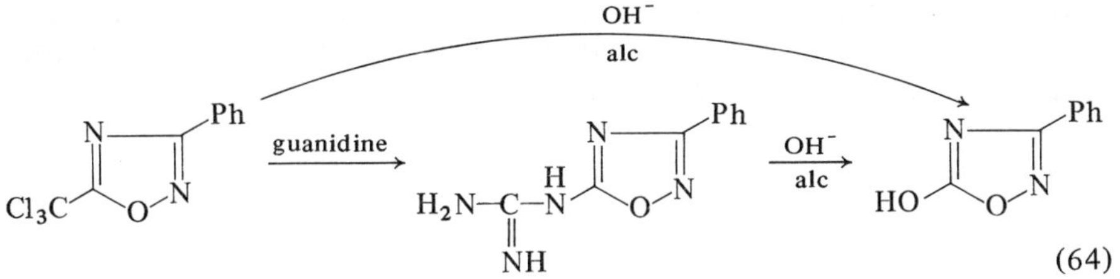

Sec. V.I]   1,2,4-OXADIAZOLES   103

[Scheme showing conversion of (87) Cl₃C-oxadiazole-Ph via guanidine to (88) H₂N−C(=NH)−NH-oxadiazole-Ph, and via OH⁻/alc to (64) HO-oxadiazole-Ph]   (64)

—CCl₃ in **87** to yield **88** and chloroform. The guanidino group in turn is displaced by —OH [Eq. (64)]. On the other hand, with the same group at C-3, nucleophilic attack occurs on the side chain, not at C-3 [Eq. (65)].[73]

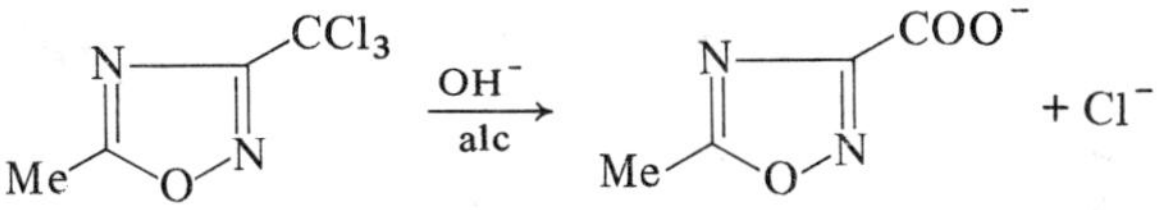

[Scheme: Me-oxadiazole-CCl₃ + OH⁻/alc → Me-oxadiazole-COO⁻ + Cl⁻]   (65)

The statement concerning ease of displacement at C-3 and C-5 was not based on competitive rate studies, and the idea deserves to be tested on a compound that possesses both groups in the same molecule, e.g. **89**.

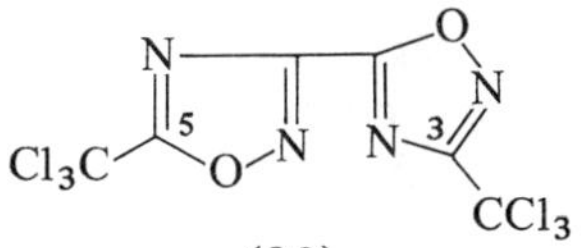

[Structure **89**: bis-oxadiazole with Cl₃C groups at positions 5 and 3]

I. COMPLEXES WITH INORGANIC COMPOUNDS

The only successful isolation of the parent compound of the oxadiazole homologs came about through complex formation with cadmium chloride[39,160] as shown in Eq. (66). Oxadiazole was then distilled away from the purified complex at 30° under reduced pressure.

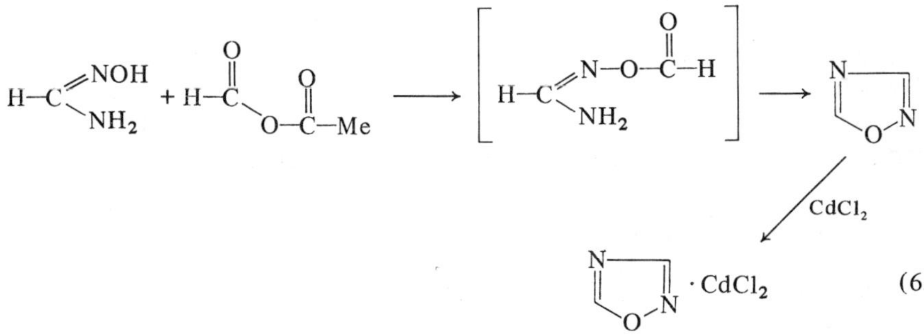

[Scheme: H-C(=NOH)(NH₂) + H-C(=O)-O-C(=O)-Me → [H-C(=N-O-C(=O)-H)(NH₂)] → oxadiazole → oxadiazole·CdCl₂]   (66)

Mercury(II) chloride also forms a 1:1 complex with 3-methyloxadiazole, which in a buffered solution forms a new complex with the electrophilic displacement product [Eq. (67)].

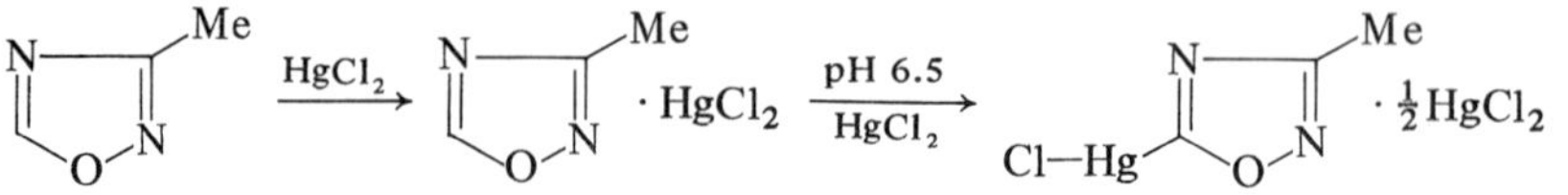

$$\underset{}{\text{N}\diagup\!\!\!\diagdown\text{Me}} \xrightarrow{\text{HgCl}_2} \underset{}{\text{N}\diagup\!\!\!\diagdown\text{Me}} \cdot \text{HgCl}_2 \xrightarrow[\text{HgCl}_2]{\text{pH 6.5}} \underset{\text{Cl-Hg}}{\text{N}\diagup\!\!\!\diagdown\text{Me}} \cdot \tfrac{1}{2}\text{HgCl}_2 \quad (67)$$

J. REARRANGEMENTS

The generalized picture of hetero-rearrangements that Katritzky and co-workers[177] pointed out in 1967 stimulated much research for other examples. In the five-membered ring (**90**), the bonding ABD gives way

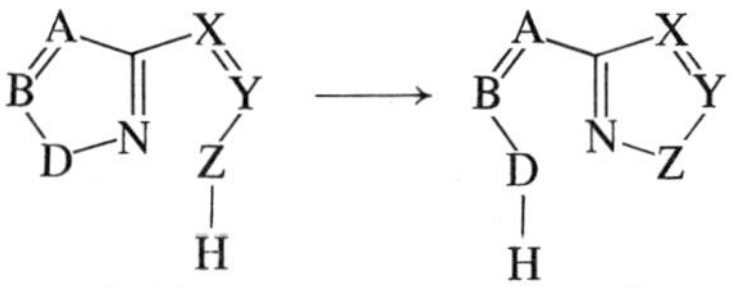

to the new bonding XYZ in **91** where C=N is common to both rings. The examples that have been recognized which involve oxadiazole rings as starting compounds rearrange to the six rings indicated in Table V.

TABLE V

REARRANGED PRODUCTS OF 1,2,4-OXADIAZOLE RINGS

XYZ	XYZ
1. CNN 1,2,3-triazole	4. NCN 1,2,4-triazole
2. CNO 1,2,5-oxadiazole	5. CCO benzisoxazole
3. NCC imidazole	6. NCS 1,2,4-thiadiazole

1. Examples of the rearrangement of 1,2,4-oxadiazoles into 1,2,3-triazoles were discovered by Ruccia and co-workers[178, 179] and earlier by Gramantieri[180] as shown in Eq. (68).[179] In the example selected, the product **93** was obtained directly from the reaction with phenyl-

---
[177] A. J. Boulton, A. R. Katritzky, and A. Majid Hamid, *J. Chem. Soc. C*, 2005 (1967); A. J. Boulton, *Lect. Heterocycl. Chem.* **1**, S-45 (1974).
[178] M. Ruccia and D. Spinelli, *Gazz. Chim. Ital.* **89**, 1654 (1959).
[179] M. Ruccia and N. Vivona, *Ann. Chim.* (*Rome*) **57**, 680 (1967).
[180] P. Gramantieri, *Gazz. Chim. Ital.* **65**, 102 (1935).

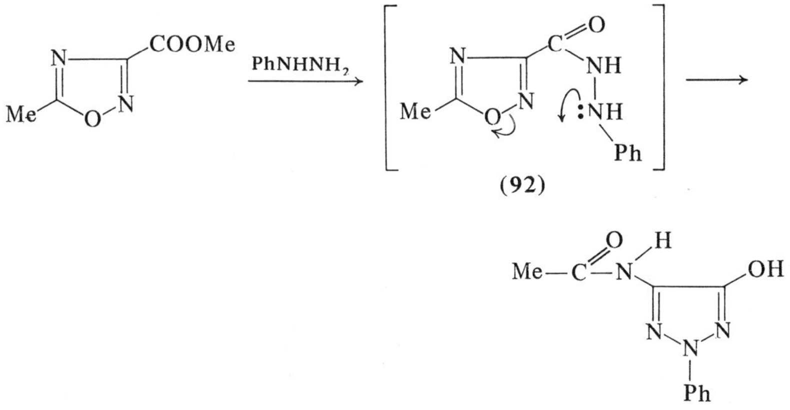

[Scheme showing conversion of methyl 1,2,4-oxadiazole-3-carboxylate with PhNHNH$_2$ via intermediate (92) to product (93), also labeled (68)]

hydrazine, but in other cases the intermediate (analogous to **92**) had to be helped along by heating in basic solution. Here the hydrazide grouping in **92** is potentially visible as X=Y—ZH if written in the zwitterionic form (**92b**).

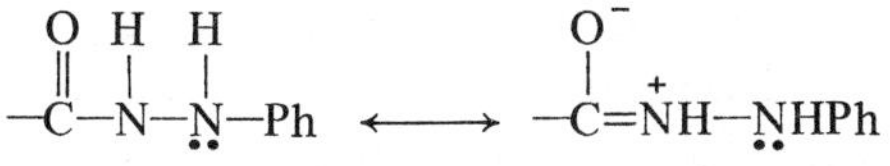

$$\underset{(92a)}{-\overset{O}{\overset{\|}{C}}-\overset{H}{\underset{|}{N}}-\overset{H}{\underset{|}{N}}-Ph} \longleftrightarrow \underset{(92b)}{-\overset{O^-}{\underset{|}{C}}=\overset{+}{N}H-\overset{..}{\underset{..}{N}}HPh}$$

2. Conversion of a 1,2,4-oxadiazole into a 1,2,5-oxadiazole is one of the earliest examples of the generalized rearrangement. An α-hydroximino group at C-3 in **94** ended as the X=Y—Z group in **95** when the 1,2,4-oxadiazole was heated [Eq. (69)].[180–185]

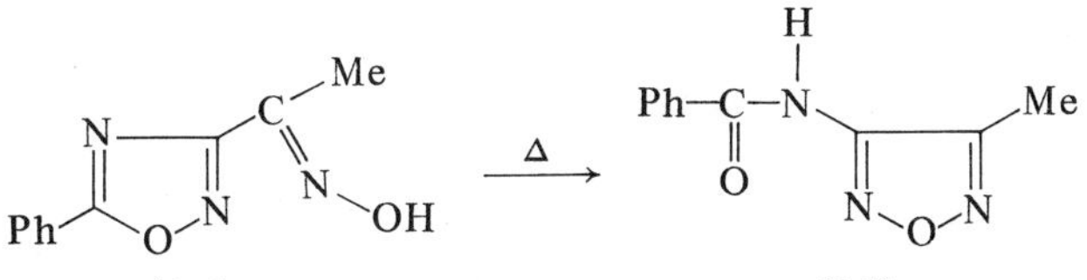

[Scheme: (94) → Δ → (95), Eq. (69)]

3. An enamine structure written for compound **96** shown in the sequence of Eq. (70) puts this rearrangement of an oxadiazole into the

---

[181] E. Durio and S. Dugone, *Gazz. Chim. Ital.* **66**, 139 (1936).
[182] G. Ponzio, *Gazz. Chim. Ital.* **61**, 138 (1931).
[183] G. Ponzio and L. Avogadro, *Gazz. Chim. Ital.* **53**, 318 (1923).
[184] G. Ponzio and G. Ruggeri, *Gazz. Chim. Ital.* **53**, 297 (1923).
[185] C. Lehmann, E. Renk, and A. Gagnaux, Swiss 498,135 (1970) [*CA* **74**, 87992 (1971)].

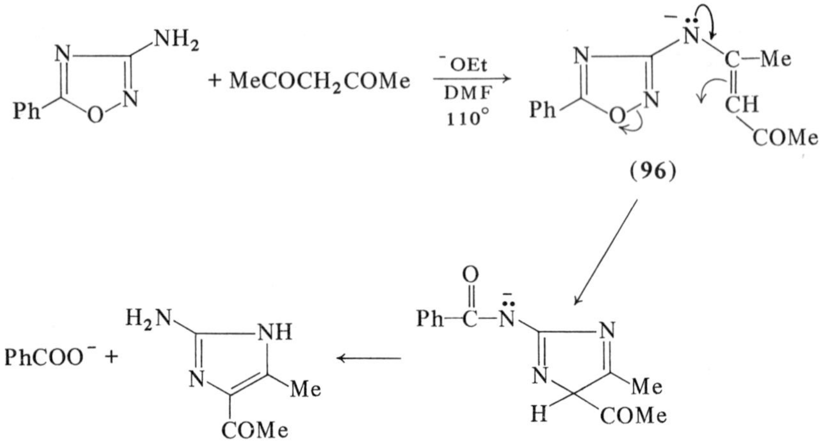

[Structures and equation (70) shown schematically, leading to compound (97)]

same classification. The 3-amino-5-phenyloxadiazole needs ethoxide ion in DMF at elevated temperature to promote this rearrangement.[186] The final product (97) is an aminoimidazole.

4. Ruccia and co-workers[187,188] also succeeded in obtaining some 1,2,4-triazoles from 3-amino-1,2,4-oxadiazoles by the series of reactions shown in Eq. (71). Isolation of the intermediate (99) could also be bypassed by heating 98 directly with the p-toluidine at 120°–170°.

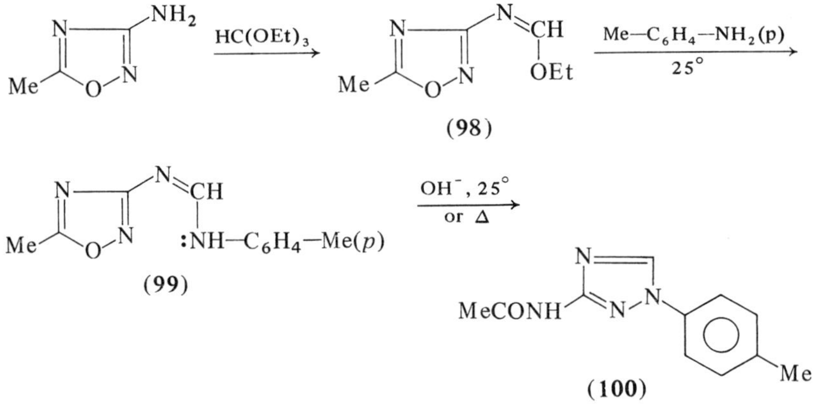

[Reaction scheme showing (98), (99), (100) leading to equation (71)]

5. Harsanyi[189] has reported a new rearrangement not described in the original postulation[177] where the X=Y—ZH group is a part of an

---

[186] M. Ruccia, N. Vivona, and G. Cusmano, *Tetrahedron Lett.*, 4959 (1972); *Tetrahedron*, **30**, 3859 (1974).
[187] M. Ruccia, N. Vivona, and G. Cusmano, *J. Heterocycl. Chem.* **8**, 137 (1971).
[188] M. Ruccia and N. Vivona, *Chem. Commun.*, 866 (1970).
[189] K. Harsanyi, *J. Heterocycl. Chem.*, **10**, 957 (1973).

aromatic ring. In one case an equilibrium between the oxadiazole **101**

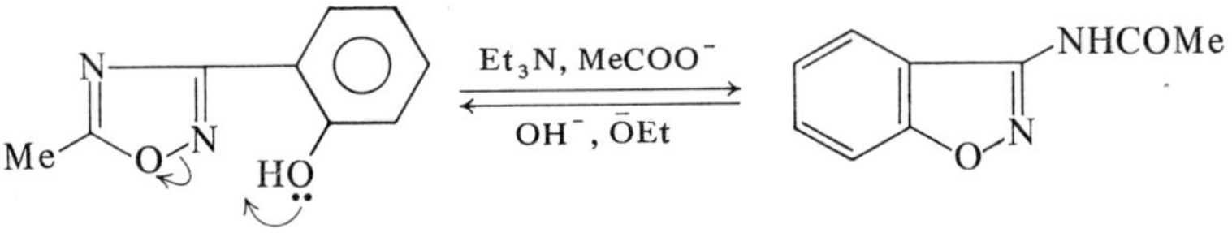

(structures 101 and 102 with reaction arrows labeled Et₃N, MeCOO⁻ and OH⁻, ŌEt)

(101)   (102)

and the benzisoxazole **102** could be controlled by changing the pH of the solution. At pH 8.4 (sodium acetate solution) **102** predominates, and in stronger base (pH ≅ 12) conversion back to **101** begins. The oxadiazoline intermediates **101a** and **101b** were suggested by Harsanyi.

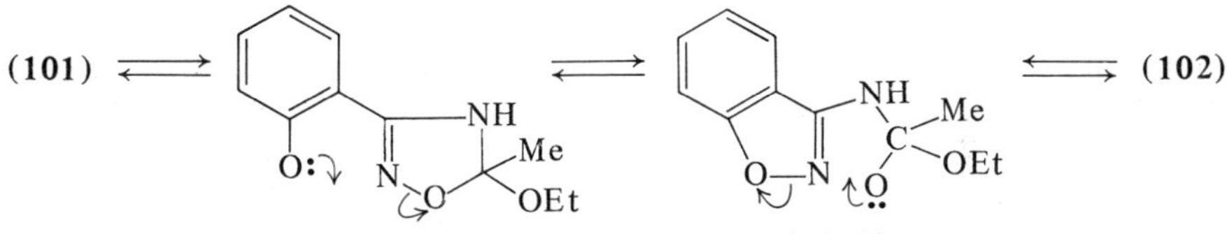

(structures showing equilibrium: 101 ⇌ 101a ⇌ 101b ⇌ 102)

(101a)   (101b)

6. Another new rearrangement, in which the final result is to replace O by S in the 1,2,4-oxadiazole, was recently reported by Ruccia and co-workers[190] [Eq. (72)].

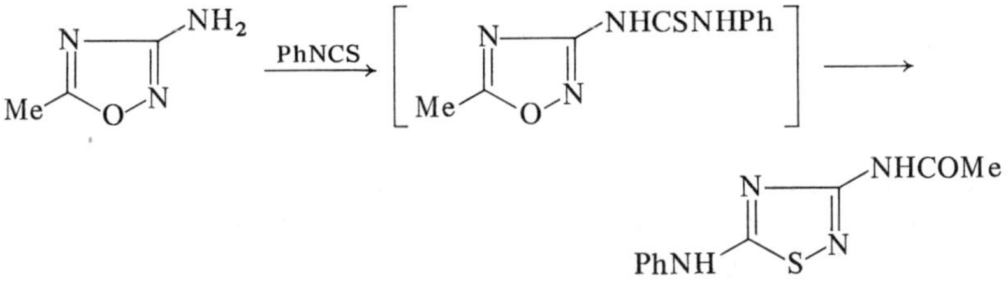

(reaction scheme showing oxadiazole-NH₂ + PhNCS → [oxadiazole-NHCSNHPh] → thiadiazole product with NHCOMe and PhNH) (72)

A rearrangement that does not fall into the pattern described in this section has been found by Cusmano and Ruccia.[191] This time a 1,2,4-oxadiazole is the final product formed by nitrosation of an imidazole. What intermediates lie between the nitroso compound (**103**) and the final product (**104**) have not been determined, but the reagent is alcoholic hydrochloric acid. Other examples of this type of rearrangement are known[191, 192] that also resemble the nitrosopyrrole to isoxazole rearrangement.[192a]

---

[190] M. Ruccia, N. Vivona, and G. Cusmano, *Chem. Commun.*, 358 (1974).
[191] S. Cusmano and M. Ruccia, *Gazz. Chim. Ital.*, **85**, 1686 (1955); **88**, 463 (1958).
[192] B. Cavalleri and G. Lancini, Ger. Offen. 2,331,058 (Jan. 1974) [*CA* **80**, 95961 (1974)].
[192a] T. Ajello and C. Petronici, *Gazz. Chim. Ital.* **72**, 333 (1942).

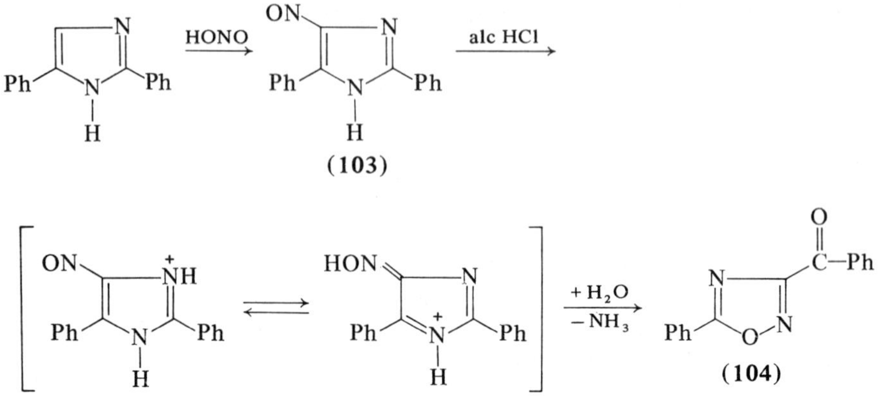

[Reaction scheme: Ph-imidazole-NH + HONO → (103) nitroso imidazole + alc HCl →]

[Bracketed tautomer equilibrium] → +H₂O, −NH₃ → (104) Ph-oxadiazole-C(=O)-Ph

From the catalytic reduction of an arylazooxadiazole (**105**), Warburton and co-workers[54] found a 1,2,4-triazole (**106**) as product.

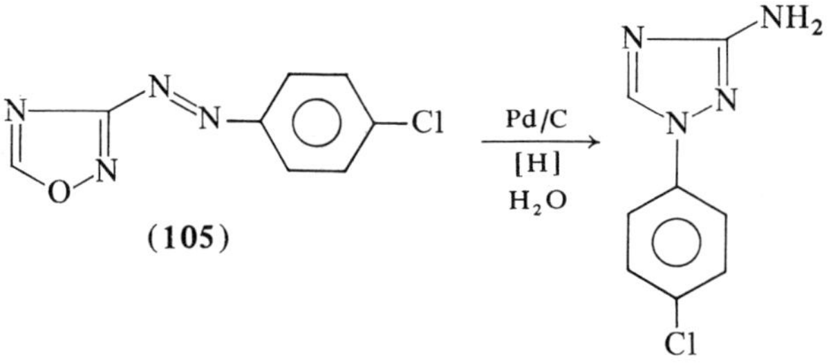

[Structures: (105) oxadiazole-N=N-C₆H₄-Cl; Pd/C, [H], H₂O → (106) triazole with NH₂ and p-chlorophenyl]

However, the explanation does not involve a rearrangement. Instead, the N—O bond was cleaved, the azo linkage was saturated, and the triazole ring was then formed by a cyclodehydration.

A reduced oxadiazolidine-5-phenylimine derivative has also been rearranged to a 1,2,4-triazolin-5-one by heating at 180° [Eq. (73)].[122]

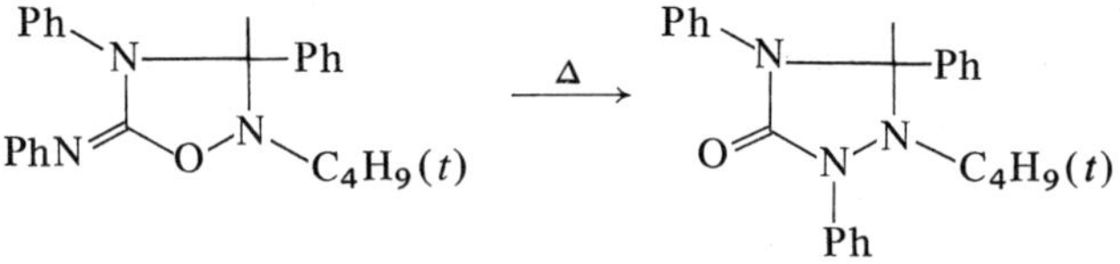

[Reaction (73): PhN—N(Ph)—O—N(C₄H₉(t))=NPh →Δ→ 1,2,4-triazolin-5-one with Ph, Ph, C₄H₉(t), Ph substituents] (73)

Hydrogenolysis with a palladium catalyst on charcoal of an oxadiazole substituted at C-3 with an enamine side chain allowed for a new synthesis of pyrimidines (**106a**) or pyrimidones (**106b**).[192b]

[192b] M. Ruccia, N. Vivona, and G. Cusmano, *J. Heterocycl. Chem.* **11**, 829 (1974).

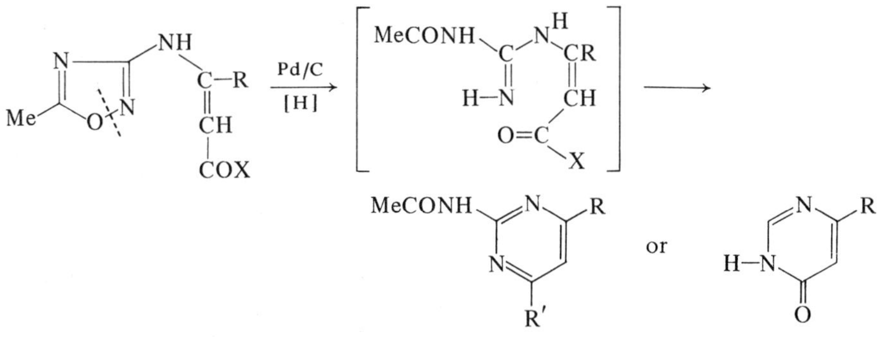

[Scheme showing Pd/C [H] reduction leading to pyrimidine tautomers (106a) and (106b)]

(106a)
X = R' = Me, Ph
R = H, Me, Ph

(106b)
X = OEt
R = H, Me, Ph

K. TAUTOMERISM AND HYDROGEN BONDING

Tautomerism in heterocyclic five-membered rings was thoroughly studied by Katritzky and co-workers beginning in 1961[193] and reviewed in 1970[194] and in the first two volumes of this series.[195] The confusing situation described by Boyer[196] has been resolved by a combination of pK_a measurements,[60, 194] the most useful single method of study, and IR, UV, and NMR spectroscopy.[60, 150–153]

The position of equilibrium in the tautomers of oxadiazoles is not the same for 3-hydroxy and 5-hydroxy derivatives although the carbons at C-3 and C-5 are both sp^2 in character. The carbon at C-3 between two nitrogens is in a less electronegative environment than that at C-5 between a nitrogen and an oxygen. Since the parent compounds are unknown, statements must be limited to 3- and 5-hydroxy, 5- and 3-substituted derivatives.

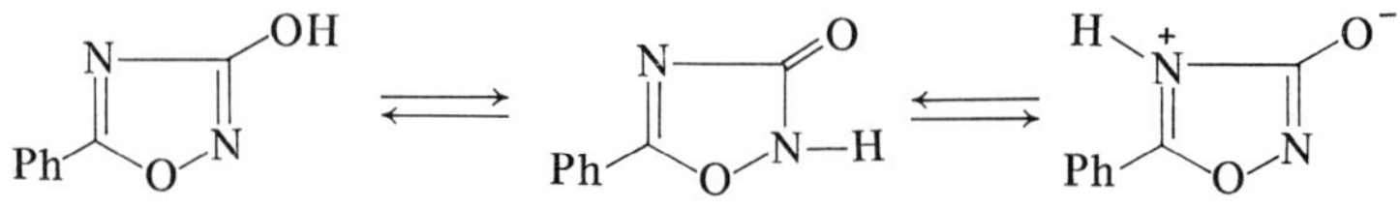

(107) (108a) (108b)

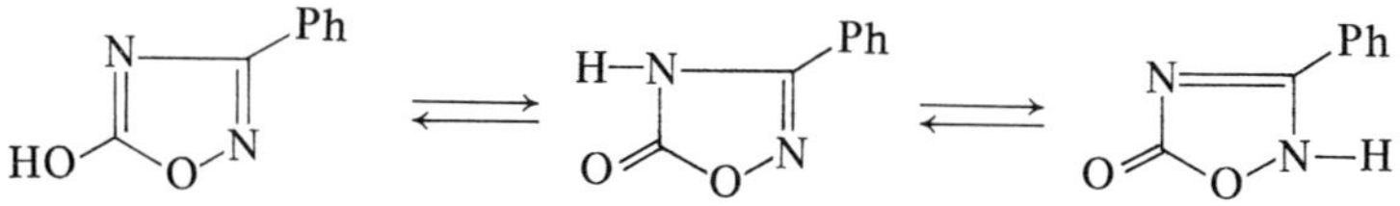

(109) (110a) (110b)

[193] A. J. Boulton and A. R. Katritzky, *Tetrahedron* **12**, 41, 51 (1961).
[194] A. R. Katritzky, *Chimia* **24**, 134 (1970).
[195] A. R. Katritzky and J. M. Lagowski, this Series, **1**, 311–437; **2**, 3–81 (1963).
[196] Boyer,[1] p. 499ff.

Spectroscopic (IR and UV, Section IV, D, E) and pK_a measurements[193] suggest that the hydroxy form (**107**) predominates in the C-3 compound, but that in the isomer the keto form (**110**) occurs. The explanation[60] is that the electronegative ring oxygen maintains a carbonyl adjacent to it (C-5) but allows the tautomeric hydroxy form when the carbon (C-3) is removed by one atom. However, the position of tautomeric equilibrium in **107, 108** is solvent dependent.[152, 153] In chloroform and in the solid state (KBr pellet), the IR spectrum indicates predominance of the keto form (**108**), but in oxygen solvents, such as ethanol, acetone, and dimethyl sulfoxide, the enol form (**107**) allows for better hydrogen bonding. The thiol derivatives exhibit the same phenomenon: no —SH band is visible in the solid state.[152] Katritzky[194] has emphasized the importance of considering zwitterionic forms (**108b**) in tautomers of heterocyclic compounds.

With diazomethane, compound **109, 110** gave a mixture of *O*-methyl and *N*-methyl derivatives[60] [Eq. (74)], but other methylating agents gave

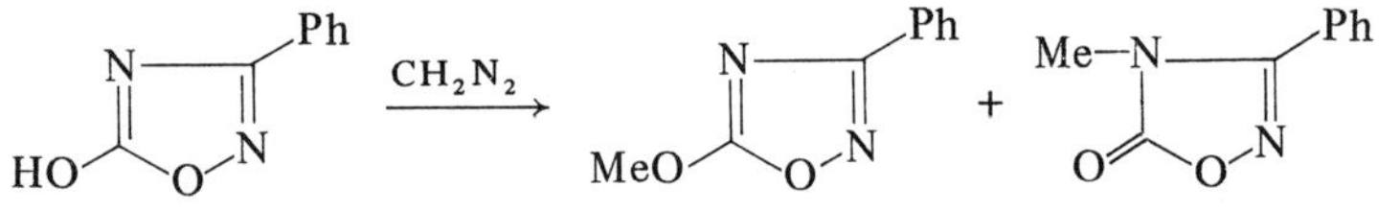

only an *N*-methyl derivative, probably substituted at N-4 (**112**). The cautious statement was made because the authors were unable to synthesize **112** by an unequivocal pathway. The results imply that diazomethane gives a kinetic result and that the *N*-methyl derivative is the thermodynamically stable product (next paragraph).

The 3-methoxy-5-phenyl isomer (**114**) was synthesized from the thio compound[59,197] (**113**), which was hydrolyzed to **115** by pyridine hydrochloride. Diazomethane, acting on 3-hydroxy-5-phenyloxadiazole (**115**)

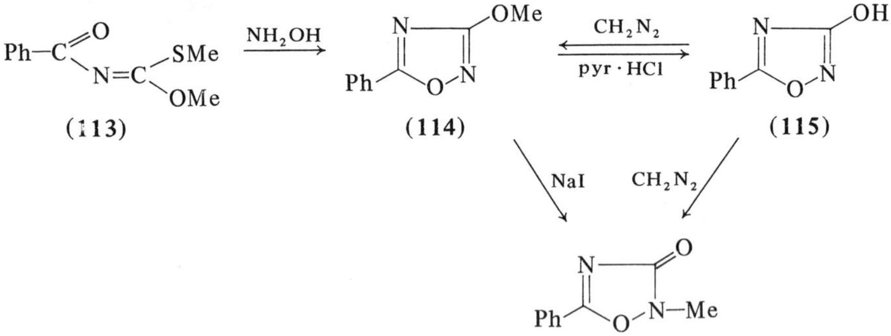

[197] T. B. Johnson and G. A. Menge, *Am. Chem. J.*, **32**, 358 (1904).

gave a mixture of *O*- and *N*-methyl derivatives (**114** and **116**). Against the 2-methyl-5-phenyl-1,2,4-oxadiazol-3-one (**116**) appears to be the thermodynamically more stable isomer since **114** was isomerized to **116** with sodium iodide as catalyst.

Only one example is known where it seems certain that both N-2 and N-4 derivatives of an oxadiazoline have been obtained[85, 86] [Eq. (29)]. Compound **22** had a chemical shift for the N—CH$_3$ group 1.03 ppm farther downfield than for the same group in **21**. (See also Scheme 7).

The tautomeric situation is simpler in the 5-amino derivatives because the equilibrium lies heavily on the side of the amino form (**117**), as is also the case in isoxazoles.[193]

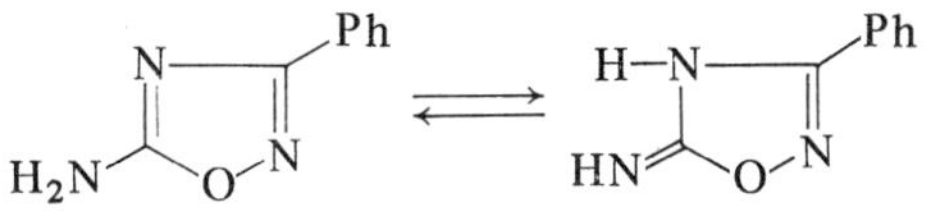

(117)   (118) 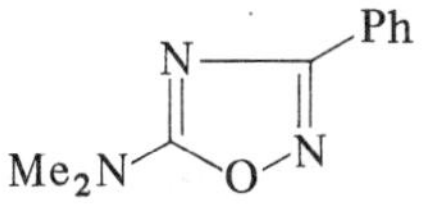   (119)

Najer and co-workers[150, 151] and Selim and Selim[152, 153] synthesized a series of 5-amino-3-phenyloxadiazoles and oxadiazolines, confirming the structures by IR and NMR spectra. The oxadiazolines **120** and **121**

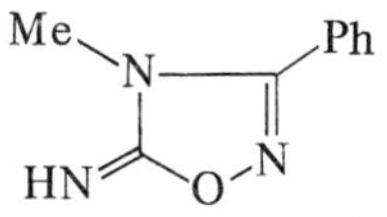

(120) 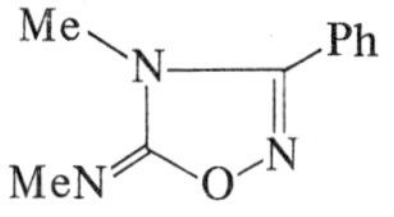   (121)

were about 6 powers of 10 stronger bases than the oxadiazoles **117** and **119**.

Hydrogen bonding in the hydroxy- and aminooxadiazoles is significant. In carbon tetrachloride, the IR spectrum[60] of **107** indicates dimer formation, well described by structure **122**. As the solution is

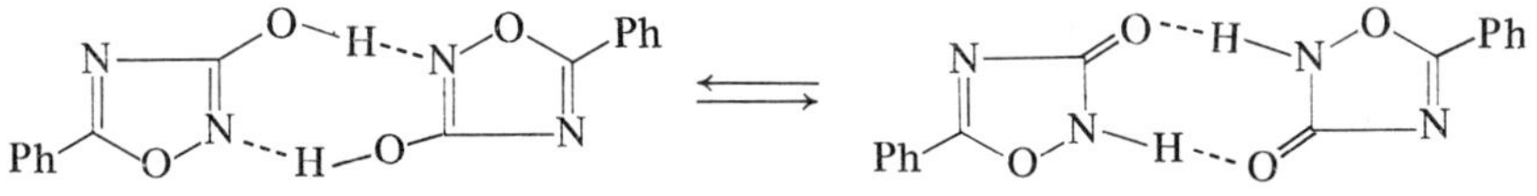

(122)   (123)

diluted, the intensity of the hydroxyl band at 3580 cm^{-1} increases. In chloroform, the IR spectrum of the same compound suggests that 5–10% of **108** may exist simultaneously with **107**. But **108** may also be hydrogen bonded, as shown in structure **123**.

VI. Uses

A. Chemotherapy

Since 1960, at least three oxadiazoles attained enough popularity as therapeutic drugs to be known by trade names: Oxolamine[198, 199] (**124**) (1960), Irrigor[199, 200] (**125**) (1964), and Libexin[27] (**126**) (1966). Oxolamine and Libexin have been used as antitussive (cough) agents, and Irrigor principally as a coronary vasodilator and local anaesthetic. These and others have been tested for a wide variety of therapeutic activities.[200–206] Silvestrini and co-workers[201–203] compared more than 100

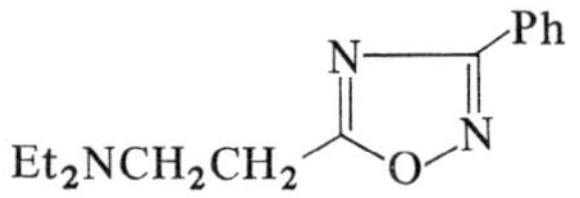

$Et_2NCH_2CH_2$—[1,3,4-oxadiazole-Ph]

(**124**)

Oxolamine (citrate)
(also Bredon, Perebron)

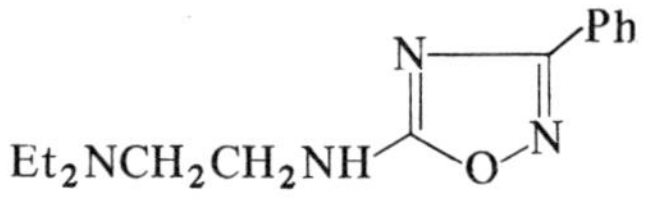

$Et_2NCH_2CH_2NH$—[1,3,4-oxadiazole-Ph]

(**125**)

Irrigor

piperidine-N—CH_2CH_2—[1,3,4-oxadiazole-CH_2CHPh_2] · HCl

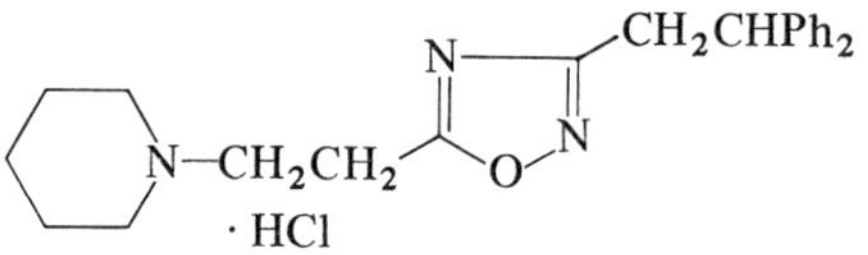

(**126**)

Libexin

amine derivatives of oxadiazoles and found highest activity for those shown in Scheme 9.

---

[198] M. De Gregorio, *Panminerva Med.* **4**, 90 (1962)
[199] F. Eloy and R. Lenaers, *Bull. Chim. Therap.*, 347 (1966).
[200] J. Sterne and C. Hirsch, *Therapie* **20**, 89, 95 (1965).
[201] G. Palazzo, M. Tavella, G. Strani, and B. Silvestrini, *J. Med. Pharm. Chem.* **4**, 351 (1961) [*CA* **56**, 1962 (1962)].
[202] B. Silvestrini and C. Pozzatti, *Arch. Int. Pharmacodyn.* **129**, 249 (1960) [*CA* **56**, 14874 (1962)].
[203] B. Silvestrini and C. Pozzatti, *Brit. J. Pharmacol.* **16**, 209 (1961) [*CA* **56**, 5356 (1962)].
[204] B. Silvestrini and G. Palazzo, *Arzneim.-Forsch.* **13**, 798 (1963) [*CA* **60**, 4665 (1964)].
[205] J. Aron-Samuel, Fr. Demande 2,148,430 (Apr. 1973) [*CA* **79**, 78810 (1973)].
[206] J. Perronnet and P. Girault, Fr. Demande 2,154,279 (June, 1973) [*CA* **79**, 92232 (1973)].

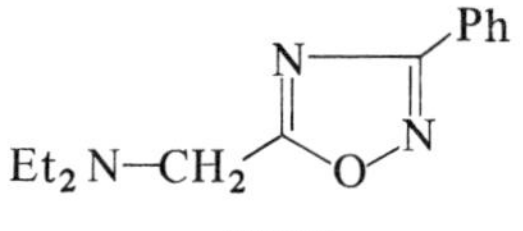

Sec. VI.A] 1,2,4-OXADIAZOLES 113

(127)

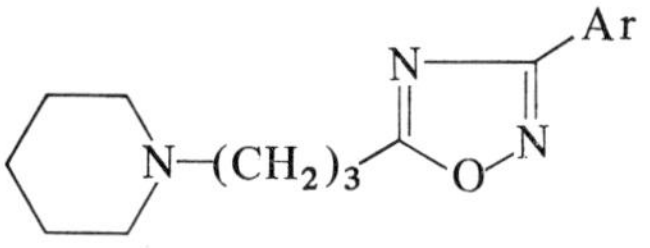

(128) Ar = Ph
(129) Ar = p-Cl—C₆H₄

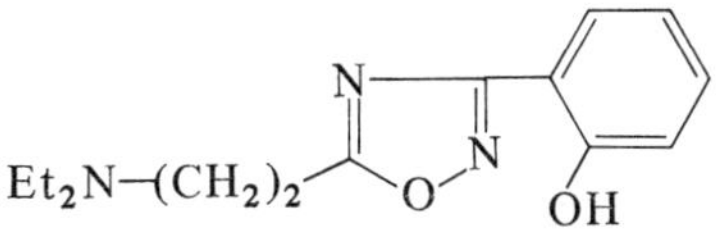

(130)

Activity (highest):
Anesthetic 127, 128
Analgesic 127, 130
Antispasmodic 130, 129
Antitussive 124
Antiinflammatory 128, 130
Toxicity (lowest): 124

SCHEME 9

However, Oxolamine was found to produce bladder irritation in rats.[207, 208] This behavior was traced to the release of diethylamine from the side chain at C-5, leaving 3-phenyl-5-vinyl-1,2,4-oxadiazole. Presumably this could occur in **130** as well.

At the clinical level Oxolamine[209] was found to give some relief of pulmonary silicosis when 18 miners were treated with it for 180 days.

Still other types of activity are claimed for oxadiazole derivatives. For example, compound **131** is an anthelmintic against *Nematospiroides dubius* in mice, dogs, and sheep.[31, 210–212] Compound **132** is a local anaesthetic and has antispasmodic and anti-inflammatory activity.[205,206,213,214] The sulfa drug **133** is said to be comparable in

---

[207] B. Catanese, G. Palazzo, C. Pozzatti, and B. Silvestrini, *Exp. Mol. Pathol., Suppl.* **2**, 28 (1963) [*CA* **60**, 16381 (1964)].

[208] B. Catanese and B. Silvestrini, *Boll. Chim. Farm.* **103**, 447 (1964) [*CA* **61**, 13761 (1964)].

[209] G. P. Nissardi, P. Cherchi, F. S. Randaccio, and P. L. Torrazza, *Clin. Terap.* **27**, 693 (1963) [*CA* **60**, 15043 (1964)].

[210] W. E. Buting and C. Ainsworth, U.S. 3,356,684 (1967) [*CA* **69**, 10441 (1968)].

[211] J. Wood and M. J. Bull, Ger. Offen. 2,310,287 (Sept. 1973) [*CA* **79**, 146504 (1973)].

[212] W. Meyer and B. Boehner, Ger. Offen. 2,312,500 (Sept. 1973) [*CA* **80**, 14933 (1974)]; M. H. Fisher and A. R. Matzuk, Ger. Offen. 2,420,779 (Nov. 1974) [*CA* **82**, 72993 (1975)].

[213] J. Sterne, M. F. Pele, and C. Hirsch, *Therapie* **24**, 735 (1969) [*CA* **71**, 79698 (1969)].

[214] J. Sterne, *Therapie* **24**, 745 (1969) [*CA* **71**, 79699 (1969)].

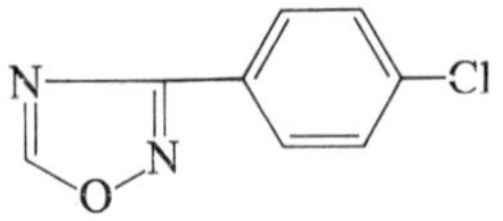

(131)

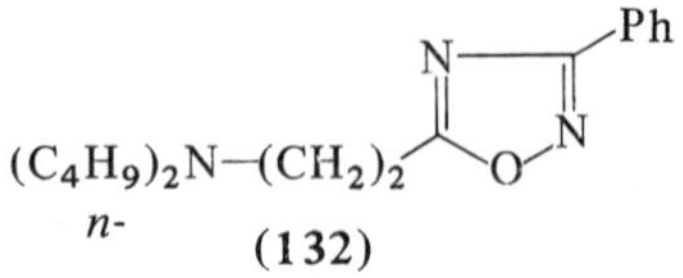

(132)

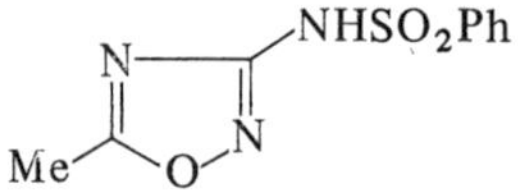

(133)

activity to sulfapyridine against streptococcus and pneumococcus infections.[215] Other sulfa drugs containing the oxadiazole ring have been tested.[216] Palazzo and Silvestrini[11] mentioned that various carbamates (**134**) have a wide spectrum of pharmaceutical activity, including anticonvulsant, analgesic, and antiinflammatory. The hydrazine derivative

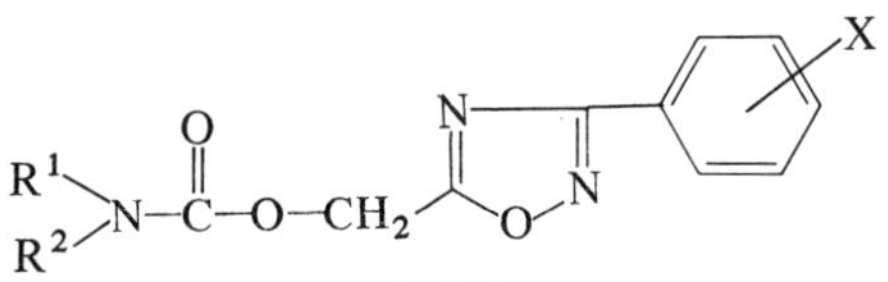

(134)

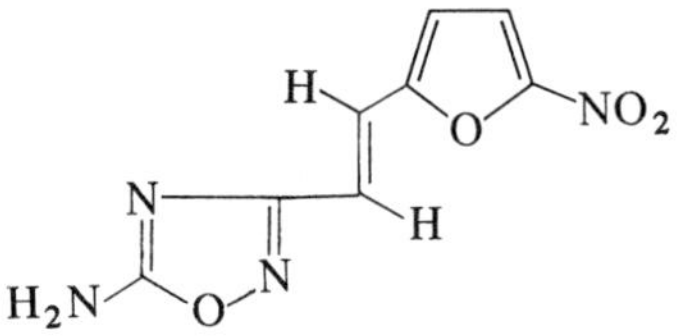

(135)

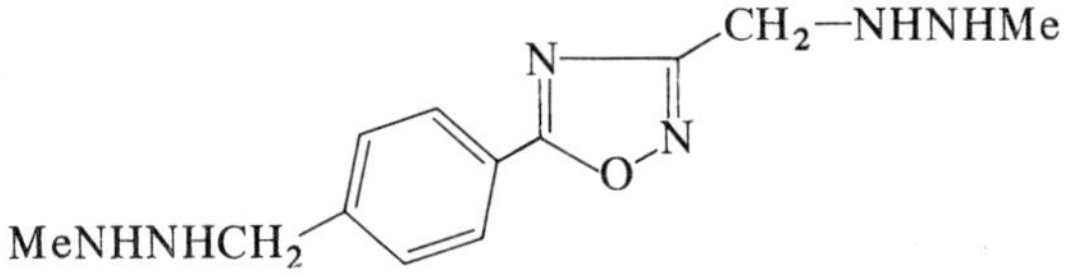

(136)

136 is said to have anticancer activity,[217] and the hydrazide **137** is useful against leprosy and tuberculosis. Anticonvulsant,[218] hypo-

---

[215] G. W. Anderson, H. E. Faith, H. W. Marson, P. Winneth, and R. Robbin, *J. Amer. Chem. Soc.* **64**, 2902 (1942).
[216] E. D. Bergmann, H. Bendas, and U. D'Avilla, *J. Org. Chem.* **18**, 64 (1953).
[217] F. Hoffmann-La Roche, Fr. Demande 1,385,856 (Jan. 1965) [*CA* **62**, 14688 (1965)].
[218] M. Nakanishi, K. Araki, T. Tahara, and M. Shiroki, Ger. Offen. 2,331,540 (Jan. 1974) [*CA* **80**, 96036 (1974)].

cholesteremic,[219-222] central depressant,[223] tranquilizing[69] and anoretic[224] activity (rats and dogs) have all been found in oxadiazoles.

The 5-nitrofuryl group[225] is promulgated as an efficacious moiety to ensure chemotherapeutic value, and the trans derivative (135) was found to be active against *Staphylococcus aureus*.[171] It is also a broad-spectrum antimicrobial agent[226] and an antischistosomal drug.[227] The latter activity is due to reduction of the phosphorylase phosphatase in the worms themselves, damaging reproduction in the female.[227, 228]

(137)

B. AGRICULTURAL AND OTHER USES

One oxazolidinedione (138), called Methazole, has been used to control yellow nutsedge in cotton,[133, 229] and a few other oxadiazole

(138)

[219] Y. Imai and K. Shimamoto, *Atherosclerosis* **17**, 121 (1973).
[220] Y. Imai, H. Matsumura, S. Tamura, and K. Shimamoto, *Atherosclerosis* **17**, 131 (1973).
[221] S. Yurugi, A. Miyake, T. Fushima, E. Imamiya, H. Matsumura, and Y. Imai, *Chem. Pharm. Bull.* **21**, 1641 (1973).
[222] S. Yurugi, A. Miyake, M. Tomimoto, H. Matsumura, and Y. Imai, *Chem. Pharm. Bull.* **21**, 1885 (1973).
[223] W. J. Fanshawe and S. R. Safir, U.S. 3,770,739 (Nov. 1973) [*CA* **80**, 48006 (1974)].
[224] R. Dalla Vedova and G. D'Alo, *Boll. Chim. Farm.* **112**, 273 (1973) [*CA* **80**, 10431 (1974)].
[225] M. C. Dodd, W. B. Stillman, M. Roys, and C. Crosby, *J. Pharmacol. Exp. Ther.* **82**, 11 (1944).
[226] H. H. Gadebusch, H. Breuer, G. J. Miraglia, H. I. Basch, and R. Semar, *Antimicrob. Agents Chemother.*, 280 (1969) [*CA* **75**, 45930 (1971)]; H. H. Gadebusch and H. I. Basch, *Antimicrob. Agents Chemother.* **6**, 263 (1974) [*CA* **82**, 68723 (1975)].
[227] C. H. Robinson, E. Bueding, and J. Fisher, *Mol. Pharmacol.* **6**, 604 (1970).
[228] S. J. Lau, I. Weliky, and E. C. Schreiber, *Xenobiotica* **3**, 97 (1973) [*CA* **79**, 49072 (1973)].
[229] P. E. Keeley, C. H. Carter, and J. H. Miller, *Weed Sci.* **21**, 327 (1973) [*CA* **79**, 112306 (1973)].

derivatives have been patented as weed killers,[131] fungicides,[230, 231] and acaricides.[211, 212]

Oxadiazole derivatives have been tested in the textile industry as antistatic agents for textile finishes,[232, 233] thermally stable polymers,[82, 234, 235] fluorescent whiteners,[236] oil and water repellents on cotton and wool,[237] and blue dyes for polyesters and polyamides.[238] Oxadiazole merocyanine dyes have been patented as photosensitizers for silver halide emulsions.[239]

---

[230] H. Hamano, K. Shimada, and S. Kuriyama, Japan. 72 03,823 [*CA* **76**, 153752 (1972)].
[231] Y. Uchiyama and Y. Hashimoto, Japan. 74 13,177 (Feb) [*CA* **80**, 133442 (1974)].
[232] H. Hagen, F. Becke, and J. Niemeyer, Ger. Offen. 2,016,692 (Oct. 1971) [*CA* **76**, 25299 (1972)].
[233] H. Hagen, F. Becke, and J. Niemeyer, U.S. 3,776,910 (Dec. 1973) [*CA* **80**, 82995 (1974)].
[234] Y. Hagiwara, M. Kuwahara, S. Toyama, N. Dogoshi, and N. Ida, Japan. 71 24,009 [*CA* **76**, 86370 (1972)].
[235] D. A. Bochvar, I. V. Stankevich, E. S. Kronganz, and V. V. Dorshak, *Dokl. Akad. Nauk SSSR* **200**, 364 (1971) *CA* **76**, 46591 (1972)].
[236] H. Davidson, K. T. Johnson, B. E. Leggeter, and A. J. Moore, Ger. Offen. 2,344,834 (Mar. 1974) [*CA* **81**, 38958 (1974)].
[237] P. L. Pacini and E. K. Kleiner, Ger. Offen. 2,065,001 (Jan. 1972) [*CA* **77**, 6019 (1972)].
[238] M. Patsch and H. Eilingsfeld, Ger. Offen. 1,949,295 (Apr. 1971) [*CA* **75**, 7417 (1971)].
[239] H. Oehlschlaeger, O. Riester, and E. Proeschal, Ger. Offen. 2,035,724 (Jan. 1972) [*CA* **76**, 155,604 (1972)].

Covalent Hydration in Nitrogen Heterocycles

ADRIEN ALBERT[1]

Research School of Chemistry, The Australian National University, Canberra, Australia

I. Introduction	117
II. Diagnosis and Location of Covalent Hydration	118
A. Qualitative Aspects	119
1. New Techniques	119
2. New Patterns of Hydration	122
B. Quantitative Aspects	127
1. Equilibrium Studies	128
2. Kinetic Studies	129
3. Discovery of Hydrated Intermediates	129
III. Time-Dependent Shifts in the Preferred Position of Hydration	131
IV. Examples of Covalent Hydration in New Ring Systems	135
V. The Isolation of Stereoisomeric Hydrates	139
VI. Covalent Hydration in Nature	140

I. Introduction

The reversible covalent hydration of a C=N bond in a nitrogen heterocycle was first reported[2] in 1952. Originally the subject was reviewed[3] in these volumes in 1965, and elsewhere[4] in 1967. The following pages are intended to bring the 1965 reviews up to date.

In that last decade, the covalent character of the hydration has been even more firmly established by mass spectroscopy, X-ray crystallography, and proton magnetic resonance (PMR). The last named has been pressed into use also as a diagnostic and locating technique. Since 1965, covalent hydrates have been found in many new ring systems, even in single rings. Some cases have been found where *two* molecules of water are added to a single nucleus. Other molecules have

[1] Present address: Department of Pharmacological Sciences, Health Sciences Center, State University of New York at Stony Brook, Stony Brook, New York 11790.
[2] A. Albert, D. J. Brown, and G. Cheeseman, *J. Chem. Soc.* 1620 (1952).
[3] (a) A. Albert and W. L. F. Armarego, this Series, **4**, 1 (1965); (b) D. D. Perrin, *ibid.* **4**, 43 (1965).
[4] A. Albert, *Angew. Chem.* **79**, 913 (1967), *Angew. Chem., Int. Ed. Engl.* **6**, 919 (1967).

been shown to add one molecule of water to a certain position, under kinetic control, whereas after the lapse of time this water molecule is found to occupy a different position under thermodynamic control. Several covalent hydrates have been reported as reaction intermediates. Stereochemical evidence has been offered of optical and of geometrical pairs of isomeric hydrates. Remarkable examples of covalent hydration have been reported in physiologically potent compounds isolated from both lower and higher forms of life.

II. Diagnosis and Location of Covalent Hydration

Of the new methods that have emerged for studying covalent hydration, PMR has been most used. With its aid added to the two most fruitful of the earlier methods—ionization constants and ultraviolet (UV) spectra—many new and often unsuspected facts have emerged. The various physical methods have been particularly revealing when a nucleus with substituents that display a range of inductive, mesomeric, and steric properties has been used. As a result, more is now known about the mechanism of hydration, and many new patterns of hydration have emerged.

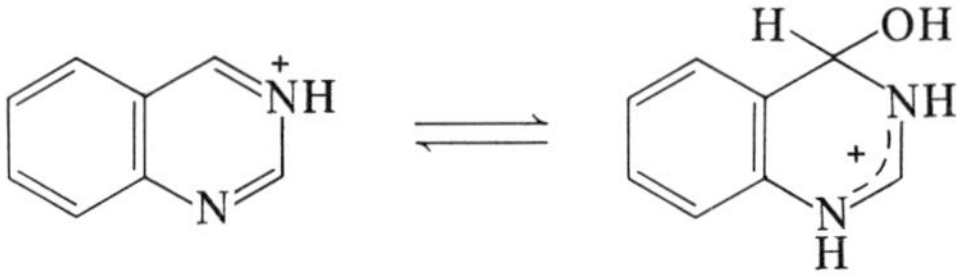

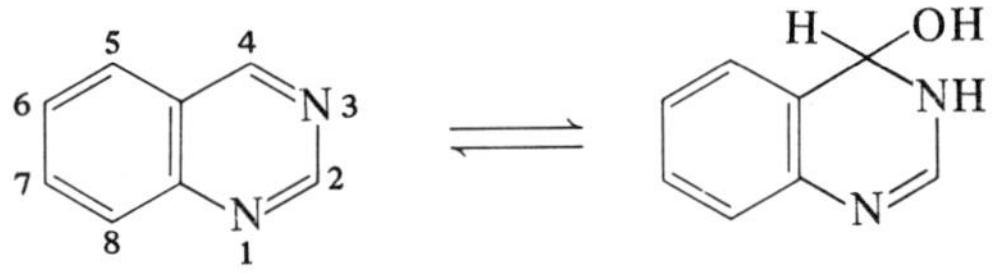

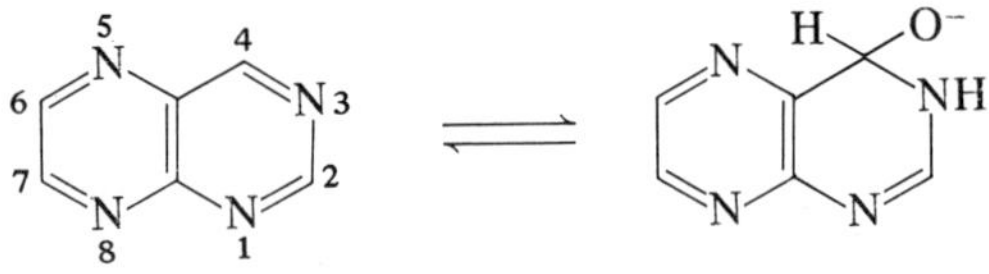

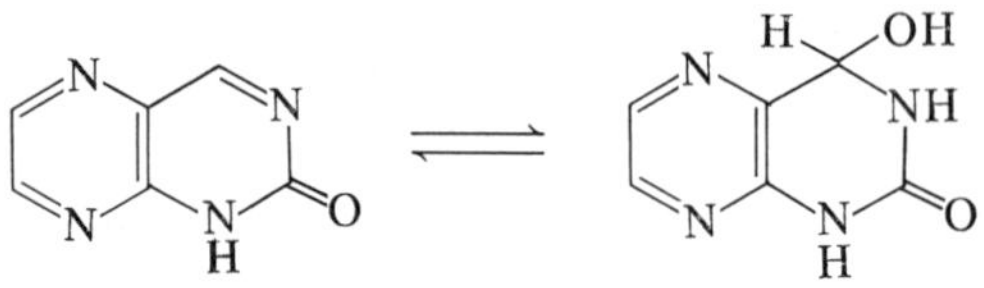

Methods and results will now be described. As before, the two essentials to produce an easily recognizable proportion of hydrated molecules (say, 2%) are (a) a nucleus whose π-layer is deprived of electrons by two or more doubly bound nitrogen atoms in the ring, or by strongly inductive ($-$I) substituents, or by both influences together, plus (b) stabilization of the hydrate by resonance. When the last reviews appeared in this series, four main patterns of hydration (**1–4**) were recognized. Type I, as exemplified by the quinazoline cation (**1**), was known to be stabilized by an amidinium-type resonance; type II, the corresponding neutral species had little stabilization, and hence it lost water fast. Type III, the anion corresponding to type II, was found only in more π-deficient nuclei,[5] such as (the anion of) pteridine (**3**). Type IV, stabilized by a urea-type resonance, was found in the 2-oxo ("2-hydroxy-") derivatives of fused pyrimidines, as in pterid-2-one (**4**); in this type, hydration was a property of the neutral species, not of the monoanion and -cation.

A. Qualitative Aspects

1. *New Techniques*

As a model for the hydrated product of the cation of 2-aminopteridine, the more stable ethanol adduct was submitted to X-ray analysis (as the hydrobromide[6]). The bromide ion, the carbon, nitrogen, and oxygen atoms, and four of the twelve hydrogen atoms were located from Patterson, electron density, and difference maps, respectively. Full-matrix least-squares refinement reduced R to 0.144 for 1968 reflexions. C-2 and the three nitrogen atoms bonded to it were seen to form a planar guanidinium system in which all four hydrogen atoms were bonded to two bromide ions. The ethoxy unit was found attached to the C-4 position. This, the first crystal structure of a heteroaromatic covalent adduct, shows the constitution to be 2-amino-4-ethoxy-3,4-dihydropteridine hydrobromide (**5**). This diffraction analysis also verified the conclusion that 3,4-pyrimidine-ring adducts are preferred to pyrazine-ring adducts because of resonance stabilization in the former by the guanidinium ion.[7,8]

Applying mass spectrometry to the hydrated and anhydrous forms of 4-trifluoromethylpteridine, which interconvert only slowly, it was found that each had its characteristic mass spectrum.[9] The fragmentation

---

[5] A. Albert, "Heterocyclic Chemistry," 2nd ed., p. 56, Athlone Press, London, and Humanities Press, New York, 1968.
[6] T. J. Batterham and J. A. Wunderlich, *J. Chem. Soc. B*, 489 (1969).
[7] A. Albert and J. J. McCormack, *J. Chem. Soc. C*, 1117 (1966).
[8] A. Albert, T. J. Batterham, and J. J. McCormack, *J. Chem. Soc. B*, 1105 (1966).
[9] J. Clark and A. E. Cunliffe, *Org. Mass. Spectrom.* **7**, 737 (1973).

(5)

(6) 8-Azapurine

pattern of the hydrate clearly showed that the water molecule was attached to the 3,4-bond, in agreement with allocation by the usual methods. Spectra obtained at lower ionizing voltages (11 eV) showed the position of added water molecules in the 5,6,7,8-dihydrated derivatives of ethyl pteridine-4-carboxylates.[9]

Apparently the first published use of PMR to study hydration was the report that the proton in the 4-position of quinazoline moved upfield (τ 0.63 $\longrightarrow$ 3.58) after the molecule was converted into the cation by aqueous acid.[10] Normally a downfield shift would be expected on ionization, and this reversal of direction afforded a further and convincing proof of the covalent character of hydration, as well as demonstrating the value of this technique in determining the position of hydration. Contemporaneous application of PMR to the hydration of 8-azapurine (6) cations established the scope and value of this technique.[11]

When using PMR, special thought must be given to the choice of solvent. Whereas some hydrates, like pterid-2-one, do not lose their covalently bound water even at 200°, others cannot exist at room temperatures in the absence of water. Hence partly nonaqueous solvents, by lowering the thermodynamic activity of the water molecules that they contain, are found to inhibit the hydration of the less susceptible substances. This makes PMR a less useful technique than UV spectra and ionization studies for detecting hydration in those neutral species that carry water-insolubilizing groups, such as $-NH_2$, :O and :S. When the neutral species is quite soluble in water, however, useful information can often be obtained by switching to a non-adding solvent. This technique can always be used for the *ionic* forms, as in the following example of cations in the 8-azapurine (6) series.

In deuterium oxide, the parent substance, 8-azapurine (neutral species), showed two peaks at τ 0.32 and 0.80 (1H each) corresponding to the two aromatic-type protons in the 6- and 2-positions, respectively (see Table I). The anhydrous cation was easily demonstrated in trifluoroacetic acid, through the downfield displacement of these signals by the usual 0.4–0.6 ppm. However in deuterium oxide–deuterium

---

[10] W. L. F. Armarego and R. E. Willette, *J. Chem. Soc.*, 1258 (1965).
[11] J. W. Bunting and D. D. Perrin, *J. Chem. Soc. B*, 433 (1966).

chloride mixture, the strong upfield shift of the signals demonstrated that the *hydrated* cation was exclusively present. Similar results were obtained with 7-methyl-8-azapurine[12] and 8-methyl-8-azapurine.[13] However the insertion of a methyl group in the 6-position of 8-azapurine effectively blocked hydration, so that even in deuterium oxide the cation was entirely anhydrous[11] (Table I).

TABLE I

COVALENT HYDRATION OF 8-AZAPURINES[a]

8-Azapurine	Solvent	Ionic species[b]	(τ) H(6)	(τ) H(2)
Unsubstituted	D$_2$O	Anh. 0	0.32	0.80
	DC1	Hyd. +	3.19	1.46
	TFA[c]	Anh. +	−0.31	+0.40
6-Me	DC1	Anh. +	Nil	0.54
7-Me	D$_2$O	Anh. 0	0.13	0.56
	DC1	Hyd. +	2.95	1.32
	TFA	Anh. +	−0.55	+0.13

[a] From Bunting and Penrin.[11] Results from proton magnetic resonance study.
[b] Anh., anhydrous; Hyd., hydrated.
[c] Trifluoroacetic acid.

It is noteworthy that purine shows the normal downfield shift when the cation is made in deuterium oxide,[11] for the removal of one doubly bound nitrogen atom (N-8) lowers depletion of the π layer, and no hydration takes place. Other attempts to produce hydration in substituted purines[14] have been unavailing, but the addition of stronger nucleophiles has been demonstrated.[15]

This addition of nucleophiles stronger than water to molecules that readily hydrate has, in recent years, grown into a well-explored topic that is worthy of a separate review. All that need be said here is that alcohols,[16] amines, and aqueous ammonia[17] readily form covalent ad-

[12] A. Albert and K. Tratt, *J. Chem. Soc. C*, 344 (1968).
[13] A. Albert, *J. Chem. Soc. C*, 2076 (1968).
[14] A. Albert, *J. Chem. Soc. B*, 438 (1966).
[15] W. Pendergast, *J. Chem. Soc. Perkin Trans. I*, 2759 (1973); 2240 (1975).
[16] A. Albert and H. Mizuno, *J. Chem. Soc. B*, 2423 (1971); A. Albert and J. J. McCormack, *J. Chem. Soc.*, 6930 (1965); A. Albert and C. F. Howell, *J. Chem. Soc.*, 1591 (1962).
[17] B. E. Evans, *J. Chem. Soc. Perkin Trans. I*, 357 (1974).

ducts and hence must be avoided as solvents in the study of covalent hydration.

2. New Patterns of Hydration

A nucleus that has already hydrated may still contain a highly polarized double bond that is only weakly conjugated with double bonds in the rest of the molecule. Such a compound may take up a second molecule of water (a phenomenon called *multiple hydration*), adjacent to the first one. This is not surprising, for the first hydration will have left an unsaturated heterocycle, and these are prone to hydration, as will be shown at the end of Section IV.

The first example of multiple hydration was discovered in the cation of 1,4,5,8-tetra-azanaphthalene. This species had already been recognized as undergoing hydration (see Albert Armarego,[3] p. 31),

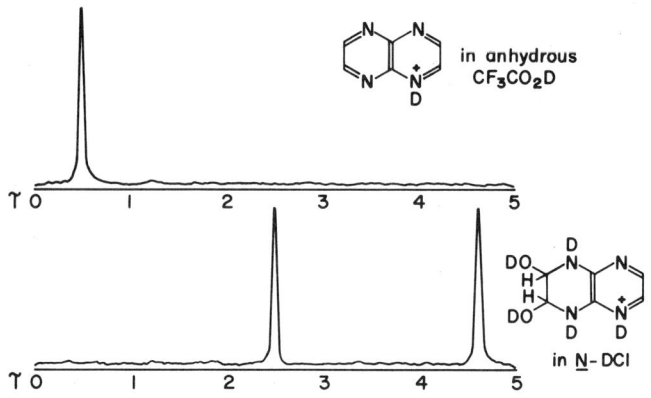

FIG. 1. The hydration of 1,4,5,8-tetra-azanaphthalene cation, as observed by proton magnetic resonance study.

but subsequent examination by PMR showed that a second molecule of water was added immediately after the first one.[18] In the absence of water, only one signal (τ 0.5) was recorded, which was a time-averaged result from the four CH groups. However, examination in dilute DCl in D_2O showed that this signal was replaced by two peaks much farther upfield (see Fig. 1). Both molecules of water must have been added in the same ring to give (**7**), otherwise the spectrum would have been more complex. Further examples of multiple hydration then came to light (see Sections III and V), and in each case a pyrazine ring was dihydrated. A

[18] T. J. Batterham, *J. Chem. Soc. C*, 999 (1966).

different pattern, known as *transmolecular hydration* will now be discussed.

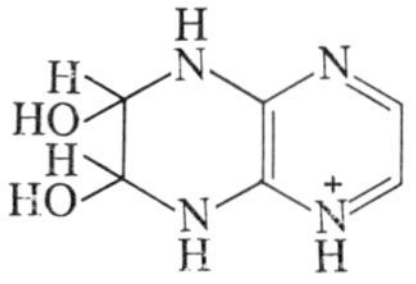

(7)

It was reported previously (Ref. 3a, p. 29) that 6,7,8-trimethylpterid-2-one (**8**) (and the isomeric -4-one) forms a sodium salt, which would not be possible without prior hydration. It has since been shown that the related 2,8-dihydro-6,7,8-trimethyl-2-methyliminopteridine (neutral species) (**9**) forms a hydrate by attaching H and OH to positions 1 and 7, respectively.[19] The oxo analog (**8**) was also found to hydrate in these

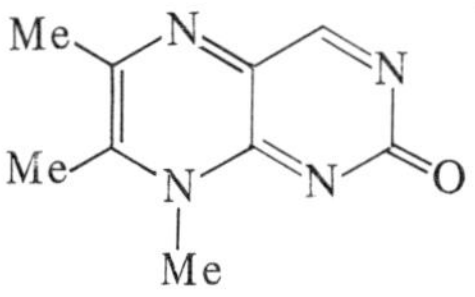

(8) 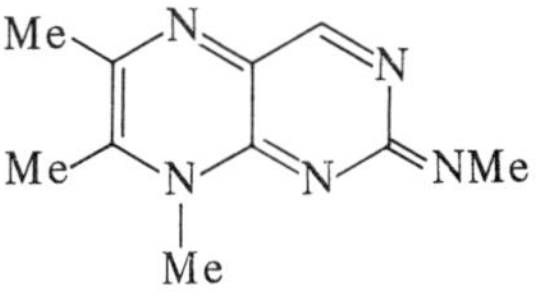   (9)

positions. These and other pteridines with a C-methyl substituent at the site of OH addition undergo facile demethylation when gently oxidized with potassium permanganate.[19]

Soon afterward, it was shown that 6,7-dimethyl-8-ribityllumazine (**10**) was an intermediate in the biosynthesis of riboflavine; it was noted that the UV spectrum of this lumazine could not be reconciled with those of

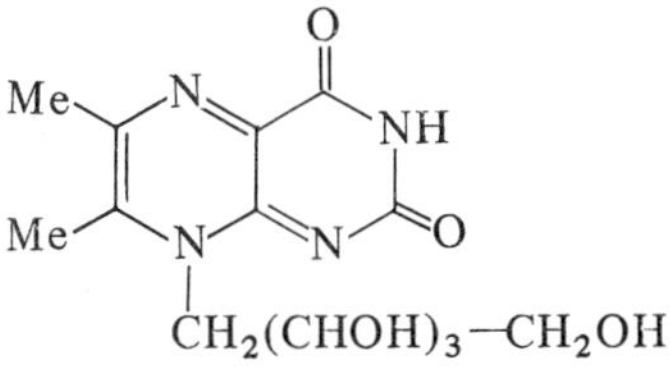

(10)

6,7,8-trimethyllumazine. This disagreement, already quite striking in the neutral species, became quite astonishing in alkaline solution, where 6,7,8-trimethyllumazine and 3,6,7,8-tetramethyllumazine, too, underwent a slow hypsochromic shift.[20] When ionization constants were

[19] N. W. Jacobsen, *J. Chem. Soc. C*, 1065 (1966).
[20] W. E. Fidler and H. C. S. Wood, *J. Chem. Soc.*, 3980 (1957); W. Pfleiderer and G. Nübel, *Chem. Ber.* **93**, 1406 (1960); C. H. Winestock and G.W. Plaut, *J. Org. Chem.* **26**, 4456 (1961).

sought, as a preliminary to assigning the UV spectra to single ionic species, it was found that these constants, too, differed widely among homologs. The apparent contradictions were finally resolved by the discovery that transmolecular hydration was occurring, to various degrees, and in some homologs more than others.[21a]

This investigation began with a study of lumazines in which the ionizable hydrogen atom on N-3 was replaced by a methyl group. It was found that 3,8-dimethyllumazine (11) (with or without further methyl groups in the 6- and 7-positions) showed no anomalies of pK_a or UV spectra, neither as the neutral species nor the cation. Alkali produced a monoanion of pK_a 6.4, which must necessarily have come from a hydrated form. This anion had a UV maximum at shorter wavelengths, corresponding to loss of a double bond (see Scheme 1). Such an anion is clearly related to that of pteridine (3), which has pK_a 11.2.

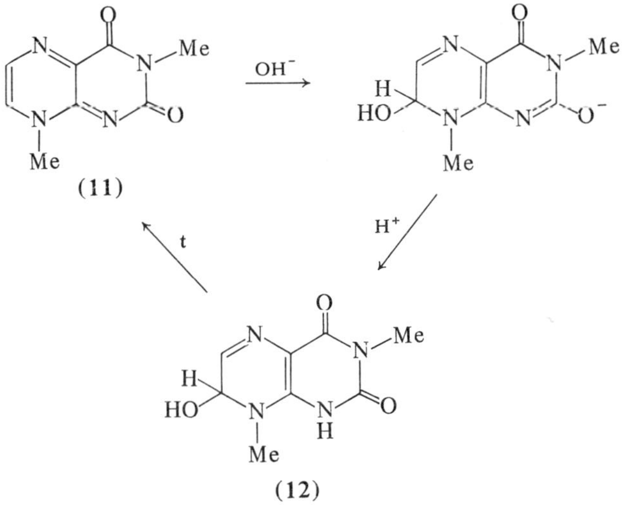

SCHEME 1. The anhydrous neutral species (11), the hydrated neutral species (12), and the hydrated anion of 3,8-dimethyllumazine; λ_{max} 394, 308, and 306 nm, respectively.

Neutralization of an alkaline solution of 3,8-dimethyllumazine produced the (unstable) hydrated form of the neutral species (12) with a UV maximum at 308 nm, quite close to that of the hydrated anion (306 nm). On standing at room temperature, this neutral form quickly became converted into the anhydrous starting material (11) which has

[21] (a) W. Pfleiderer, J. W. Bunting, D. D. Perrin, and G. Nübel, *Chem. Ber.* **99**, 3503 (1966); (b) W. Pfleiderer, J. W. Bunting, D. D. Perrin, and G. Nübel, *ibid.* **101**, 1072 (1968).

its maximum at 394 nm. By rapid-reaction apparatus technique,[3b] it was shown that there were about 5400 molecules of the *anhydrous* neutral species at equilibrium with one molecule of the *hydrated* neutral species. That the anion was not dihydrated was confirmed by PMR.

The next, and more diffcult, part of the investigations used 8-alkyllumazines with a hydrogen atom on N-3. Here again, the neutral species and cation proved to be normally anhydrous, but the UV spectra of alkaline solutions were wildly sensitive to small changes in the nature of alkyl groups in the 6-, 7-, or 8-positions. This variation was traced to the formation of equilibrium mixtures of two monoanions (one of them hydrated) and a dianion. According to the PMR signals and the UV spectra, this dianion corresponds to the monoanion in Scheme 1 but represents a much weaker acidic group (pK_a 12.4); this weakness derives from the operation of Coulomb's law. The two monoanions of 8-methyllumazine were found to exist in approximately equal proportions at equilibrium, and each anion is stabilized by an effective resonance.[4] The anhydrous monoanion (pK_a 10.1) was obtained pure in solution by basification of a solution of the neutral species in the rapid-reaction apparatus. The hydrated monoanion (pK_a 6.4) was similarly isolated by slightly lowering the pH of a solution of the dianion.

The proportion of hydrate in the neutral species of 8-methyllumazine was found to be 1 part in 5700. The presence of two isopropyl groups in the 6- and 7-positions raised the hydrated contribution to 1 in 86, an effect attributed to mutual steric interference of these groups leading to distortion of the pyrazine ring, and relief of the steric strain through hydration.[21a] Two phenyl groups give rise to a similar, although smaller, effect. Such 6,7-steric effects had already been noted by Jacobsen.[19] Kinetic studies showed that the neutral species of 6,7-diisopropyl-8-methyllumazine became hydrated about 20,000 times more slowly than the neutral species of 8-methyllumazine, although the former contained a much higher proportion of hydrate once equilibrium was established.

Lumazines with a β-hydroxyethyl group on N-8, of which the riboflavine precursor (**10**) is an example, were found to have anhydrous forms in equilibrium, not with a hydrate, but with a form in which the β-hydroxy group had attacked the 7-position. Understandably, these internal adducts had UV spectra similar to those of the hydrates, even though they were themselves anhydrous. Prolonged exposure to alkali led to ring opening of all the 8-alkyllumazines discussed so far; ring-opening and reclosure proved to depend on the ionic strength of the solution.[21a]

Solving the long-standing problems posed by the 8-alkyllumazines led the same team of workers to examine the 8-alkylpterins (8-alkyl-2-aminopterid-4-ones).[21b] The results differ somewhat from those obtained in the earlier studies[19] of 6,7,8-trimethyl-2-methylimino-2,8-dihydropteridine (9). The neutral species of the latter was anhydrous and gave the same UV spectrum in cyclohexane as in water, with a prominent peak at 350 nm. The anhydrous cation formed in dilute acid had a very similar spectrum, but this changed completely in 20 minutes as follows. New peaks appeared at lower wavelengths, although not more than 75% of the height of the original peaks was lost at equilibrium. It was established that the new cation was hydrated (—OH on C-7) (see Ref. 4 for the stabilizing resonances). When this mixture of cations was made alkaline, the anhydrous neutral species was totally regenerated in 20 minutes. Ring opening was excluded in this study.[19]

The cations of the 8-alkylpterins were found[21b] to be stronger bases than those of the lumazines and were anhydrous. The neutral species were, in general, anhydrous; but an unstable hydrated neutral species could be made by acidifying the anion (13), which is necessarily hydrated, because these pterins, unlike the 8-alkyllumazines, have no mobile hydrogen atom on N-3. As with the 8-alkyllumazines, the hydrated species showed a UV peak near 300 nm, whereas the anhydrous species had its peak about 400 nm. The kinetics of hydration showed first-order reactions, catalyzed by both hydrogen ions and hydroxide ions. The kinetics of ring opening, also studied, showed catalysis by salts.[21b]

(13)

(14)
Dianion of hydrated
pterid-2-one

Hydrated dianions have also been encountered in a different context. The monoanions that pterid-2-one and pterid-4-one form in weakly alkaline solution are almost completely anhydrous. However, at higher pH values, further ionization takes place to give hydrated dianions. By the use of rapid-reaction techniques,[3b] the pK_a values of 13.02 and 12.70, respectively, were found for this second ionization (cf. 10.15 and 8.55, respectively, for pK_a^{equil} of the monoanions respectively).[21a] Because the monoanions lack an ionizable proton, the dianions must be formed from a hydrate, hence pterid-2-one dianion should have the structure 14.

Other patterns of hydration will be described for various natural products in Section VI.

Pterid-6-thione ("6-mercaptopteridine") was found to add a molecule of water across the 7,8 double bond in the cation and neutral species whereas the monoanion was anhydrous.[22b]

The effect of substituents in the 5-, 6-, 7-, or 8-position of quinazoline was summed up in the earlier review.[3a] In general, (−I) substituents promote hydration of the 3,4-bond by lowering the electron density on C-4. Later it was found that a (−I) substituent in the 2-position had the opposite effect. The addition of the negatively charged pole of a water molecule to C-4 is favored by the polarization of the 3,4-bond in this sense: $-C^{\delta+}=N-^{\delta-}$. But a (−I) group in the 2-position can oppose this polarization. In a study of twenty 2-substituted quinazolines,[23] it was found that hydration was helped by (+I) substituents, not greatly affected by (+M), and much diminished by (−I) substituents. The pH rate profile (first-order kinetics) for the hydration of 2-aminoquinazoline, measured from pH 2 to 10, was parabolic,[23] typical of molecules that undergo reverse covalent hydration.[3b]

The effect of the various substituents in the 2-position of pteridine has been investigated.[24] Here the (−I) substituents, which disfavour hydration of the 3,4-position, do not interfere with its occurrence in the alternative site for the pteridine nucleus, namely in the pyrazine ring.

B. Quantitative Aspects

The possibilities for employing rapid reaction techniques in the study of hydration have been widened by introduction[11] of a more sensitive, miniaturized version of the Britton Chance apparatus. A commercial manual UV spectrometer was rebuilt so that quartz cylindrical lenses focused the light onto a quartz capillary tube used as the optical cell and rendered the emergent beam parallel before it fell onto the photomultiplier tube. The electrical output of this tube was applied, through a cathode follower, as a single sweep to a cathode-ray oscilloscope fitted with a camera. The time between mixing the reactant solutions and arrival of the mixture in the cells was about 3 msec. This apparatus was satisfactorily used by the stopped-flow technique, where the change of optical density with time was repeatedly measured for a stationary mixture, at a carefully selected wavelength, and extrapolated back to zero time.

---

[22] (a) J. W. Bunting and D. D. Perrin, *Aust. J. Chem.* **19**, 337 (1966); (b) A. Albert and J. Clark, *J. Chem. Soc.*, 27 (1965).
[23] W. L. F. Armarego and J. I. C. Smith, *J. Chem. Soc. C*, 234 (1966).
[24] J. Clark and W. Pendergast, *J. Chem. Soc. C*, 1124 (1968).

1. Equilibrium Studies

The pK_a of quinazoline, as commonly measured, is 3.51; this represents mainly the equilibrium between the two most stable species, namely, the hydrated cation and the anhydrous neutral species. The true anhydrous pK_a (i.e., for the instantaneous equilibrium between anhydrous cation and anhydrous neutral species) was obtained[25] for quinazoline, twelve substituted quinazolines, and triazanaphthalenes in the rapid-reaction apparatus just described. The true anhydrous pK_a of quinazoline turned out to be 1.95. The true hydrated pK_a of quinazoline has already been reported[26] as 7.77, the slower rate of hydration permitting its determination in the usual rapid-reaction apparatus. Thus, in general, three pK_a values exist for each hydrating base, and the equilibrium between the totally hydrated species furnishes the strongest basic properties.

Although there is yet no method for directly detecting less than 2% of a hydrated species in the presence of its anhydrous counterpart, yet the ratio of hydrated to anhydrous species can be determined, even when it is as low as 1 in 100,000, by Eq. (1) where R is the ratio of hydrated (hyd) to anhydrous (anhyd) neutral species.[3b,27]

$$R = \frac{K_a^{hyd}(K_a^{anhyd} - K_a^{equil})}{K_a^{anhyd}(K_a^{equil} - K_a^{hyd})} \tag{1}$$

When the value of pK_a^{equil} differs from that of pK_a^{anhyd} by more than 1.5, a simplified Eq. (2) gives a close approximation:

$$R = K_a^{hyd}/K_a^{equil} \tag{2}$$

In this way, it was found that quinazoline (neutral species) has one hydrated molecule for every 5500 anhydrous ones. Altogether twenty-eight variously substituted quinazolines were examined,[27] and the ratio was seen to vary from 10^{-2} to 10^{-5}. It was concluded that $(-I)$ substituents favor hydration, and, in general, the relative effects of various substituents on the ratio was very similar to that which they had earlier been found to exert on the cations, although much less intense.

Equilibrium ratios for the hydrated–anhydrous cationic species of 8-azapurine (**6**) and its 2-methyl and 2-amino derivatives were similarly determined, and also for the neutral species of 2-amino-8-azapurine and 8-azapurin-2-one, the latter being substantially hydrated as the neutral

---

[25] J. W. Bunting and D. D. Perrin, *J. Chem. Soc. C*, 436 (1966).
[26] A. Albert, W. L. F. Armarego, and E. Spinner, *J. Chem. Soc.*, 5267 (1961).
[27] W. L. F. Armarego and J. I. C. Smith, *J. Chem. Soc. B*, 449 (1967).

species.[11] This work was extended to the cations of 7- and 8-methyl-8-azapurine[12,13] using

$$\log(1 + R) = pK_a^{equil} - pK_a^{anhyd} \qquad (3)$$

To summarize: the cations of 8-azapurine and its 7- and 8-methyl derivatives are almost completely hydrated at equilibrium, whereas that of the 9-methyl isomer shows little evidence of hydration. None of the corresponding neutral species is appreciably hydrated.[13]

A linear relation was found[28] to exist between the logarithm of the equilibrium constant for the addition of water and other nucleophiles to quinazoline and the y-value of Sander and Jencks. The latter is an equilibrium measure of the avidity of a nucleophile to add across the carbonyl bond in aldehydes.[29]

2. Kinetic Studies

A rapid-reaction technique was used to study the pH dependence of the reversible addition of water across the 3,4-double bond of eighteen quinazolines and four triazanaphthalenes. The pH range of 0–13 was covered, at 20°. When the rate constants for hydration were plotted against pH, a paraboloid curve was obtained with the minimum rate near neutrality. It was calculated that there is a strong acceleration of hydration in acidic solution due to the successive formation of mono- and dications (the attacking species is the water molecule). The increasing rate of hydration in alkaline solution was seen as the catalytic effect of the hydroxyl ion on the neutral species.[30] The kinetics of dehydration in neutral solution proved to be 10^5 times faster than those for hydration. For quinazoline, the two curves crossed at pH 3.5, below which hydration ran much the faster. Substituent and positional effects, particularly the slowing effect of a substituent in the 4-position, were quantified.[30]

First-order rate constants for hydration and dehydration at 20° were obtained[11] for 8-azapurine and its 2-methyl-, 2-amino-, and 2-oxo-derivatives in the pH range 2–7. Strong acceleration of hydration was observed in the cation.

3. Discovery of Hydrated Intermediates

It has often been found, during exploration of a reaction mechanism, that covalent hydration of a C=N bond plays an important early part in

---

[28] M. J. Cho and I. H. Pitman, *J. Amer. Chem. Soc.* **96**, 1843 (1974).
[29] E. G. Sander and W. P. Jencks, *J. Amer. Chem. Soc.* **90**, 6154 (1968).
[30] J. W. Bunting and D. D. Perrin, *J. Chem. Soc. B,* 950 (1967).

the reaction. Thus, in the rearrangement of 1,2-dihydro-2-imino-1-methylpyrimidine (15) to give 2-methylaminopyrimidine (a typical Dimroth rearrangement,[31]) it was found that the starting material exists in equilibrium with a small proportion of the carbinolamine (16) formed from it by covalent addition of a molecule of water.[32] The carbinolamine, characterized by its UV spectrum, underwent reversible ring-fission to yield the corresponding guanidino-aldehyde (17), which could be recycled to give either the starting material (15) or 2-methylaminopyrimidine (18). Kinetic studies showed that the rearrangement, which hung fire in anhydrous tetrahydrofuran (at 20°), gathered pace in proportion as water was added to the mixture.[32]

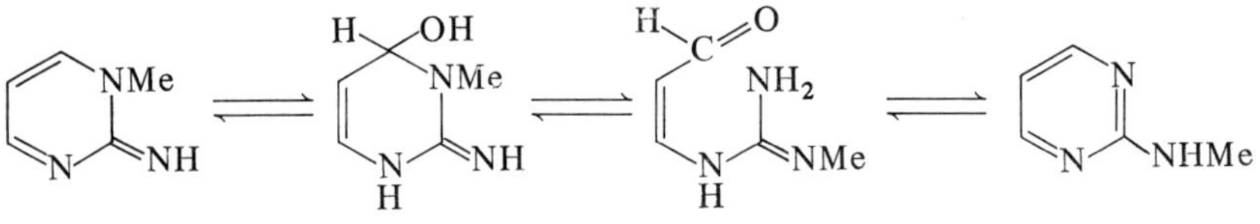

Although the substitution of a doubly bound nitrogen atom for a CH group in a heteroaromatic ring usually greatly decreases the rate of electrophilic substitution, it was found that pyrimid-2-one (19) unexpectedly exchanged H(5), for deuterium, 10^4 times faster than the H(5) exchanged in pyrid-2-one. It was concluded[33] that exchange with deuterium took place on a small equilibrated proportion of the covalent deuterio-hydrate (20). The similar behavior of 1,2-dihydro-1,3-dimethyl-2-oxopyridinium salts confirmed this mechanism and excluded exchange by an ylid mechanism.

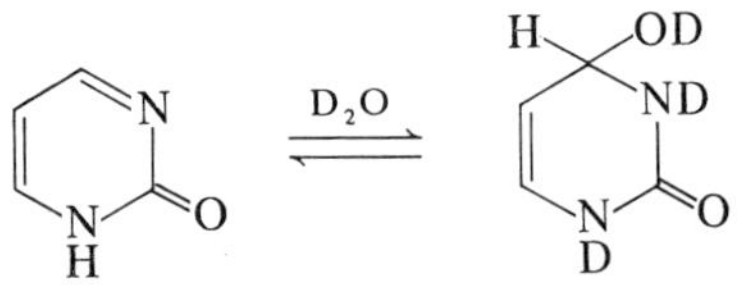

A rare opportunity to follow the growth and decay of a tetrahedral intermediate occurred in the acidic (pH 2.6) hydrolysis of the N,O-trimethylenephthalimidium ion (21). The rapid disappearance of the UV spectrum of the starting material, and its replacement by one compatible

---

[31] D. J. Brown, in "Mechanisms of Molecular Migrations" (B. S. Thyagarajan, ed.), Vol. 1, p. 209. Wiley (Interscience), New York, 1968.
[32] D. D. Perrin and I. H. Pitman, J. Chem. Soc., 7071 (1965).
[33] A. R. Katritzky, M. Kingsland, and O. S. Tee, Chem. Commun., 289 (1968).

with the covalent hydrate (22), was followed by a slow replacement of the latter spectrum by that of the product (23), *N*-hydroxyethylphthalimide.[34] All stages were found to be reversible.

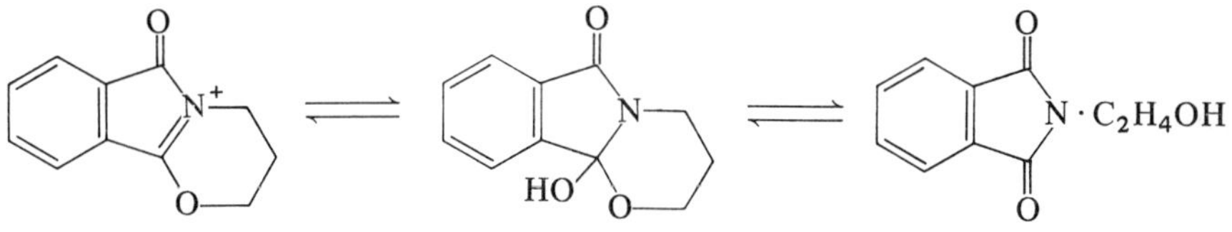

(21)   (22)   (23)

The rearrangement of the triazolo[4,3-*c*]pyrimidines, e.g., **24**, to their [2,3-*c*) isomers has been hypothesized as going through a neutral covalent hydrate followed by ring fission.[35] Kinetic evidence has been

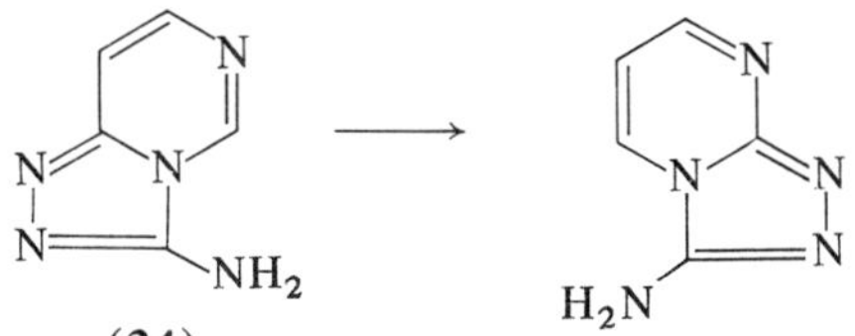

(24)

put forward in favor of a covalently hydrated intermediate in the acid-catalyzed rearrangement of *s*-triazolo[4,3-*a*]pyrazines, e.g., **25**, to imidazo[2,1-*c*]-*s*-triazoles.[36]

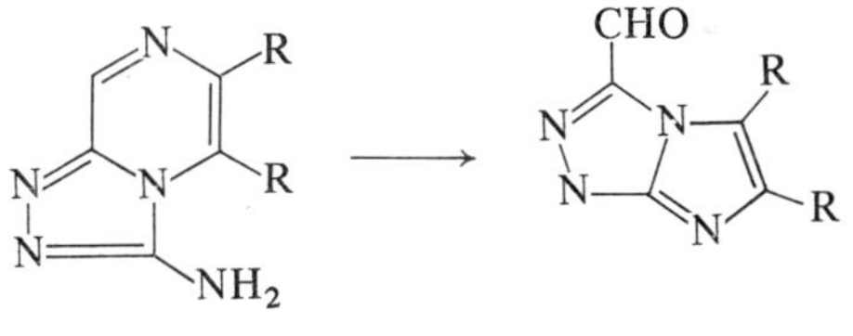

(25)

III. Time-Dependent Shifts in the Preferred Position of Hydration

In 1963, it was discovered that the neutral species of pterid-2,6-dione (26), which initially adds a molecule of water to the 3,4-bond, becomes (in the course of about an hour) hydrated exclusively on the 7,8-bond. This behavior was explained as follows: the 3,4-bond has only a small

[34] N. Gravitz and W. P. Jencks, *J. Amer. Chem. Soc.* **96**, 489 (1974).
[35] G. W. Miller and F. L. Rose, *J. Chem. Soc.*, 5642 (1963).
[36] S. Nicholson, G. J. Stacey, and P. J. Taylor, *J. Chem. Soc., Perkin Trans. II*, 4 (1972).

affinity for water but a low energy barrier, so that water is added to it by a kinetic process; however, the 7,8-position has a high affinity for water but a high energy barrier, so that the water is only slowly transferred there, namely by a thermodynamically controlled process.[37] It is noteworthy that each hydrate retains one "aromatic" ring.

Three years later, studies of pteridine with PMR showed that the cation underwent a time-dependent shift of water molecules. To follow this, it may be helpful first to look at the spectrum of the neutral species in chloroform, a solvent in which it is anhydrous. This shows four signals: two singlets at τ 0.20 and 0.35 assigned to the 4- and 2-protons, respectively, and a typical AB system (J 3.4 Hz) with doublets at 0.67 and 0.85 assigned, collectively, to the 6- and 7-positions.[38] The assignments were made with the help of a set of C-methyl derivatives (the methyl group shifted the remaining signals upfield from 0.06 to 0.22 ppm). A freshly prepared solution in deuterium oxide showed a similar spectrum (upon standing, a small upfield peak appeared corresponding to equilibration with 20% of the 3,4-monohydrate).

(26)   (27)

Acidification of a freshly prepared solution (in D_2O) of pteridine showed, momentarily, a UV spectrum similar to that of the neutral species.[39] This spectrum was rapidly replaced by that of the cation of the 3,4-hydrate, which was observed to come slowly to equilibrium with a new species, originally thought to be the product of ring-opening, but now known to be the dihydrate (27). The course of events was more clearly revealed by a PMR study,[8] as follows.

As can be seen from Fig 2, the isolated H-4 signal at τ 0.11 in the monohydrate cation has moved strikingly upfield to τ 3.40, indicating that this proton has taken on a more aliphatic character because the 3=4 bond has become saturated. That this was truly the cation of 3,4-dihydro-4-hydroxypteridine was confirmed by measuring a set of C-methylpteridines. In this monohydrate, the complex multiplet signal presented by the H(2), H(6), and H(7) atoms had moved slightly upfield with no loss of coherence. When this solution of the monocation was

[37] A. Albert, Y. Inoue, and D. D. Perrin, J. Chem. Soc., 5151 (1963).
[38] S. Matsuura and T. Goto, J. Chem. Soc., 1773 (1963).
[39] D. D. Perrin, J. Chem. Soc., 645 (1962).

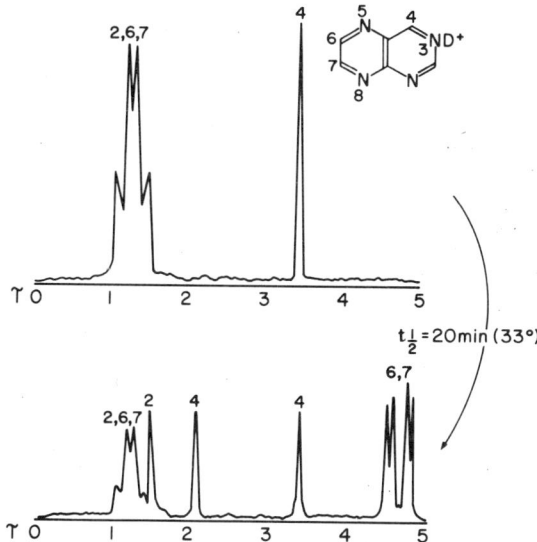

FIG. 2. The hydration of the cation of pteridine, as observed in 10% solution by proton magnetic resonance study, at 33° and pH 2 (DCl/D$_2$O). The upper spectrum was recorded 3 (and the lower 20) minutes after dissolution of the pteridine.

maintained at 33°, a change occurred with a half-time of 20 minutes. The new spectrum, while still showing all the signals for the monohydrate, then displayed a new double doublet near τ 4.7 assigned to the H(6) and H(7) in a new cation doubly hydrated at the 5,6- and 7,8-bonds (27). In this new cation, the H(4) signal had moved downfield to 2.12, whereas H(2) now gave a singlet signal at 1.53. It was concluded that the monohydrate was formed under kinetic, and the dihydrate under thermodynamic, control.[8]

Because of the similar energy of the two possible resonances that can ensure hydration of the pteridine cation (namely, 4-aminopyridinium stabilization of 7,8-hydration and amidinium stabilization of 3,4-hydration[3a]), a study of the effects of substituents on this delicately balanced system was undertaken. An initial survey showed that the majority of pteridine cations were subject neither to dihydration nor to shifting hydration. Only four examples were found in which 5,6,7,8-dihydration was preceded by 3,4-hydration, namely 2-methyl-, 6-methyl-, 7-methyl-, and 2-amino-4-methylpteridine, whereas the related cations of 6,7-dimethyl-, 2-amino-, 2-amino-6-methyl-, and 2-amino-7-methylpteridine remained stably hydrated in the 3,4-position.[8,40] Further examples, including a reversed direction of shift, came to light from a

[40] A. Albert and K. Ohta, *J. Chem. Soc. C*, 1540 (1970).

more intensive study of substituents in the 4-position, which will now be discussed.

The special, and somewhat complex, role played by a substituent in the 4-position was first observed in the quinazoline series. The carbon atom that received the —OH group during hydration was located by preparing all six C-methylquinazolines. Only when inserted in the 4-position did a methyl group prevent the addition of water.[41,23] This inhibition has both an inductive (+I) and a steric component. A kinetic study of quinazoline-4-carboxamide showed that the carbamoyl substituent, because of its steric effect, slowed the attainment of hydration from less than a second to 2–4 hours. However, this substituent has the

TABLE II

VARIATIONS IN THE NUCLEOPHILIC ATTACK OF WATER MOLECULES ON THE CATIONS OF PTERIDINES

Substituent in 4-position	Effect of substituent on attack at 4-position		Site of hydration	
	Steric	Electronic	Initial	Final
H—	Nil	Nil	3,4	5,6–7,8
Me—	Hinders	Hinders	5,6–7,8	5,6–7,8
F_3C—	Hinders	Attracts	5,6–7,8	3,4
EtO·CO—	Prevents	Attracts	5,6–7,8	5,6–7,8

opposite (—I) inductive effect to that of the methyl group, and hence is ultimately very favorable to hydration. A 4-carboxy group behaved similarly.[27] Similar effects operate in the 4-position of pteridine cations (see Table II).

It can be seen from the examples in Table II that both (+I) and (—I) substituents in the 4-position, as exemplified by methyl and trifluoromethyl groups, respectively, can sterically hinder the attack of water molecules on C(4), so that attack occurs in the 5,6–7,8 positions. After the lapse of some hours, however, the hydration of 4-trifluoromethyl-pteridine cation has shifted to the 3,4-position, which is thermodynamically favored by the (—I) effect of the substituent.[42] When a (—I) group is not merely hindering but, from its bulk, actually preventive, the water molecules are permanently excluded from the 3,4-position even when the electronic effects are favorable; the 4-ethoxycarbonyl-

---

[41] A. Albert, W. L. F. Armarego, and E. Spinner, *J. Chem. Soc.*, 2689 (1961).

[42] J. Clark and W. Pendergast, *J. Chem. Soc. C*, 1751 (1969); J. Clark and F. S. Yates, *J. Chem. Soc. C*, 2278 (1971).

pteridine cation provides an example.[43] The *neutral species* of 4-trifluoromethylpteridine forms a 3,4-hydrate, apparently directly, which is stable as a solid.[42] However, the neutral species of 4-ethoxycarbonylpteridine, which is 5,6–7,8 hydrated, does not transfer water to the 3,4-position.[43]

A steric inhibitory effect can, in turn, be inhibited, as in the cation of 4,5-dimethylquinazoline, which, unlike that of 4-methylquinazoline, is highly hydrated, the water being added across the 3,4-double bond. Apparently, overcrowding in the C-4 and C-5 area, which is relieved by hydration, encourages this process.[44]

A further and most unusual example of shifting hydration has been reported in the 7-azapteridine series. (The parent substance, **28**, is also called pyrimido[5,4-*e*]-*as*-triazine but is numbered here as a pteridine.)

(28)

7-Azapteridine

The neutral species of this parent is completely hydrated in aqueous solutions[45] whereas pteridine is only 20% hydrated under the same circumstances in accord with its less electron-depleted π-layer. Hydrated derivatives of 7-azapteridine have since been noted.[46] It has been claimed that 2-chloro-4-ethoxycarbonyl-7-azapteridine forms a 1,2-monohydrate in which the hydroxyl group and the chlorine are attached to C-2. This very unusual type of hydrate spontaneously isomerizes to the 3,4-monohydrate.[47]

IV. Examples of Covalent Hydration in New Ring Systems

This section will deal with nuclei not known to undergo hydration when the previous review[3a] was written.

Not until 1967 was covalent hydration found in a single-ring aromatic nucleus, but several examples are now known. The cation of 5-nitropyrimidine (also its 2-methyl and 2-benzyl derivatives) adds a molecule

[43] J. Clark, *J. Chem. Soc. C*, 1543 (1967).
[44] W. L. F. Armarego and J. I. C. Smith, *J. Chem. Soc.*, 5360 (1965).
[45] M. E. C. Biffin, D. J. Brown, and T. Sugimoto, *J. Chem. Soc. C*, 139 (1970).
[46] J. Clark and F. S. Yates, *J. Chem. Soc. C*, 2475 (1971).
[47] F. S. Yates, *J. Chem. Soc. C*, 2475 (1971).

of water across the 3,4-position,[48] as signaled by a large upfield shift in PMR. Stabilization may occur through amidinium-type resonance (1) as in the quinazoline cation, but it is puzzling why 1,3,5-triazine, electronically so similar, should be instantly hydrolyzed (to formamide) by cold water. Perhaps 5-nitropyrimidine makes use of a *p*-nitroaniline-type resonance (29)? 5-Methylsulfinyl- and 5-methylsulfonylpyrimidine cations also form hydrates (both substituents are strongly electron-attracting, as seems necessary for the hydration of pyrimidines).[49] 5-Diazouracil readily forms a 1,6-hydrate in water.[49a]

(29)   (30)

The acidic properties of 1-methyl-4-dimethylamino-5-nitro-2-oxopyrimidine (pK_a 9.04), a molecule with no ionizable proton, have been traced to the structure 30, which is the anion of a hydrate,[50] The acidic properties (pK_a 10.57) of the corresponding primary amine, 5-nitrocytosine,[51] probably depend on a similar hydration. In solution, 1,3-dimethyl-5-nitrouracil adds water across the 5,6 (C=C) double bond at pH 9.5, but not at neutrality.[52]

1,2,4-Triazine furnishes the species 31 when a solution in trifluoroacetic acid is titrated with water. The new upfield signals thus produced in PMR spectra are not seen when the hydrogen atom on C-5 is replaced by a methyl group. The hydration is reversible by basification.[53] Acidification of an aqueous solution of the anhydrous sodium salt of 3-hydroxy-1,2,4-triazine precipitated a hydrate of the neutral species, and elemental analysis combined with PMR (τ_5 4.62) indicated structure 32. This substance was not dehydrated during sublimation or heating in distilling toluene.[54]

8-Azapurine (6) is the parent of a series in which most members with a free 6-position undergo 1,6-hydration (see Table I, and pp. 120, 128). No shift of the position of the added water molecule has ever been

---

[48] M. E. C. Biffin, D. J. Brown, and T. C. Lee, *J. Chem. Soc. C*, 573 (1967).
[49] D. J. Brown, P. W. Ford, and M. N. Paddon-Row, *J. Chem. Soc. C*, 1452 (1968).
[49a] T. C. Thurber and L. B. Townsend, *J. Heterocycl. Chem.* **9**, 629 (1972).
[50] H. U. Blank, I. Wempen, and J. J. Fox, *J. Org. Chem.* **35**, 1131 (1970).
[51] D. J. Brown, *J. Appl. Chem.* **9**, 203 (1959).
[52] I. H. Pitman, M. J. Cho, and G. S. Rork, *J. Amer. Chem. Soc.* **96**, 1840 (1974).
[53] W. W. Paudler and T.-K. Chen, *J. Heterocycl. Chem.* **7**, 767 (1970).
[54] W. W. Paudler and J. Lee, *J. Org. Chem.* **36**, 3921 (1971).

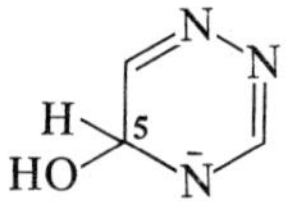

(31) 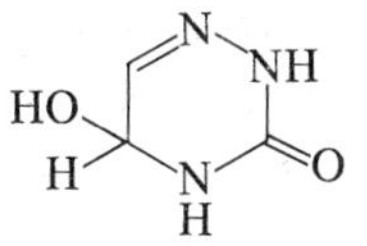   (32)

observed. In their π-deficient nature, 8-azapurines fall between the purines and the pteridines.[55] Corresponding to this assessment, water molecules tend to be less strongly bound in 8-azapurines than in the corresponding pteridines, but hydration could not be detected at all in the purine series.[14]

The hydration of 4-methylthiopyrimido[4,5-d]pyrimidine across the 5,6-bond to give **33** has been detected by PMR.[56] The neutral species of tetrazolo[1,5-c]pyrimidine absorbs moisture from the air to form 5-hydroxy-5,6-dihydrotetrazolo[1,5-c]pyrimidine (**34**), but this reaction is not easily reversible.[57]

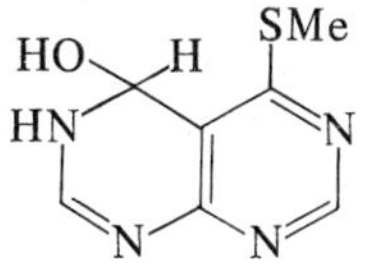

(33) 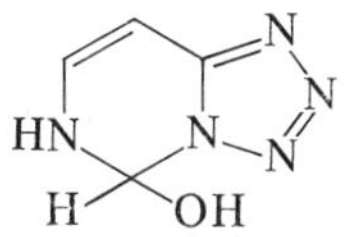   (34)

Alkali converts 4,6-dinitrobenzofuroxan (**35**), which is usually considered to be a heteroaromatic substance, into the hydrated anion **36**.[58]

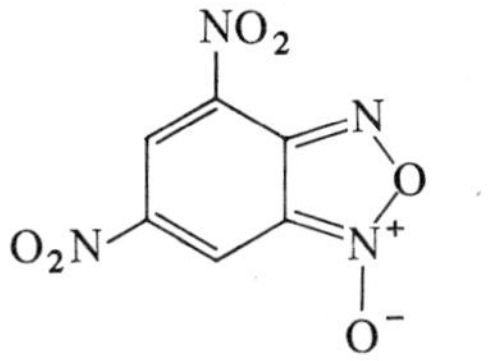

(35) 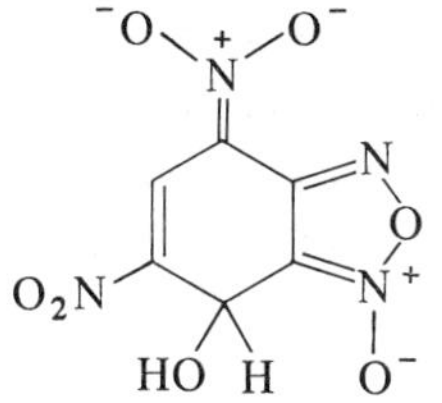   (36)

Fairly recently, the first five-membered ring to undergo hydration was reported.[59] 1-Oxo-2,3-dihydro-3-hydroxy-3phenyl-1H-isoindole (**37**,

---

[55] A. Albert and W. Pendergast, *J. Chem. Soc., Perkin Trans. I*, 457 (1972).
[56] E. C. Taylor, in "Pteridine Chemistry" (W. Pfleiderer and E. C. Taylor, eds.), p. 126. Pergamon, Oxford, 1964.
[57] C. Temple, R. McKee, and J. Montgomery, *J. Org. Chem.* **30**, 829 (1965).
[58] N. E. Brown and R. T. Keyes, *J. Org. Chem.* **30**, 2452 (1965); W. P. Norris and J. Osmundsen, *ibid.* 30, 2407 (1965); A. J. Boulton and D. P. Clifford, *J. Chem. Soc.* 5414 (1965).
[59] H.-J. W. Vollmann, K. Bredereck, and H. Bredereck, *Chem. Ber.* **105**, 2933 (1972).

R=H) gave off water at 220°, thus forming 1-oxo-3-phenyl-1H-isoindole (**38**) in 90% yield. The reverse hydration was not mentioned in this paper, but the anhydrous product (**38**) took up methanol (overnight at room temperature) to give a good yield of the adduct (**37**, R = Me). The phthalimidinium ion (**22**), discussed in the foregoing, provides another example of the hydration of a five-membered ring.

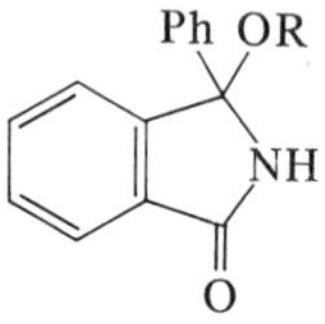

So far, heteroaromatic substances have dominated the discussion, although the addition of a second molecule of water to a hydrate (see p. 133) is an example of the hydration of an unsaturated heterocycle. In other examples, the addition of water occurs at an ethylenic site (C=C). This location usually requires irradiation to effect a hydration in pyrimidines such as uracil,[60] but it has been observed in a few heteroethylenic compounds under very mild conditions (see Section VI).

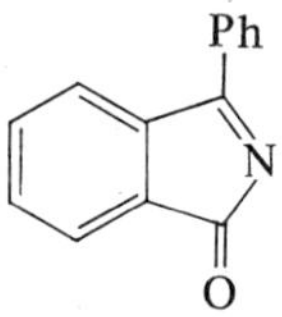

The reduced form of the respiratory coenzyme diphosphopyridine nucleotide (DPNH), which is vitally important for all cellular metabolism, is changed by dilute acid to a substance that, in the light of the following model experiments, may be a covalent hydrate. The 1,4- and 1,6-dihydro derivatives of 1-benzylpyridine-3-carboxamide furnish a single substance when dissolved in dilute acid at 20° and the solution is basified. The stable yellow product, which has a prominent peak at 292 nm, was assigned the constitution 1-benzyl-6-hydroxy-1,4,5,6-tetrahydropyridine-3-carboxamide[61] (see Scheme 2). 3-Acetyl-1-benzyl-1,4-

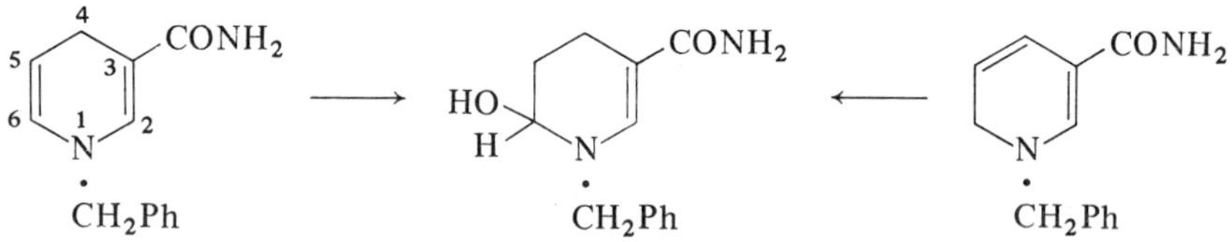

SCHEME 2. Common product from hydration of 1,4- and 1,6-dihydropyridines.

[60] A. M. Moore, *Can. J. Chem.* **36**, 281 (1958).
[61] R. Segal and G. Stein, *J. Chem. Soc.*, 5254 (1960); R. M. Burton and N. O. Kaplan, *Arch. Biochem. Biophys.* **101**, 139 (1963).

dihydropyridine similarly forms the hydrate, 3-acetyl-1-benzyl-6-hydroxy-1,4,5,6-tetrahydropyridine.[62] The kinetics have been studied (by UV spectrometry) of the acid-catalyzed hydration of DPNH and of the model substance N-propyl-1,4-dihydronicotinamide.[63]

When the much used antibacterial drug, sulfadiazine (2-4'-aminobenzenesulfonamidopyrimidine), is brominated, one double bond in the pyrimidine ring becomes saturated by the addition of hypobromous acid, and the product at once adds a molecule of water across the adjacent C=N bond.[64]

V. The Isolation of Stereoisomeric Hydrates

Each act of covalent hydration introduces a center of asymmetry, and a double hydration can provide the opportunity for geometrical isomerism.

Pteridine is changed by the enzyme *adenosine deaminase* to the levorotatory form of the hydrate, 3,4-dihydro-4-hydroxypteridine. When this reaction was approached from the other direction, the same enzyme dehydrated this enantiomer most rapidly, leaving a net positive optical rotation at equilibrium.[65] Specimens of the enzyme from both mammalian and fungal sources, although of very different molecular weight, were found to catalyze the stereospecific hydration of pteridine; also pteridine inhibited the deamination of adenosine by both enzyme specimens. Assuming that the first step in the deamination of adenosine is the formation of a tetrahedral hydrated intermediate, it was argued that both the hydration of pteridine and the deamination of adenosine were parallel phenomena.[65]

(39)

4-Ethoxycarbonyl-5,6,7,8-tetrahydro-6,7-dihydroxypteridine (**39**), which is the hydrated form of 4-ethoxycarbonylpteridine, has been

---

[62] A. G. Anderson and G. Berkelhammer, *J. Amer. Chem. Soc.* **80**, 992 (1958).
[63] C. C. Johnston, J. L. Gardner, C. H. Suelter, and D. E. Metzler, *Biochemistry* **2**, 689 (1963).
[64] W. Barbieri, L. Bernardi, F. Luini, and G. Palamidessi, *Farmaco, Ed. Sci.* **24**, 561 (1969).
[65] B. E. Evans and R. V. Wolfenden, *Biochemistry* **12**, 392 (1973).

prepared in two geometrically isomeric forms, of which the lower-melting (112°) and more water-soluble isomer was made quite simply by shaking ethyl pteridine-4-carboxylate with water. The more stable higher-melting (181°) isomer arose by direct synthesis, or by rapidly neutralizing a solution of 4-ethoxycarbonylpteridine in dilute acid. Both compounds are clearly hydrated across the 5,6 and 7,8-bonds as shown by PMR and UV spectroscopy. The lower-melting isomer had less thermal stability during mass spectrometry.[66] No conclusion was reached as to which isomer was the *cis*-diol and which the *trans*-diol.

VI. Covalent Hydration in Nature

Increased interest is added to the topic of this review by the discovery of several examples of covalent hydration in nature; although no systematic investigation has been made, an increasing alertness to its possible occurrence can be detected among natural product workers. Some naturally occurring pteridines, such as xanthopterin,[67] are covalently hydrated. The hydration of uracil and of the coenzyme diphosphopyridine nucleotide (DPNH) were mentioned in Section IV.

Aflatoxins, e.g., **40**, comprise a family of related products that are elaborated by the fungus *Aspergillus flavus* and end by contaminating the food, such as peanuts, upon which the fungus grows. Several aflatoxins, notably B_1 and G_1, become covalently hydrated in the mammalian liver to give the true toxin **41**, which is a hemiacetal. This change

(40)
Aflatoxin B_1

(41)
Relevant portion of aflatoxin B_1 after hydration

can be brought about *in vitro* by dilute acid. It is suggested that the hydrate exerts its poisonous effects by attacking amino groups in key enzymes. The nonspecific, severe necrosis which it brings about in the liver often precedes cancer. Other aflatoxins, lacking the 2,3-bond of the

[66] J. Clark, *J. Chem. Soc. C*, 313 (1968).
[67] A. Albert and F. Reich, *J. Chem. Soc.*, 127 (1961); Y. Inoue and D. D. Perrin, *ibid.*, 2600 (1962).

vinyl ether in **40**, cannot form a hydrate; these types have little toxicity.[68]

Anthramycin (**42**) is a pyrrolobenzodiazepine antibiotic with powerful antibacterial and carcinostatic properties. When recrystallized from acetone, water is lost from the 10–11 bond, and λ_{max} increases from 333

(**42**)
Anthramycin

to 352 nm because of the new, conjugated double bond. Exposure to moisture at ambient temperatures brings about rehydration.[69] Biochemical studies showed that it became slowly but covalently bound to DNA, but not by intercalation, and that it prevented the synthesis of both DNA and RNA. (It binds most strongly to DNAs that are richer in G-C than in A-T.) Biological action disappears if the hydrogen atoms in positions 9 or 10 are replaced.[70] It seems likely that the delayed but powerful binding to DNA depends on shedding water from the 10–11 bond and attack of a nucleophile group (say NH_2 or $-O^-$) at the 11-position.

Hortiamine, which can lower blood pressure, is a red alkaloid and the major hypotensive component of the bark of the Brazilian plant *Hortia arborea* (Rutaceae). Chemically, it is 10-methoxy-14-methyl-5-oxo-5,7,8,14-tetrahydroindolo[2′,3:3,4]pyrido[2,1-*b*]quinazoline.[43] When boiled with moist benzene it forms a yellow product, hydrated as shown in **44**, and λ_{max} falls from 411 nm to 375 nm. The anhydrous form is regenerated on gentle drying.[71]

Sunlight effects the covalent hydration of the lysergic acid portion (**45**) of ergotamine and other ergot alkaloids.[72] The reaction, carried out

---

[68] D. Patterson and B. Roberts, *Food Cosmet. Toxicol.* **10**, 501 (1972).
[69] W. Leimgruber, V. Stefanović, A. Karr, and J. Berger, *J. Amer. Chem. Soc.* **87**, 5791 (1965).
[70] V. Stefanović, *Biochem. Pharmacol.* **17**, 315 (1968); I. H. Goldberg and P. A. Friedman, *Annu. Rev. Biochem.* **40**, 775 (1971).
[71] I. J. Pachter, R. F. Raffauf, G. E. Ullyot, and O. Ribeiro, *J. Amer. Chem. Soc.* **82**, 5187 (1960).
[72] A. Stoll and W. Schlientz, *Helv. Chim. Acta* **38**, 585 (1955); H. Hellberg, *Acta Chem. Scand.* **13**, 1106 (1959).

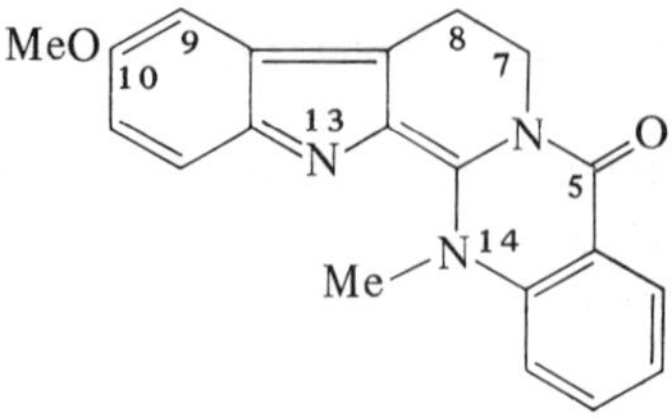

(43)
Hortiamine

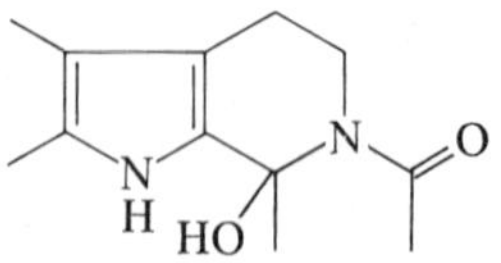

(44)
Relevant portion of
hydrated form of hortiamine

in aqueous solution at 30°, was catalyzed by anions (bicarbonate or phosphate). Addition was across the 9,10-double bond in "ring D." By its attachment at position 10, the hydroxyl group created a new center of

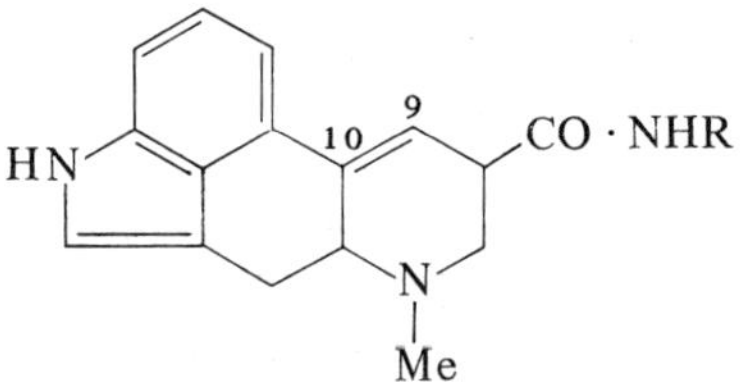

(45)

asymmetry, and both isomers were isolated. This hydration lowered λ_{max} from 318 nm to 293 nm, and the spectrum of the hydrated product was identical with that of 9,10-dihydroergotamine. The hydrated products turned out to be less physiologically active than the starting material. The reaction is important as causing loss of activity in dispensed preparations of ergot alkaloids.

Tetrodotoxin, an extremely poisonous heterocycle present in Japanese puffer fish (various species of *Spheroides*), exerts its action by blocking the sodium ion channels in the peripheral nervous system. Chemically it has an adamantane-like structure, namely octahydro-12-(hydroxymethyl)-2-imino-5,7,9,10a-dimethano-10aH-[1,3]-dioxocino-[6,5-d]pyrimidine-4,7,10,11,12-pentol. When chemically dehydrated, it rehydrates in dilute acid.[73] Methanol, and ammonia, can add similarly. Because 2-amino-1,6-dihydropyrimidine forms the central part

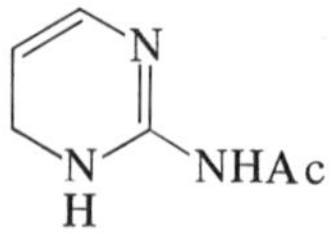

(46) 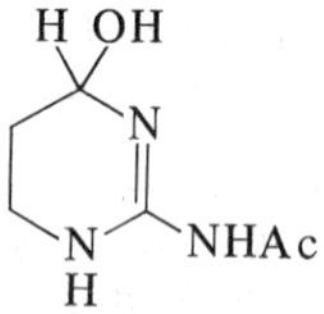   (47)

[73] T. Goto, Y. Kishi, S. Takahashi, and Y. Hirata, *Tetrahedron* **21**, 2059 (1965).

of these structures, the more stable 2-acetamido-1,6-dihydropyrimidine (**46**) was examined[74] and found to form a hydrate (**47**) in dilute acid (24 hours at 80°); the added water was lost by heating for 18 hours in a vacuum at 125°.

Acknowledgments

I thank Drs. W. L. F. Armarego and D. D. Perrin for reading this manuscript and making helpful suggestions.

---

[74] E. Suzuki, S. Inoue, and T. Goto, *Chem. Pharm. Bull.* **16**, 933 (1968).

1,2,3,4-Thiatriazoles

ARNE HOLM

The H. C. Ørsted Institute, Chemical Laboratory II (General and Organic Chemistry) University of Copenhagen, Copenhagen, Denmark

I. Introduction	145
II. Physicochemical Properties of 1,2,3,4-Thiatriazoles	146
A. Quantum Mechanical Calculations	146
B. Infrared Spectra	147
C. Mass Spectrometry	148
D. Nuclear Magnetic Resonance Spectra	149
III. Chemical Properties of 1,2,3,4-Thiatriazoles	149
A. Thermal Reactions	149
B. Electrophilic and Nucleophilic Reactions	152
1. Alkylation	152
2. Oxidation	155
3. Reactions of Thiatriazole with Trivalent Phosphorus	158
4. Complex Formation	159
C. Photochemical Reactions	159
IV. 1,2,3,4-Thiatriazoles Substituted with C-Radicals	162
V. 1,2,3,4-Thiatriazole-5-thiol and Its Derivatives	163
VI. 5-Alkoxy and 5-Aryloxy 1,2,3,4-Thiatriazoles	166
VII. N-Substituted 5-Amino-1,2,3,4-thiatriazoles	167
VIII. Synthesis and Chemical Properties of 1,2,3,4-Thiatriazolines and Alleged 1,5-Dihydro-1,2,3,4-thia(S^{IV})triazoles	168
A. Thiatriazolines	168
B. Alleged 1,5-Dihydro-1,2,3,4-thia(S^{IV})triazoles	173
Notes Added in Proof	173

I. Introduction

This survey of the chemistry of the heteroaromatic thiatriazoles is a continuation of the 1964 review by K. A. Jensen and C. Pedersen.[1] In the meantime more work on this type of compound has appeared, and some interesting developments have taken place that justify a presentation of the results from the past decade. In particular the facile decomposition of thiatriazoles and thiatriazolines has led to several types of hitherto unknown or little known compounds. In this review, references preceded by an asterisk are taken from the article by Jensen and Pedersen. Since only

[1] K. A. Jensen and C. Pedersen, this Series **3**, 263 (1964).

1,2,3,4-thiatriazoles are considered, the shorter term thiatriazole is often used.

II. Physicochemical Properties of 1,2,3,4-Thiatriazoles

A. Quantum Mechanical Calculations

A series of CNDO (complete neglect of differential overlap) calculations have been performed on the structure, formation, and nucleophilic reactivity of thiatriazole.[2] Figures 1 and 2 indicate the lowest energy structures, with optimal geometry and electron distribution, for thiatriazole (2) and the related 1,3,4-thiadiazole (1).

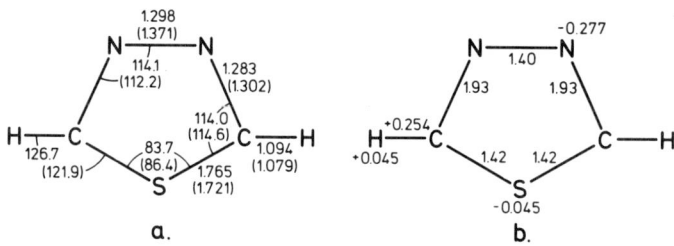

a.   b.

FIG. 1. Calculated values: (a) bond length (Å), bond angles (degrees); (b) bond orders and charge densities for 1,3,4-thiadiazole. Experimental values[3] are given in parentheses.

For the thiadiazole, bond angles and bond lengths are in good agreement with microwave[3] and electron diffraction[4] determinations. Although no structural measurements have been made for thiatriazole, the predicted structure seems reasonable by comparison. Of interest is the change in electron distribution from **1** to **2** upon introduction of the third nitrogen at position 2. Sulfur and the nitrogen at position 3 become positively charged. Moreover, whereas the 1,2 and 1,5 C—S bond orders for thiadiazole indicate moderate delocalization around the ring, the 1,2 bond in thiatriazole is suggested to be essentially a single bond. Thus 1,3,4-thiadiazole is symmetrical and cyclically delocalized, and thiatriazole is predicted to be only partially conjugated. These differences should be reflected in the chemistry of the two hetero systems.

[2] J. P. Snyder (University of Copenhagen, H. C. Ørsted Institute), private communication, 1975.
[3] L. Nygaard, R. Lykke Hansen, and G. O. Sørensen, *J. Mol. Struct.* **9**, 163 (1971).
[4] P. Markov and R. Stølevik, *Acta Chem. Scand.* **24**, 2525 (1970).

As pointed out below (Section III, A), phenylthiatriazole can be prepared by the thermal ring closure of thiobenzoyl azide. CNDO potential-surface calculations[2] suggest the reaction to be thermally "allowed" with an energy barrier of 63 kJ/mole. The heat of the reaction, ΔH, is negative by 54 kJ/mole. The lack of a HOMO-LUMO[4a] crossing is unusual in view of the fact that the terminal nitrogen and sulfur atoms of the open-chain azide bear only a lone electron pair. Similar ring closures to three-membered rings are consistently predicted to be thermally "forbidden."[5]

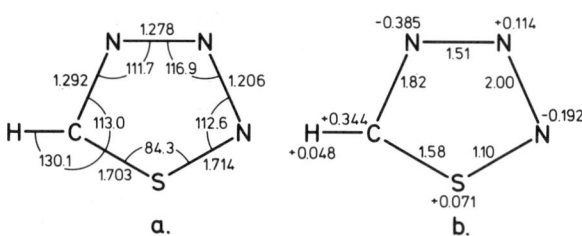

Fig. 2. Calculated values: (a) bond length (Å), bond angles (degrees); (b) bond orders and charge densities for 1,2,3,4-thiatriazole.

A PMO-CNDO[4a] analysis[6] of thiatriazole and several 5-substituted derivatives (R = Ph—, MeS—, NH_2—, CH_2=CH—) leads to the prediction that alkylation at position 4 is favored, the relative order being N-4 $\geqslant$ N-2 > N-3 (see also Section III, B). Companion potential surface calculations suggest that both H^+ and CH_3^+ prefer in-plane attack on the nitrogen σ-lone pairs.[2] Position 3 is again the least-favored attack site, in direct contradiction to experiment.[6] Either the calculations are misleading or the complexity of thermodynamic versus kinetic control in the thiatriazole system has yet to be unraveled.

B. Infrared Spectra

Previously it was reported that infrared (IR) spectra of all N-mono and N-disubstituted 5-aminothiatriazoles exhibit strong bands in the 1540–1590 cm^{-1} range,[8–10] which are believed to be due to the C=N

[4a] HOMO, highest occupied molecular orbital; LUMO, lowest unoccupied molecular orbital; PMO, perturbation molecular orbital.

[5] B. Schilling and J. P. Snyder, *J. Amer. Chem. Soc.* **97**, 4422 (1975).

[6] A. Holm, K. Schaumburg, N. Dahlberg, C. Christophersen, and J. P. Snyder, *J. Org. Chem.* **40**, 431 (1975).

[7] C. N. R. Rao and R. Venkataraghavan, *Can. J. Chem.* **42**, 43 (1964).

*[8] E. Lieber, D. R. Levering, and L. J. Patterson, *Anal. Chem.* **23**, 1594 (1951).

*[9] M. Kuhn and R. Mecke, *Z. Anal. Chem.* **181**, 487 (1961).

*[10] K. A. Jensen, A. Holm, and C. T. Pedersen, *Acta Chem. Scand.* **18**, 566 (1964).

and N=N stretching vibrations of the heteroaromatic ring system.[1] Rao and Venkataraghavan[7] examined 5-amino-, 5-mercapto-, and 5-phenylthiatriazoles (25 compounds) and what they believed to be 4-substituted 1,2,3,4-thiatriazoline-5-thiones (9 compounds). As before, all types of thiatriazoles were reported to exhibit absorptions around 1600 cm^{-1} assigned to ring-skeletal vibrations. Surprisingly, however, it was later found that 5-(methylthio)thiatriazole is devoid of absorptions around 1600 cm^{-1}.[11]

The compounds considered to be thiatriazoline-5-thiones were later shown in fact to be 5-substituted thiatriazoles (Section V). Therefore the assignments by Rao et al.[7] of ring breathing and CH out-of-plane deformations to the supposed two different systems of 4- and 5-substituted thiatriazoles all correspond to 5-substituted thiatriazoles.[11] The ring breathing absorptions are found at 1320–1260, 1238–1215, 1105–1080, and 1030–1000 cm^{-1} and the CH out-of-plane deformations at 923–880 cm^{-1}. However, most of the bands are rather weak and have little diagnostic value.[11]

Solanki and Trivedi[12] synthesized a series of 5-benzylamino-, 5-benzylthio-, and 5-benzyloxythiatriazoles for spectroscopic studies. Spectral details or spectra are not given, but the authors assigned the regions at 960–930 and 1160–1115 cm^{-1} as characteristic absorptions for the thiatriazole ring system. As will be noted, these figures are in only fair agreement with those previously reported. According to the authors, there is no specific effect of isosterism (—NH—, —O—, and —S—) on the spectra of the thiatriazoles studied.

C. Mass Spectrometry

In connection with a study of the electron impact-induced fragmentations of 1,2,3-thiadiazoles, the mass spectrum of 5-phenylthiatriazole has been scrutinized.[13] Jensen et al.[14] have undertaken a detailed investigation including 5-aryl-, 5-amino-, and 5-alkylthiothiatriazoles. The electron impact-induced decompositions resemble the pyrolytic loss of N_2S (Section III, A). In all cases the M—N_2S ion together with its fragmentation is responsible for the major part of the total ion current. A detailed discussion of the spectra is outside the scope of this review.

---

[11] C. Christophersen and A. Holm, *Acta Chem. Scand.* **25**, 2015 (1971).
[12] M. S. Solanki and J. P. Trivedi, *J. Indian Chem. Soc.* **48**, 843 (1971).
[13] K.-P. Zeller, H. Meier, and E. Müller, *Tetrahedron* **28**, 1353 (1972).
[14] K. A. Jensen, S. Treppendahl, C. Christophersen, and G. Schroll, *Acta Chem. Scand., Ser.* B **28**, 97 (1974).

D. Nuclear Magnetic Resonance Spectra

Nuclear magnetic resonance (NMR) spectral data of thiatriazoles and thiatriazolium salts are recorded in Section III, B, 1.

III. Chemical Properties of 1,2,3,4-Thiatriazoles

A. Thermal Reactions

The characteristic thermal reaction of thiatriazoles—decomposition upon slight heating into nitrogen, sulfur, and an organic fragment—has been submitted to mechanistic investigations. Jensen and Holm[15] examined which of the individual nitrogen atoms are lost on decomposition of 2-[^{15}N]-5-isobutoxythiatriazole. The labeled compound was prepared by the usual procedure from Na^{15}NO$_2$ (97.4% ^{15}N) and isobutoxythiocarbonylhydrazine [Eq. (1)].

$$i\text{-}C_4H_9O\text{-}CS\text{-}NHNH_2 + {}^{15}NO_2^- + H^+ \longrightarrow i\text{-}C_4H_9O-\underset{S}{\overset{N=N}{\underset{\diagdown}{\diagup}}}{}^{15}N \quad (1)$$

The nitrogen formed was analyzed by mass spectrometry and found to contain 96.4% of ^{15}N^{14}N, corresponding to 98.8% loss of nitrogen from the N-2 and N-3 positions [Eq. (2)].

$$i\text{-}C_4H_9O-\underset{S}{\overset{N=N}{\underset{\diagdown}{\diagup}}}{}^{15}N \xrightarrow{\Delta} i\text{-}C_4H_9OCN + {}^{15}N^{14}N + S \quad (2)$$

Possible participation of an intermediate, such as a three-membered ring compound (thiazirine), was briefly discussed.

The kinetics of the decomposition of 5-alkoxythiatriazoles in dibutyl phthalate solution was studied by Jensen *et al.*[16] The reaction was followed by manometric measurement of the nitrogen evolved and follows first-order kinetics. Rates of decomposition and activation parameters are shown in Table I.

The activation energies for the different alkoxythiatriazoles differ very little, in agreement with the idea that the reaction proceeds via the same mechanism in all cases. The decomposition of alkoxythiatriazoles is analogous to the decomposition of 5-alkylthiatriazoles,[17] in which

---

[15] K. A. Jensen and A. Holm, *Acta Chem. Scand.* **23**, 2183 (1969).
[16] K. A. Jensen, S. Burmester, and T. A. Bak, *Acta Chem. Scand.* **21**, 2792 (1967).
[17] H. E. Wijers, L. Brandsma, and J. F. Arens, *Rec. Trav. Chim. Pays-Bas* **86**, 670 (1967).

TABLE I

RATES OF DECOMPOSITION OF 5-ALKOXY-1,2,3,4-THIATRIAZOLES, RO—CSN_3^a

R	$k_1(25°)$ (10^{-5} sec^{-1})	$t_{1/2}$ (hours)	A (10^{14} sec^{-1})	$E\ddagger$ (kcal.mol^{-1})	$\Delta S\ddagger$ (cal.mol^{-1} K^{-1})
Me	12.71 ± 0.09	1.52 ± 0.01	1.4 ± 0.3	24.6 ± 0.1	4.2 ± 0.4
Et	8.95 ± 0.05	2.15 ± 0.01	1.3 ± 0.3	24.8 ± 0.1	4.1 ± 0.4
n-C_3H_7	9.48 ± 0.04	2.03 ± 0.01	1.8 ± 0.3	24.9 ± 0.1	4.7 ± 0.3
n-C_7H_{15}	9.21 ± 0.10	2.09 ± 0.02	1.2 ± 0.4	24.7 ± 0.2	3.9 ± 0.6
Ph	13.13 ± 0.18	1.47 ± 0.02	0.9 ± 0.3	24.3 ± 0.2	3.2 ± 0.6

$^a k_1$ = rate constant (25°), A = Arrhenius factor, $E\ddagger$ = activation energy, $S\ddagger$ = activation entropy.

isothiocyanates have been observed as by-products. Jensen et al.[16] assume that this shows that nitrogen is split off first with the formation of an intermediate, R'C(=S)N, which may either lose sulfur and form a nitrile or rearrange to isothiocyanate. An isothiocyanate as by-product is not observed in the decomposition of 5-alkoxythiatriazoles. A scheme for the decomposition of 5-alkoxythiatriazoles involving primary ring opening is accordingly as shown in Eq. (3).

$$\text{RO}-\underset{S}{\overset{N=N}{\underset{|}{C}}}\!\!\!{\overset{}{N}} \longrightarrow \text{RO}-C\!\!\begin{array}{c}\nearrow N_3\\ \searrow S\end{array} \longrightarrow \text{RO}-C\!\!\begin{array}{c}\nearrow \bar{N}|\\ \searrow S\end{array} + N_2$$

$$\text{RO}-C\!\!\begin{array}{c}\nearrow \bar{N}|\\ \searrow S\end{array} \longrightarrow \text{ROCN} + S \tag{3}$$

The thioazide must decompose rapidly since the characteristic azide band near 2130 cm^{-1} was not observed in the IR spectrum of a decomposing thiatriazole.

Attracted by these results, Pilgram et al.[18] tried to intercept possible intermediates, such as RO—C(=S)N$_3$, by trivalent phosphorus compounds. With phosphorous triamides 2,2-dihydro-1,3,4,5,2-thiatriazaphosphorines are formed, but it seems impossible to distinguish between attack of the reactant directly on the heterocyclic ring and attack on a thioacyl azide formed by ring opening (see Section III, B, 3).

In this laboratory a number of experiments have been conducted to trap or obtain evidence for the hitherto unknown thioacyl azides, thioacyl nitrenes, and thiazirines during thermally[19] or photochemically

---

[18] K. Pilgram, F. Görgen, and G. Pollard, *J. Heterocycl. Chem.* **8**, 951 (1971).
[19] A. Holm and L. Carlsen, unpublished results.

induced[20] thiatriazole decomposition (for the latter see Section III, C). In the presence of 1,3-dipolarophiles, such as tetracyanoethylene, dimethyl acetylenedicarboxylate, and other reagents, a high yield of benzonitrile (90–100%) from thermally decomposing 5-phenylthiatriazole is invariably obtained.[19] Since nitrenes normally are extremely reactive species, this result renders the participation of a thioacyl nitrene, proposed by Jensen et al.[16] rather unlikely, and suggests that sulfur and nitrogen are lost simultaneously.

If thioacyl azides are intermediates in the decomposition [Eq. (3)], then the first step (thiatriazole to thioacyl azide) must be reversible, since thioacyl azides are certainly intermediates when thiatriazoles are formed from, for example, a thioacyl chloride and azide ions. Therefore a steady-state concentration of thioacyl azide should be present in a decomposing thiatriazole. On this basis, an equimolar amount of Na^{15}N$_3$ (95% isotopically labeled on one nitrogen) was added to decomposing 5-phenylthiatriazole in ethanol, the rationale being that the powerfully nucleophilic azide ion should attack the thioacyl azide, resulting in isotope exchange.[19] After approximately 50% of the original thiatriazole had decomposed, the remaining thiatriazole was recovered and analyzed by mass spectrometry, but only the natural abundance of ^{15}N was present. When benzoyl azide is treated similarly, an isotope exchange is observed.[19] Hence there is no reason to suppose that thioacyl azides are intermediates in the thermal decomposition of thiatriazoles.

An attempt has been made to obtain the elusive thiobenzoyl azide at low temperature.[19] Ethanolic solutions of thiobenzoyl chloride and tetrabutylammonium azide were mixed at $-115°$ directly in a low-temperature UV cell. On slight warming (to ca. $-100°$) a reaction took place with formation of thiatriazole, but no new absorptions due to thiocarbonyl chromophore were observed. This most probably means that the ring closure to thiatriazole, not unexpectedly, is more rapid than the bimolecular formation of thiocarbonyl azide.

Jensen et al.[16] also studied the decomposition of 5-ethoxythiatriazole in dibutyl phthalate solution in the presence of trichloroacetic acid, tripentylamine, 4-benzylpyridine, anhydrous aluminum chloride, or trinitrobenzene, but practically no effect on the reaction rate was observed. However, later experiments have shown that the decomposition can indeed by enhanced catalytically by Lewis acids under conditions where the catalyst is not sequestered by complex formation with the solvent.[19] Thus the addition of aluminum chloride to 5-phenylthiatriazole in benzene causes a brisk evolution of nitrogen at room temperature, and if instead boron tribromide is added, a rather violent reaction sets in. However, when esters or ethers are used as

[20] A. Holm, N. Harrit, and N. H. Toubro, J. Amer. Chem. Soc. **97**, 6197 (1975).

solvents, complexes are formed between the solvent and the catalyst and catalytic rate enhancement of the thiatriazole decomposition is no longer observed, explaining the negative result of Jensen *et al.*[16] The reaction with aluminum chloride in benzene was originally studied in the hope of obtaining a complex with the postulated thioacyl azide or thioacyl nitrene, but the reaction mixture, after cessation of nitrogen evolution, constituted a rather complex mixture.[19] Thus, diphenyl sulfide and disulfide and benzyl phenyl sulfide are formed, proving that this reaction proceeds in a completely different manner from the uncatalyzed thermal reaction of thiatriazoles (see also Section III, B, 4).

The sulfur extruded in these reactions is probably primarily eliminated as atoms, which would be expected to react as "active sulfur." This was borne out by addition of triphenylphosphine, triphenylarsine, or triphenylstibine to decomposing 5-ethoxythiatriazole in ether.[21] From the solution the corresponding sulfides could be isolated in quantitative yields. It is remarkable that triphenylarsine sulfide is formed so easily, since it could not be prepared directly from the arsine and sulfur.[21] (See also Section III, B, 3 for the possibility of a ring expansion in these reactions.) Sulfur thermally generated is not reactive enough to react with C—H or C=C bonds.[22] However, triplet sulfur atoms appear to be formed on irradiation of thiatriazoles with UV light.[22] In the presence of alkenes, episulfides are formed. This difference in reactivity is suggested to be explicable in terms of low excess kinetic (translational) energy of sulfur atoms generated in the thermal reaction[22] (see Section III, C).

B. Electrophilic and Nucleophilic Reactions

Thiatriazoles with an aromatic substituent have previously been regarded as extremely stable toward oxidizing agents as well as electrophilic attack in general.[1] By the use of potent alkylating and oxidizing agents, it has been possible to attack the heteroaromatic ring and obtain thermally stable derivatives.

1. *Alkylation*

5-Phenyl-, 5-methylthio-, and 5-ethylthiothiatriazole are alkylated with Et_3OBF_4 to form monomeric, crystalline salts.[6] Under similar alkylating conditions, 5-ethoxythiatriazole decomposes entirely to

---

[21] K. A. Jensen, A. Holm, and E. Huge-Jensen, *Acta Chem. Scand.* **23**, 2919 (1969).
[22] R. Okazaki, K. Okawa, S. Wajiki, and N. Inamoto, *Bull. Chem. Soc. Jap.* **44**, 3167 (1971).

TABLE II
^{1}H NMR Values of Thiatriazoles and Thiatriazolium Salts (δ, TMS)

Compound	$-SCH_3$	$-SCH_2CH_3$	$-SCH_2CH_3$	$-NCH_2CH_3$	$-NCH_2CH_3$
1a	—	3.43[a]	1.57	—	—
2a	—	3.57[b]	1.51	5.26	1.83
1b	2.88[a]	—	—	—	—
2b	3.02[b]	—	—	5.38	1.86
2c	—	—	—	5.34[c]	1.94

[a] Solvent, CCl_4.
[b] Solvent, d_6-acetone.
[c] Solvent, $CD_3OD/CDCl_3$ (1:1).

nitrogen, sulfur, and ethyl cyanate in the usual thermal decomposition reaction of 5-alkoxythiatriazoles. Apparently the inductive effect of the ethoxy moiety reduces the electron density in the ring so that alkylation cannot compete with fragmentation.

5-Substituted thiatriazoles possess four different ring heteroatoms to which an alkylating agent may be delivered. Location of the ethyl group in the ring at position 3 was accomplished by means of ^{1}H, ^{15}N, and ^{13}C NMR spectroscopy[6] [Eq. (4)]. Spectral data are given in Tables II–IV.

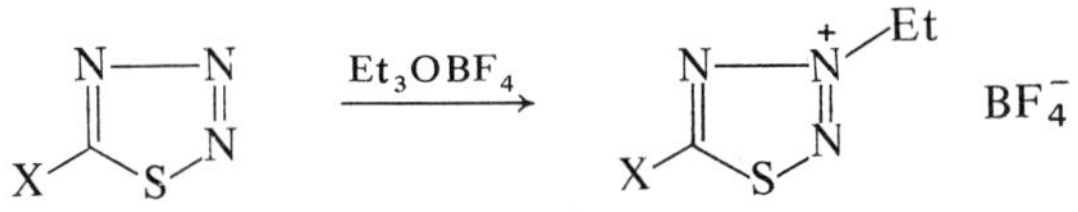

$$\underset{\substack{(1a)\ X = SEt \\ (1b)\ X = SMe \\ (1c)\ X = Ph}}{\text{thiatriazole}} \xrightarrow{Et_3OBF_4} \underset{\substack{(2a)\ X = SEt \\ (2b)\ X = SMe \\ (2c)\ X = Ph}}{\text{thiatriazolium}} BF_4^- \quad (4)$$

TABLE III
^{15}N Chemical Shifts and ^{15}NH Coupling Constants for Thiatriazolium Salt 2c Labeled with ^{15}N

^{15}N position	$\delta(^{15}N)$[a] (ppm)	$J(^{15}N-CH_2)$ (Hz)	$J(^{15}N-CH_3)$ (Hz)
2	-31 ± 2	2.43 ± 0.05	<0.2[b]
3	-91 ± 2	<1.20[b]	3.74 ± 0.1
4	-73 ± 2	1.64 ± 0.05	<0.2[b]

[a] Position relative to CH_3NO_2; $\delta(NH_4^+) = -360$ ppm.
[b] Coupling not observed. The values listed represent the observable halfwidth limit.

TABLE IV
^{13}C Chemical Shifts for Phenylthiatriazole 1c and Thiatriazolium Salt 2c[a]

Compound	C_1	C_{ortho}	C_{meta}	C_{para}	C-5	CH_2	CH_3
1c	125.84	129.14	129.14	132.63	178.46	—	—
2c	122.79	129.07	129.52	135.48	186.42	60.65	49.58
$\Delta\delta$	−3.05	−0.07	0.38	2.85	7.96	—	—

[a] Lines are measured relative to internal TMS. δ (CDCl$_3$) has been used as secondary standard. Positive δ values correspond to low field shifts.

Both nitrogen chemical shifts and coupling constants (J_{NH}, J_{NC}) were used to locate the alkylated nitrogen. All the ^{15}N-labeled thiatriazoles were prepared as outlined in Scheme 1. For a detailed discussion, the reader is referred to the original paper.[6]

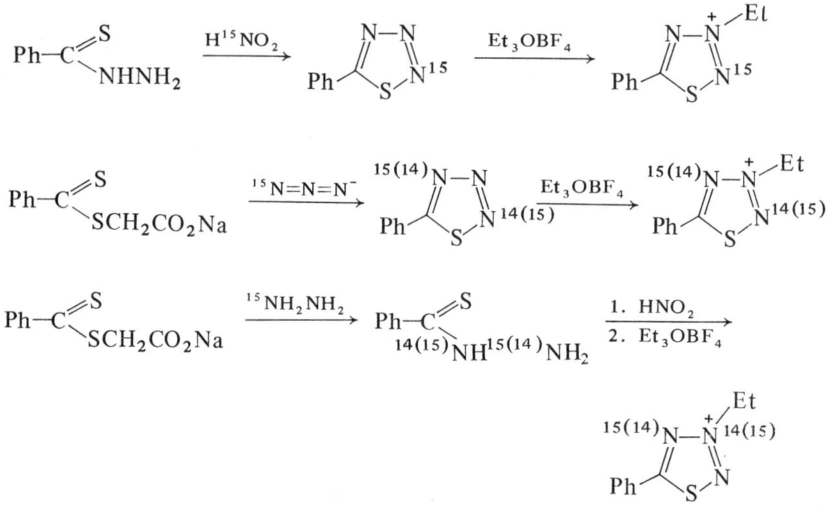

SCHEME 1

It is surprising that the ethyl group is located on N-3 of the thiatriazolium salts. Comparison with other multisite heterocycles for which the relative nitrogen nucleophilicity is known suggested the reactivity order N-4 ≥ N-3 ≫ N-2.[6] The authors attempted to derive a reactivity rationale for the assignment from self-consistent field (SCF)-MO-CNDO calculations. Total charge densities and the frontier orbital distribution were obtained for the energy-geometry optimized heterocycle (see also Section II, A). These theoretical quantities have been used as indices of positional reactivity when the reaction is charge or orbital controlled, respectively. Both criteria predict the reactivity order N-4 ⩾ N-2 > N-3. No satisfactory explanation was obtained for this apparent disagreement. The possibility that a 4-ethyl salt is generated in a kinetically controlled step followed by dealkylation and realkylation at N-3 to give a thermodynamically favored isomer could not be substantiated. The reaction was monitored by NMR at ordinary temperature during its early stages, but the only new signals appearing in the spectrum of the freshly prepared reaction mixture were those arising from the 3-ethyl salt. A second explanation may be steric but seems unattractive since 5-(N-monosubstituted-amino)thiatriazoles are readily alkylated by diazomethane in the 4-position (Section VIII). A 1,5-sigmatropic shift has been suggested as a possible alternative rearrangement pathway.[6]

In all the alkylations, the products are stable up to 180–200° in the solid state. This is consistent with alkylation at the 3-position, which clearly blocks the ability of the system to eject nitrogen under mild conditions.

2. Oxidation

By the action of peroxytrifluoroacetic acid on 5-phenylthiatriazole, a 3-oxide is obtained[23] [Eq. (8)].

$$\underset{Ph}{\overset{N-N}{\underset{S}{\diagdown}}\overset{\diagup}{\underset{N}{\diagup}}} \xrightarrow{CF_3COOH} \underset{Ph}{\overset{N-\overset{+}{N}\diagup O^-}{\underset{S}{\diagdown}}\overset{\diagup}{\underset{N}{\diagup}}} \qquad (8)$$

The structure assignment problem, which offers difficulties similar to those discussed for the alkylation products, was solved by means of mass spectrometry in combination with isotope labeling at position 2 with ^{15}N. ESCA (X-ray photoelectron) spectroscopy and IR spectra are in agreement with the assignment as a 3-oxide.[23]

Like the 3-ethylated thiatriazoles, the 3-oxide is thermally quite stable, decomposing at around 185°.

The oxide may be deoxygenated to the starting thiatriazole, but the oxygen is released rather slowly, even with the potent deoxygenating agent hexachlorodisilane.

---

[23] A. Holm, L. Carlsen, S.-O. Lawesson, and H. Kolind-Andersen, *Tetrahedron* **31**, 1783 (1975).
[23a] Two compounds: 2-^{15}N- and 4-^{15}-labeled, respectively.
[23b] Two compounds: 3-^{15}N- and 4-^{15}-labeled, respectively.

A quaternary salt is formed on treatment of the oxide with triethyloxonium tetrafluoroborate [Eq. (9)]. The salt is soluble in water, but is

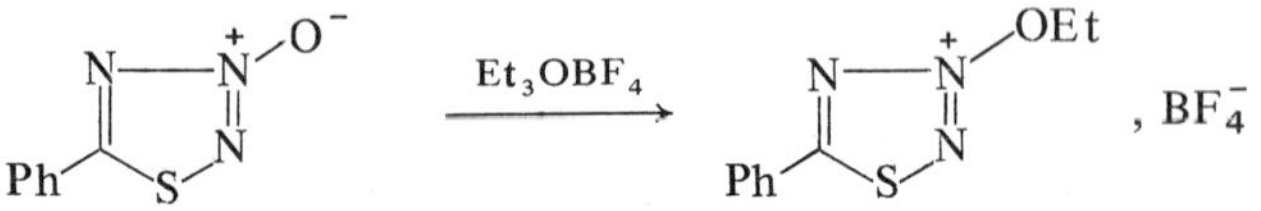

$$\underset{Ph}{\overset{N-N-O^-}{\underset{S}{\parallel}\underset{N}{\overset{+}{\mid}}}} \xrightarrow{Et_3OBF_4} \underset{Ph}{\overset{N-N-OEt}{\underset{S}{\parallel}\underset{N}{\overset{+}{\mid}}}} , BF_4^- \quad (9)$$

rapidly transformed into a mixture of thiatriazole and thiatriazole 3-oxide [Eq. (10)]. The first of these reactions is also known with other alkoxyammonium salts although it ordinarily occurs under basic conditions.[23]

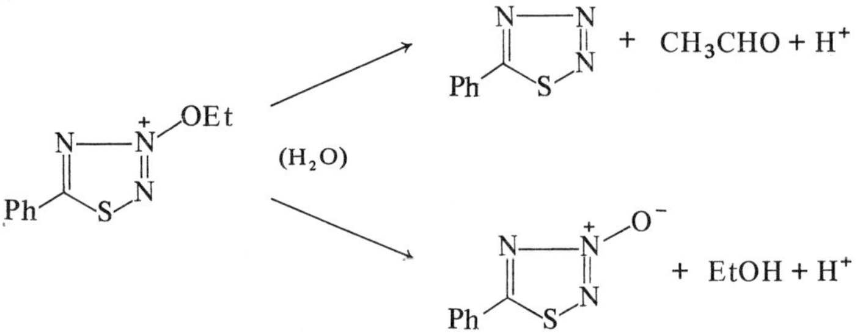

$$(10)$$

The UV spectrum of the 3-oxide in ethanol exhibits maxima at 254 nm ($\epsilon = 19300$) and 320 nm ($\epsilon = 3000$), the latter appearing as a shoulder. On irradiation ($\lambda = 250$ nm, in methylene chloride), benzonitrile and phenyl isothiocyanate are formed in 65% and 3% yields, respectively. A similar result is obtained on photolysis of 5-phenylthiatriazole (Section III, C). The major part of the products was shown to be formed from the singlet excited state.[23]

Attempts have been made to obtain 5-phenylthiatriazole-1-oxide (the S-oxide) for thermal and photochemical studies.[24] A nonoxidative route was investigated by mixing ethanolic solutions of either cis- or trans-thiobenzoyl chloride S-oxide and ammonium or tetrabutylammonium azide. At room temperature a rapid evolution of gas takes place, and besides nitrogen, both sulfur and sulfur dioxide are formed in amounts approximately according to Eqs. (11) and (12).

$$PhC(=S=O)Cl + N_3^- \longrightarrow PhCN + SO + N_2 + Cl^- \quad (11)$$

$$SO \longrightarrow \tfrac{1}{2}SO_2 + \tfrac{1}{2}S \quad (12)$$

At $-80°$ a yellow precipitate is obtained, which on heating evolves the gases. An infrared spectrum of a 1% solution in methylene chloride at $-80°$ indicates, in striking contrast to the formation of ordinary

[24] A. Holm and L. Carlsen, *Tetrahedron Lett.*, 3203 (1973).

thiatriazoles, that the aromatic thiatriazole *S*-oxide is not formed, but that the compound has an open structure, i.e., is thiobenzoyl azide *S*-oxide (3).

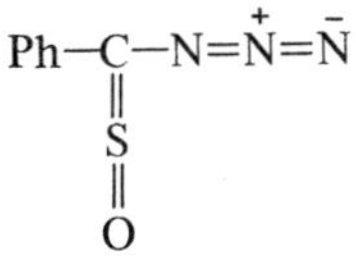

$$\text{Ph}-\underset{\underset{\text{O}}{\overset{\|}{\text{S}}}}{\overset{\|}{\text{C}}}-\text{N}=\overset{+}{\text{N}}=\overset{-}{\text{N}}$$

(3)

This is inferred from the intense IR absorption at 2120 cm^{-1}, in the region where organic azides generally absorb. A strong absorption at 1110 cm^{-1} is ascribed to the C=S=O group. It is not known whether the oxide is a cis or a trans form or a mixture of both forms. The same infrared observations are made whether *cis*- or *trans*-thiobenzoyl chloride *S*-oxide is used as starting material. On heating, the absorptions at 2120 and 1110 cm^{-1} begin to disappear around $-40°$ to $-30°$. At room temperature both bands have vanished, confirming their assignment to the yellow unstable compound.

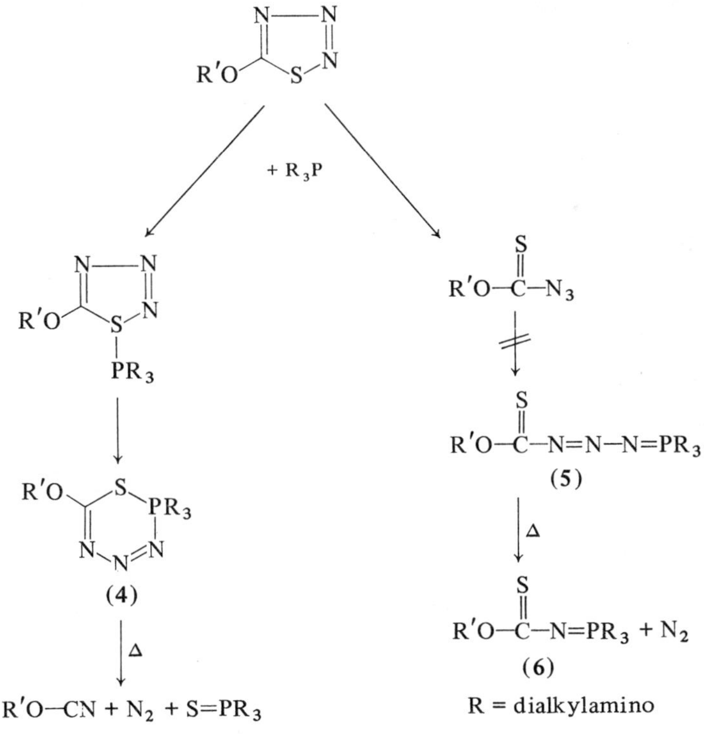

SCHEME 2

No conclusions with regard to the thermal stability of a true thiatriazole S-oxide can be drawn with safety from the above results. They may indicate only that decomposition to nitrogen, sulfur monoxide, and benzonitrile is much faster than ring closure.

3. *Reactions of Thiatriazole with Trivalent Phosphorus*

Pilgram et al.[18] investigated the reaction between 5-alkoxythiatriazoles and phosphorous triamides. Orange to orange-red 1:1 reaction products are formed in an exothermic reaction at $-15°$ to $-20°$, which, according to the authors, are 2,2-dihydro-1,3,4,5,2-thiatriazaphosphorines (**4**) (Scheme 2). This assignment is based on IR spectroscopy (lack of azide band around 2100 cm^{-1}) and mass spectrometry. The thiatriazaphosphorines are stable at room temperature. On heating in benzene or toluene, nitrogen is evolved and triamidophosphorothioates and aryl cyanates are formed. This is in contrast to the thermal properties of the reaction products between trivalent phosphorus compounds and acyl azides. In the latter case "phosphazides" are produced that lose nitrogen on heating to produce N-benzoylaminophosphoranes [Eq. (13)].

$$R_3P + N_3-\underset{\underset{O}{\|}}{C}-(O)_n-Ph \longrightarrow$$

$$R_3P=N-N=N-\underset{\underset{O}{\|}}{C}-(O)_n-Ph \xrightarrow{\Delta} R_3P=N-\underset{\underset{O}{\|}}{C}-(O)_n-Ph + N_2 \quad (13)$$

$n = 0, 1$
$R = $ dialkylamino

Had the phosphazide **5** been formed in the reaction of the thiatriazoles, then loss of nitrogen to form compound **6** would be the logical consequence. The authors therefore favor the direct insertion route without intervention of the (unknown) thioacyl azide, but no decisive choice can be made between alternative reaction mechanisms, since, if a compound **5** is actually formed from an open-chain thioacyl azide, it may rapidly cyclize to form the thermodynamically stable **4**.

Reaction between 5-alkoxythiatriazole and triphenylphosphine, which is a weaker nucleophile, proceeds more sluggishly.[18] At 25° triphenylphosphine reacted exothermically with 5-(4-chlorophenyl)thiatriazole in ether; the products were those of total fragmentation, i.e., 4-chlorophenyl cyanate and triphenylphosphine sulfide, in addition to nitrogen. Since thiatriazaphosphorines normally are stable at room temperature the facile decomposition may indicate a catalytic action on the thiatriazole instead of thiatriazaphosphorine formation. This question, however, is not discussed by the authors cited (see also Section III, A).

4. *Complex Formation*

In Section III, A the catalytic action of $AlCl_3$ and BBr_3 on the thermal decomposition of thiatriazoles was mentioned. This effect is evidently connected with complex formation between a thiatriazole and a Lewis acid since the catalytic activity is lost on addition of compounds that complex more effectively with the Lewis acid.[19] It is remarkable that titanium tetrachloride, in contrast to this, does not catalyze decomposition, but instead forms a thermally stable, orange 1:1 complex with 5-phenylthiatriazole.[19] The complex is sensitive to atmospheric moisture and is hydrolyzed in high yield to the starting thiatriazole on addition of water.

C. Photochemical Reactions

Photolysis of thiatriazoles was first described by Kirmse,[25] who found that aryl cyanides, aryl isothiocyanates, and sulfur are formed on irradition of 5-arylthiatriazoles with UV light. Recently Okazaki *et al.*[22] reported the formation of sulfur atoms on irradiation of thiatriazoles. Photolysis of 5-phenylthiatriazole ($\lambda_{max} = 280$ nm, $\epsilon = 10800$)[25] in the presence of cyclohexene yielded cyclohexene episulfide. Cyclohexene thiols were not formed; this was taken as an indication that the reactive species is a triplet sulfur atom. Tetramethylethylene behaves similarly.[22]

The potential use of the reaction was examined, but the episulfides decompose partially during the process, and the yield of cyclohexene episulfide under optimum conditions was 20%. Since episulfides may be obtained in much higher yields from the photodecomposition of COS to give triplet sulfur atoms in gas-phase reactions,[26] the method hardly has synthetic importance. It was assumed that the low yield is due to deactivation of the sulfur atoms by collision with solvent molecules, causing them to polymerize to molecular sulfur instead of reacting with cyclohexene.[22]

Okazaki *et al.* found that the thermal decomposition of 5-phenylthiatriazole in the presence of alkenes gives no episulfides and suggested that this is explicable in terms of low excess kinetic (translational) energy of sulfur atoms generated in the thermal reaction.[22]

*[25] W. Kirmse, *Chem. Ber.* **93**, 2353 (1960).

Since quenching experiments with piperylene and naphthalene showed no effect, it was suggested that the triplet state of the thiatriazole is not involved in the reactions.[22] It was later shown that the photoreactions leading to nitrile, isothiocyanate, sulfur, and nitrogen actually take place from a singlet excited state of the thiatriazole.[20] Thus the photoreduction of benzophenone in isopropyl alcohol is effectively quenched by addition of 5-phenylthiatriazole, but analysis revealed that all thiatriazole could be recovered from the photolysis mixture, indicating the lack of photoreactivity of the triplet state.

Okazaki et al.[22] proposed for the production of triplet sulfur a decomposition scheme that involves the hitherto unknown thiobenzoylnitrene, although attempts to trap it were unsuccessful.

The question of transient intermediates in the photolysis of thiatriazoles was investigated in detail by Holm et al.[20] The nitrogen formed during irradiation of 2-^{15}N-5-phenylthiatriazole was analyzed by high-resolution mass spectrometry. It was found that it originated exclusively from the N-2 and N-3 positions [Eq. (14)].

$$\text{Ph-thiatriazole} \xrightarrow[\text{(no scrambling)}]{h\nu} \text{Ph-CN} + \text{PhNCS} + {}^{15}\text{N}{\equiv}\text{N} \qquad (14)$$

Scrambling of the nitrogen atoms during irradiation does not take place. This was substantiated by irradiating 2-^{15}N-5-phenylthiatriazole to give partial transformation and analyzing the remaining material by means of ^{15}N NMR spectroscopy.

When 5-phenylthiatriazole is irradiated in neat dimethyl acetylenedicarboxylate, compound 7 is obtained in 9% yield.[20] The same compound is also obtained upon photolysis of 4-phenyl-1,3,2-oxathiazolylio-5-oxide (8)[20,27] or 5-phenyl-1,3,4-oxathiozol-2-one[20] in the same solvent in 9% and 22% yield, respectively. The formation of 7 is explained by the intervention of benzonitrile sulfide as a common intermediate (Scheme 3).

Benzonitrile sulfide has previously been trapped by the same solvent as a product of the thermal decomposition of 5-phenyl-1,3,4-oxathiazol-2-one.[28] Spectral studies of this elusive substance can be made when photolyzing either thiatriazole or the other precursors mentioned in an EPA glass at 85°K.[20]

[26] E. M. Lown, E. L. Dedio, O. P. Strausz, and H. E. Gunning, *J. Amer. Chem. Soc.* **89**, 1056 (1967).
[27] H. Gotthardt, *Chem. Ber.* **105**, 188 (1972).
[28] J. E. Franz and L. L. Black, *Tetrahedron Lett.*, 1381 (1970); R. K. Howe and J. E. Franz, *J. Chem. Soc., Chem. Commun.*, 524 (1973).

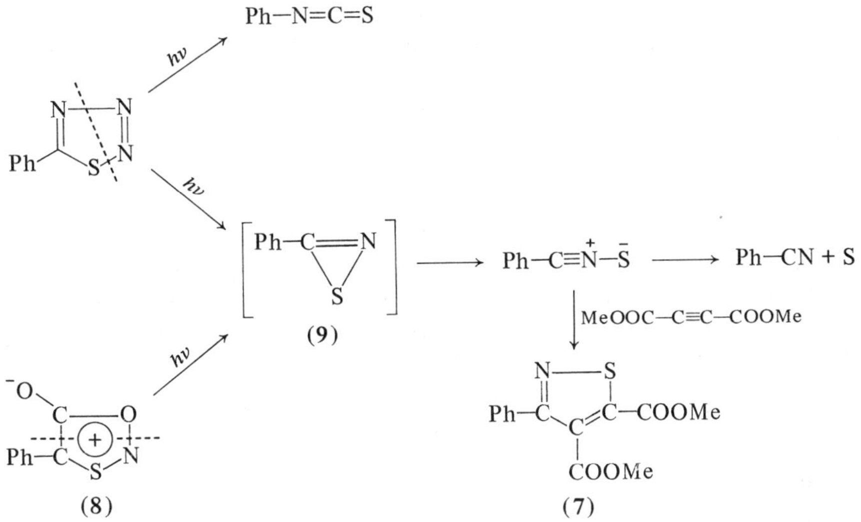

SCHEME 3

Phenylthiazirine (9) could be a yet undetected precursor for the nitrile sulfide when generated from the thiatriazole or 8. Its possible formation from the thiatriazole system has attracted other investigators[15,29] with regard to certain thermal reactions. However, as a cyclic conjugated 4-π electron-system it should be antiaromatic,[30] and therefore thermally unstable. It could possibly be generated in photochemical reactions.

The amount of isothiocyanate formed on irradiation of 5-phenylthiatriazole is remarkably independent of the conditions employed.[20] In a number of solvents, yields of 6.0–7.6% were found at room temperature. At 193°K the yield was 4.6–4.8% in methylene chloride. 5-(2-Methylphenyl)-1,2,3,4-thiatriazole behaved similarly, giving rise to 7.6% isothiocyanate. This insensitivity of the yield of isothiocyanate under various photolytic conditions is believed virtually to rule out nitrene intermediates. Furthermore, characteristic nitrene products were sought for in the photolysis mixture but not found. ESR spectroscopy also failed to provide evidence for the intermediacy of a thioacyl nitrene (10).

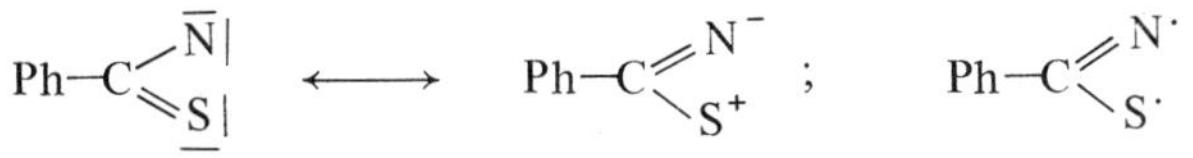

[29] R. Neidlein and J. Tauber, *Arch. Pharm. (Weinheim)* **304**, 687 (1971).
[30] R. Breslow, *Accounts Chem. Res.* **6**, 393 (1973).

Since nitrene intermediates appear to be ruled out, the photochemical generation of isothiocyanate could either take place via other types of intermediates or directly from a singlet excited state of the thiatriazole. As possible intermediates thiobenzoyl azide, benzonitrile sulfide, and phenylthiazirine were considered.[20] Since isothiocyanate is formed from the thiatriazole at 85°K but not from **8** at 85°K where benzonitrile sulfide is stable, the latter was ruled out as a precursor for phenyl isothiocyanate. Thiobenzoyl azide is assumed to absorb in the visible but could not be observed when the thiatriazole was irradiated in an EPA glass at 85°K, and thus no evidence for its intermediacy was obtained. In conclusion it appears that the isothiocyanate is formed directly from the singlet excited state of the thiatriazole, although the intermediacy of phenylthiazirine cannot be directly excluded.

IV. 1,2,3,4-Thiatriazoles Substituted with C-Radicals

Whereas a number of 5-aryl thiatriazoles have been reported, the only previously known true aliphatic and alicyclic representatives were 5-*tert*-butyl[31] and cyclohexyl thiatriazole.[32] These are unstable oils that decompose at 0° with nitrogen evolution and formation of sulfur. They are prepared from the corresponding thioacylhydrazides and nitrous acid, but the method is not generally applicable because of difficulties in obtaining the required aliphatic thioacylhydrazides.[1] Wijers *et al.*[17] have found that aliphatic thiatriazoles can be prepared from 1-acetylthio-1-alkynes. Thus a substance believed to be 5-pentylthiatriazole was isolated from the reaction between 1-acetylthio-1-hexyne and ammonium azide. It is an oil that solidifies at about −16° and could not be analyzed because of its explosive character and poor stability at room temperature. Its formation is explained by the following scheme [Eq. (15)].

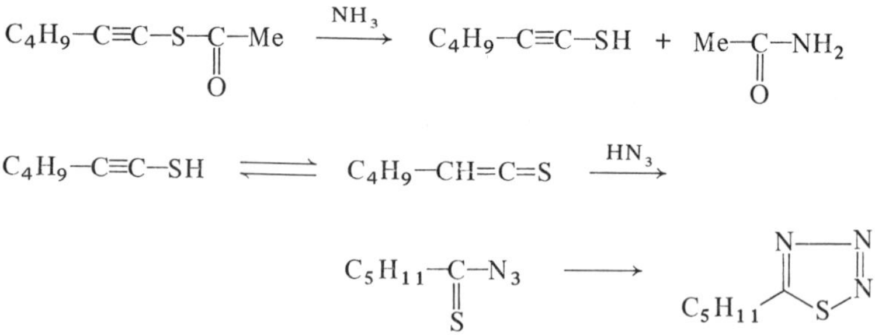

$$C_4H_9-C\equiv C-S-\underset{O}{\overset{\parallel}{C}}-Me \xrightarrow{NH_3} C_4H_9-C\equiv C-SH + Me-\underset{O}{\overset{\parallel}{C}}-NH_2$$

$$C_4H_9-C\equiv C-SH \rightleftharpoons C_4H_9-CH=C=S \xrightarrow{HN_3}$$

$$C_5H_{11}-\underset{S}{\overset{\parallel}{C}}-N_3 \longrightarrow \underset{C_5H_{11}}{\overset{N=N}{\underset{S}{\bigvee}}}N \quad (15)$$

---
[31] K. A. Jensen and C. Pedersen, *Acta Chem. Scand.* **15**, 1104 (1961).
[32] P. A. S. Smith and D. H. Kenny, *J. Org. Chem.* **26**, 5221 (1961).

The structure is in part inferred from the observation that in boiling tetrahydrofuran the substance loses nitrogen and sulfur and gives pentyl cyanide. The absence of absorption in the neighbourhood of 2150 cm^{-1} indicative of an azide group excludes the thioacyl azide structure.[17]

It is remarkable that thermal decomposition in polar solvents (dimethylformamide, tetrahydrofuran) gives pentyl cyanide whereas in a nonpolar solvent (decane, 150°) pentyl cyanide is formed together with pentyl isothiocyanate in a 4:1 molar ratio. The same ratio is obtained on injecting 5-pentylthiatriazole into glass wool heated at 350° in an evacuated flask.

Raasch[33] has prepared 5-[2,2,2-trifluoro-1-(trifluoromethyl)ethyl]-thiatriazole from bis(trifluoromethyl)thioketene and hydrogen azide [Eq. (16)]. This highly volatile solid (m.p. 60.5–61°) is interesting in that it is thermally stable at 25°. At 100° it decomposes with formation of sulfur, but the other products were not determined.

$$(CF_3)_2C=C=S \xrightarrow{HN_3} (CF_3)_2CH-\underset{S}{\overset{N-N}{\underset{}{\bigtriangleup}}}N \qquad (16)$$

V. 1,2,3,4-Thiatriazole-5-thiol and Its Derivatives

Lieber et al.[34] discussed whether the compound obtained from the sodium salt of 5-mercaptothiatriazole and methyl iodide is an N- or an S-derivative. By unequivocal synthesis it was substantiated that it is the S-derivative.[34] However, acyl derivatives were formulated as 4-substituted thiatriazoline-5-thiones[34] because isothiocyanates were obtained on thermal decomposition. Jensen and Pedersen[1] challenged this argument, as acyl thiocyanates are very unstable and may easily rearrange to the corresponding acyl isothiocyanates during the thermal degradation experiment. Christophersen and Holm[11] then reinvestigated the reaction products obtained by treating sodium thiatriazole-5-thiolate with benzoyl chloride as well as with diphenylmethyl and triphenylmethyl chlorides because the reaction products with the latter compounds had also been formulated as 4-substituted thiatriazoline-5-thiones.[35] 5-(Diphenylmethylthio)thiatriazole was obtained by unequivocal synthesis as shown in Eq. (17).[11]

---

[33] M. S. Raasch, *J. Org. Chem.* **35**, 3470 (1970).
*[34] E. Lieber, E. Oftedahl, S. Grenda, and R. D. Hites, *Chem. Ind. (London)* 893 (1958).
*[35] E. Lieber, E. Oftedahl, and C. N. R. Rao, *J. Org. Chem.* **28**, 194 (1963); E. N. Oftedahl, M.S. Thesis, De Paul University, Chicago, Illinois, 1960.

$$(Ph_2CHS-\underset{\underset{S}{\|}}{C}-S)_2 + NH_2NH_2 \longrightarrow$$

$$S + Ph_2CHS-\underset{\underset{S}{\|}}{C}-NHNH_2 + Ph_2CHS-\underset{\underset{S}{\|}}{C}-S^-, NH_2NH_3^+$$

$$Ph_2CHS-\underset{\underset{S}{\|}}{C}-NHNH_2 + HNO_2 \xrightarrow{-H_2O} Ph_2CHS\underset{S}{\overset{N=N}{\diagdown\diagup}}N \quad (17)$$

The substance so obtained was shown by melting point and IR and ^{1}H NMR spectroscopy to be identical with the substance designated as a 4-substituted thiatriazoline-5-thione by Lieber *et al.* By comparison with authentic diphenylmethyl thiocyanate and isothiocyanate, it was also clearly demonstrated that the thermal degradation product from the so-called thiatriazoline-5-thione actually is the thiocyanate as expected.[11] When the benzoylated so-called thiatriazoline-5-thione was subjected to thermal degradation in solution (CHCl$_3$), a very sharp band appeared in the infrared spectrum at 2170 cm^{-1}. As the decomposition proceeded, a very strong and complex broad band developed at around 1981 cm^{-1}. After a sufficient length of time, the band at 2170 cm^{-1} disappeared, and the band at 1981 cm^{-1} became the most intense in the spectrum. It was proposed that the 2170 cm^{-1} band is due to the thiocyanate group in benzoyl thiocyanate and the 1981 cm^{-1} band to benzoyl isothiocyanate[11] (authentic benzoyl isothiocyanate in CHCl$_3$ solution absorbs near 1981 cm^{-1}). L'abbé and co-workers[36] later compared the ^{13}C NMR spectra (C-5 in the ring, CDCl$_3$) of authentic 5-benzylthiothiatriazole (δ 179.8) with the spectra of the products obtained from acylation of sodium thiatriazole-5-thiolate with benzoyl (δ 171.5) and phenacyl (δ 179) chlorides. Although the position of the ^{13}C chemical shift of C-5 in a thiatriazoline is not known, the narrow range in which the signals are observed favours acylation at S rather than at N. The ^{13}C chemical shift of CO (δ 185.2) in the benzoylated sodium thiatriazole-5-thiolate is comparable with that found for 1-benzyl-5-benzoylthio-1,2,3,4-tetrazole (COS, δ 184.4), which also indicates attachment of the benzoyl group to sulfur. In conclusion, these data support the results obtained by Christophersen and Holm, and the thiatriazoline-5-thione structure for the acylation and alkylation products is decisively eliminated. Interestingly, it has been found that 5-aminothiatriazole actually can be alkylated in the ring by diazomethane to give 4-substituted thiatriazolines (Section VIII).

The formation of hitherto unknown or difficultly accessible types of thiocyanates by thermal degradation of S-substituted 5-mercapto-

---

[36] G. L'abbé, S. Toppet, G. Verhelst, and C. Martens, *J. Org. Chem.* **39**, 3770 (1974).

thiatriazoles has been explored. The known N,N-diphenylcarbamoyl thiocyanate can be prepared by this method from 5-(N,N-diphenylcarbamoylthio)thiatriazole.[11] A number of 5-acylthiothiatriazoles have been prepared, and from these aroyl, aralkanoyl, alkanoyl, and 2-alkenoyl thiocyanates were obtained.[37-39]

By analogy with the well known reaction of sodium azide and carbon disulfide, which gives sodium thiatriazole-5-thiolate,[40] carbon disulfide was condensed with $[\pi\text{-}C_5H_5Ni(PBu_3)_2]^+N_3^-$ to yield brown crystals with the composition $\pi\text{-}C_5H_5NiPBu_3CS_2N_3$.[41] This compound reacted at room temperature with benzoyl chloride in acetone to give what was believed to be 4-benzoylthiatriazoline-5-thione in accordance with the incorrect structure assignment of Lieber et al.[34] Pyrolysis at 80° gave $\pi\text{-}C_5H_5NiPBu_3NCS$. However, as discussed above, the benzoyl derivative is actually 5-(benzoylthio)thiatriazole, and consequently the formation of this compound does not support the structure assigned to the Ni compound. Because of the well known stability of thiol compounds, it would also seem much more plausible that Ni should be bound to S rather than to N.

Tributyltin azide reacts with carbon disulfide in solution to give what is supposed to be 4-(tributyltin)thiatriazoline-5-thione because, on evaporation of the solvent, tributyltin isothiocyanate is formed.[42] However, isolation of this compound does not show whether tin is coordinated to S or N in the thiatriazole since rearrangements may have taken place.

Organometallic azides of the type $[(Ph)_3P]_2M(N_3)_2$ (M = Pt, Pd) react with CS_2, probably with formation of 5-mercaptothiatriazole complexes.[43]

Extremely shock-sensitive alkylene bis[5-thiatriazolyl]sulfides were prepared by Pilgram et al.[44,45] by treating sodium thiatriazole-5-thiolate with dihaloalkane. Dichloromaleimide reacted analogously.[45]

The synthesis of some new 5-benzylthiothiatriazoles by standard methods have been described.[12] Nilsson et al.[46] reported on the forma-

[37] C. Christophersen, Acta Chem. Scand. **25**, 1160 (1971).
[38] C. Christophersen, Acta Chem. Scand. **25**, 1162 (1971).
[39] C. Christophersen and P. Carlsen, Tetrahedron, in press (1976).
*[40] F. Sommer, Ber. **48**, 1833 (1915).
[41] F. Sato, M. Etoh, and M. Sato, J. Organometal. Chem. **37**, C51 (1972).
[42] P. Dunn and D. Oldfield, Aust. J. Chem. **24**, 645 (1971).
[43] P. Kreutzer, C. Weis, H. Boehme, T. Kemmerich, W. Beck, C. Spencer, and R. Mason, Z. Naturforsch. B **27**, 745 (1972); W. Beck and W. P. Fehlhammer, Angew. Chem., Int. Ed. Engl. **6**, 169 (1967).
[44] K. Pilgram and F. Korte, Angew. Chem. **77**, 348 (1965).
[45] K. Pilgram and F. Görgen, J. Heterocycl. Chem. **8**, 899 (1971).
[46] N. H. Nilsson, C. Jacobsen, and A. Senning, J. Chem. Soc., Chem. Commun., 314 (1971).

tion of S-substituted 5-mercaptothiatriazoles from trithiocarbonate S,S-dioxides [Eq. (18)]. The method hardly has practical use since the thiatriazoles may be prepared directly from the chlorodithioformates.

$$\text{MeSO}_2^- + \text{Cl}-\underset{\underset{S}{\|}}{\overset{O}{\overset{\|}{C}}}-\text{SR} \longrightarrow \text{MeS}-\underset{\underset{S}{\|}}{\overset{O}{\overset{\|}{C}}}-\text{SR} \xrightarrow[\text{MeSO}_2^-]{\text{N}_3^-} \underset{RS}{\overset{N-N}{\underset{S}{\diagdown}\diagup}}\text{N} \quad (18)$$

Sodium thiatriazole-5-thiolate gives colored precipitates with copper(II) or bismuth salts. This reaction may be used as a test for soluble inorganic azides by way of their reaction with CS_2, which yields thiatriazole-5-thiolates.[47]

VI. 5-Alkoxy and 5-Aryloxy 1,2,3,4-Thiatriazoles

The first synthesis of ethyl cyanate[48] from 5-ethoxythiatriazole[49] was briefly mentioned in the previous review of thiatriazoles.[1] The thiatriazole decomposes smoothly in ethereal solution at 20° with formation of nitrogen, sulfur, and ethyl cyanate. Although several other methods are now available for the preparation of both alkyl and aryl cyanates the thiatriazole method prevails as the most generally applicable for the thermally and chemically sensitive alkyl cyanates, since the only by-products are inert—sulfur and nitrogen. Thus a number of 5-alkoxy,[50–54] 5-aryloxy,[55–57] and 5-aralkoxythiatriazoles[12,53] have been prepared.

The 5-(2-alkenyloxy)thiatriazoles on thermal degradation gave alkenyl isocyanates by a presumed allylic rearrangement of the expected 2-alkenyl cyanates.[54] Although the data do not exclude decomposition via intramolecular cyclic attack at the 4-position in the ring followed by very rapid decomposition of the 4-substituted thiatriazolines, these

---

[47] G. S. Johar, *Talanta* **19**, 1461 (1972).
[48] K. A. Jensen and A. Holm, *Acta Chem. Scand.* **18**, 826 (1964).
*[49] K. A. Jensen, A. Holm, and B. Thorkilsen, *Acta Chem. Scand.* **18**, 825 (1964).
[50] D. Martin, *Tetrahedron Lett.*, 2829 (1964).
[51] K. A. Jensen, M. Due, and A. Holm, *Acta Chem. Scand.* **19**, 438 (1965); K. A. Jensen, A. Holm, C. Wentrup, and J. Møller, *ibid.* **20**, 2107 (1966).
[52] D. Martin and W. Mucke, *Chem. Ber.* **98**, 2059 (1965).
[53] K. A. Jensen, A. Holm, and J. Wolff-Jensen, *Acta Chem. Scand.* **23**, 1567 (1969).
[54] C. Christophersen and A. Holm, *Acta Chem. Scand.* **24**, 1512 (1970).
[55] D. Martin, *Chem. Ber.* **97**, 2689 (1964); *Angew. Chem.* **76**, 303 (1964).
[56] M. Hedayatullah and L. Denivelle, *C. R. Acad. Sci.* **260**, 2839 (1965).
[57] M. Hedayatullah, *Bull. Soc. Chim. Fr.*, 422 (1967).

were later prepared and found to be stable at ambient temperature (Section VIII).

The thermal degradation of isotopically labeled 5-isobutoxy-thiatriazole[15] and the kinetics of the decomposition of 5-alkoxy thiatriazoles have been studied[16] (Section III, A).

VII. N-Substituted 5-Amino-1,2,3,4-thiatriazoles

The rearrangement of 5-phenylaminothiatriazole to 1-phenyl-5-mercaptotetrazole in basic solution[1Δ58] has been shown to be reversible[59] (Scheme 4). The mercaptotetrazole (**11**) rearranges rapidly in refluxing

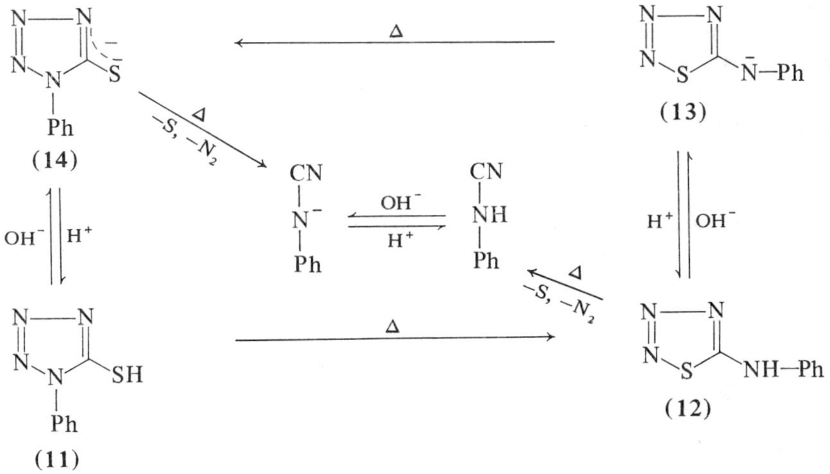

SCHEME 4

benzene to the thiatriazole (**12**). When **12** is dissolved in cold base and the solution is immediately acidified, unchanged **12** precipitates. However, when the basic solution is warmed, rearrangement of the anion **13** to **14** occurs, and on acidification of the solution only **11** is obtained. On prolonged heating of **11** in benzene or in aqueous base, nitrogen and sulfur are lost and a cyanamide is formed.

Acylation of 5-(phenylamino)thiatriazole may in principle give rise to several isomeric products. From ^{1}H NMR data Lippmann et al.[60] concluded that only 5-(N-acylphenylamino)thiatriazoles are formed. This is

---

[*58] E. Lieber, C. N. Pillai, and R. D. Hites, *Can. J. Chem.* **35**, 832 (1957).
[59] J. C. Kauer and W. A. Sheppard, *J. Org. Chem.* **32**, 3580 (1967).
[60] E. Lippmann, D. Reifegerste, and E. Kleinpeter, *Z. Chem.* **13**, 134 (1973).

in analogy to the acylation reaction of 5-mercaptothiatriazole, which leads to 5-(acylthio)thiatriazoles (Section V). However, Neidlein and Tauber[29] succeeded in obtaining 4-methyl-5-(arylimino)thiatriazolines (Section VIII) by methylation of 5-(arylamino)thiatriazoles with diazomethane. Methylation with dimethyl sulfate yielded only 5-(N,N-arylmethylamino)thiatriazoles. In contrast to this, 5-mercaptothiatriazole gave 5-(methylthio)thiatriazole with both diazomethane and dimethyl sulfate.[6] On treatment of 5-(arylsulfonylamino)thiatriazoles with diazomethane, 4-substituted thiatriazolines are similarly obtained.[61]

Some new 5-(benzylamino)thiatriazoles have been prepared by standard methods.[12] 5-(Tetra-O-acetyl-β-D-glucopyranosylamino)-thiatriazole was prepared by the action of HNO_2 on 4-(tetra-O-acetyl-β-D-glucopyranosyl)thiosemicarbazide as well as from hydrogen azide and tetra-O-acetyl-β-D-glucopyranosyl isothiocyanate.[62] Substituted 5-aminothiatriazoles may be obtained from C-sulfonylthioformamides on reaction with azide ion.[63]

Substituent parameters for the 5-aminothiatriazole group have been measured (σ_m, 0.30; σ_p, 0.19; σ_I, 0.42; σ_R or σ_R^0, −0.23).[59]

VIII. Synthesis and Chemical Properties of 1,2,3,4-Thiatriazolines and Alleged 1,5-Dihydro-1,2,3,4-thia(S^{IV})triazoles

A. Thiatriazolines

As already mentioned, the previously reported 4-substituted thiatriazolines have been shown in fact to be 5-substituted thiatriazoles (Section V). However, Neidlein and co-workers[29,61] and L'abbé and co-workers[64,65] have recently, by different methods, isolated representatives of this type of compound which have interesting chemical properties. Neidlein and Tauber[29] found that 5-(arylamino)thiatriazoles can be alkylated to yield either 5-(alkylarylamino)thiatriazoles and/or 4-alkyl-5-(arylimino)thiatriazolines depending on the alkylating agent and the aryl group (Scheme 5). With dimethyl sulfate in alkaline solution, all compounds investigated gave 5-substituted thiatriazoles (**17**), while alkyla-

---

[61] R. Neidlein and K. Salzmann, *Synthesis*, 52 (1975).
[62] R. Bognár, L. Somogyi, L. Szilágyi, and Z. Györgydeák, *Carbohyd. Res.* **5**, 320 (1967).
[63] N. H. Nilsson, A. Senning, S. Karlsson, and J. Sandström, *Synthesis*, 314 (1972).
[64] E. Van Loock, J.-M. Vandensavel, G. L'abbé, and G. Smets, *J. Org. Chem.* **38**, 2916 (1973).
[65] G. L'abbé, E. Van Loock, R. Albert, S. Toppet, G. Verhelst, and G. Smets, *J. Amer. Chem. Soc.* **96**, 3973 (1974).

tion with diazomethane gave either compound **16** or a mixture of **16** and **17**. Thus 5-phenylamino and 5-(*p*-methylphenylamino)thiatriazoles gave compound **16** exclusively with diazomethane whereas *p*-chloro and *m*-

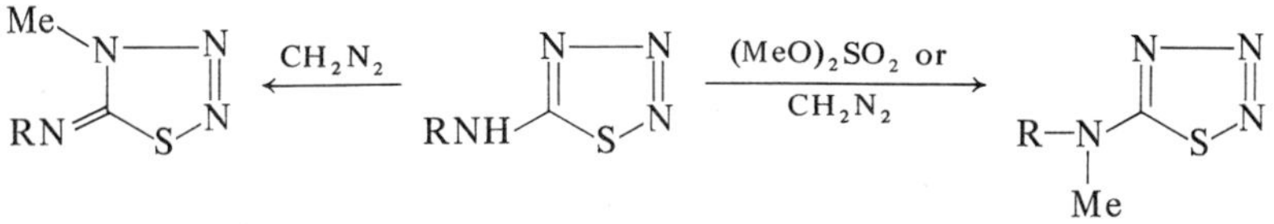

(16)	(15)	(17)
m.p. °C	R	m.p. °C
68–69	Ph	56–57
25–26	*p*-MeC$_6$H$_4$—	56–58
oil	*p*-Cl—C$_6$H$_4$—	70–72
52–54	*m*-Cl—C$_6$H$_4$—	74–76

SCHEME 5

chlorophenyl thiatriazoles gave a mixture of compounds **16** and **17** which was separated by thin-layer chromatography.

The structure assignment of the diazomethane products as 4-substituted thiatriazolines (**16**) is based on their thermolysis. On heating in toluene at 120–130° for several hours, this type of compound loses nitrogen with formation of benzothiazoles[29] [Eq. (19)] in contrast to ordinary 5-aminothiatriazoles, which yield cyanamides, sulfur, and nitrogen.

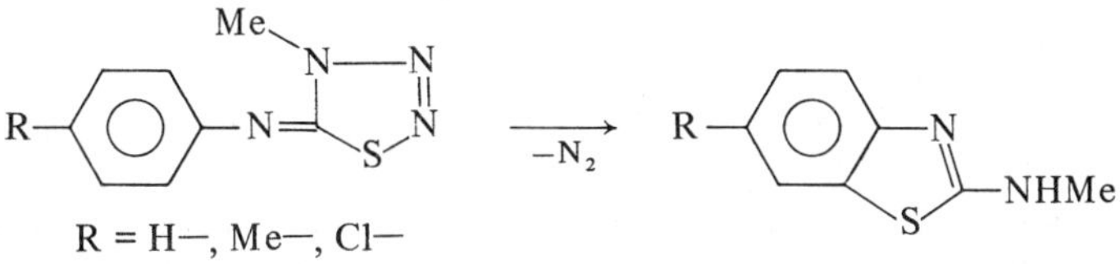

R = H—, Me—, Cl—

With the *m*-chloro compound two isomeric benzothiazoles are formed that result from attack either ortho or para to chlorine.[29] The mechanism of this process is not discussed but appears to be the result of electrophilic attack of a 1,3-dipole formed by loss of nitrogen (see later).

4-Substituted thiatriazoline-5-thiones are probably formed in the reaction between alkyl azides and phenyl isothiocyanate.[66] In this case the reaction is conducted at 80–100° in the absence of solvent, and the primary products cannot be obtained. With two equivalents of phenyl isothiocyanate, 4-alkyl-3,5-bis(phenylimino)-1,2,4-dithiazolidines are formed in 29–65% yield [Eq. (20)].

[66] D. M. Revitt, *J. Chem. Soc., Chem. Commun.*, 24 (1975).

$$RN_3 + 2Ph-N=C=S \longrightarrow \underset{(18)}{R-N\begin{array}{c}\overset{N-Ph}{\underset{\|}{C}}-S\\ |\\ \underset{\|}{C}-S\\ N-Ph\end{array}} \qquad (20)$$

This reaction may involve stepwise addition of the two equivalents of PhNCS via a 3-membered ring intermediate, $R\overline{N-S-C}$=NPh.[66] A thiaziridine was also suggested by Borsche[67] to explain the formation of 4-phenyl-5-phenylimino-1,2,4-dithiazolidine-3-thione from the aluminum chloride-catalyzed decomposition of phenyl azide in carbon disulfide. If a 4-substituted thiatriazoline is formed from phenyl isothiocyanate and the alkyl azide, the reaction may then be formulated as indicated in Eq. (21). This scheme is supported by the recent findings of Neidlein and

$$Ph-N=C=S + R-N_3 \longrightarrow \begin{array}{c}R-N\underset{}{}N\\ \underset{Ph-N}{\|}\underset{S}{\diagup}\underset{}{N}\end{array} \xrightarrow{-N_2}$$

$$\begin{array}{c}Ph-N=C-S\\ \diagdown\diagup\\ N\\ |\\ R\end{array} \xrightarrow{Ph-N=C=S} (18) \qquad (21)$$

Salzmann[61] (see Scheme 6 for the reaction between thiatriazolines and isothiocyanates).

When the more reactive sulfonyl isothiocyanates are used as 1,3-dipolarophiles 4-alkyl-5-sulfonylimino-Δ^2-1,2,3,4-thiatriazolines are readily prepared at room temperature by reaction with alkyl azides.[64,65] The thiatriazolines are obtained in 50–75% yield. Their structures are deduced from NMR, IR, and mass spectral data and degradation experiments. Thus thermal decomposition at moderate temperature (45°–80°) in inert solvents (dry toluene, CCl_4, acetone) furnished the corresponding carbodiimides [Eq. (22)]. These were identified by their

$$RN_3 + p\text{-}X\text{-}C_6H_4\text{-}SO_2\text{-}N=C=S \longrightarrow \begin{array}{c}N\text{------}N-R\\ \||\\ N\diagdown_S\diagup C=N-SO_2-C_6H_4-X\text{-}p\end{array} \xrightarrow{\Delta}$$

$$RN=C=N-SO_2-C_6H_4-X\text{-}p + N_2 + S \qquad (22)$$

[67] W. Borsche, *Chem. Ber.* **75**, 1312 (1942).

IR spectra or trapped by oxalyl chloride or water. Furthermore, the isomeric 1,4-substituted Δ^2-tetrazoline-5-thiones, which would have been formed by C=N instead of C=S addition, were definitively excluded by independent synthesis.[64]

Of particular interest is the demonstration that a 1,3-dipole or a three-membered ring compound is a first-formed intermediate in the thermal decomposition. Thus, when 4-benzyl-5-tosyliminothiatriazoline was decomposed in the presence of enamines, adducts were obtained as shown in Eq. (23).[64,65]

$$\underset{N\diagdown S}{\overset{N\text{———}N\text{—}CH_2Ph}{\|}} \diagup C=N-SO_2-C_6H_4Me\text{-}p \quad \xrightarrow[-N_2]{\Delta, \; Me_2N\diagup C=C\diagdown R^3} $$

$$\begin{array}{c} \overset{R^1}{\downarrow} \quad \overset{R^2}{\downarrow} \\ Me_2N\text{---}C\text{———}C\text{---}R^3 \\ PhCH_2-N\diagdown C\diagup S \\ \| \\ N \\ | \\ SO_2-C_6H_4Me\text{-}p \end{array} \quad (23)$$

The indicated stereochemistry of the adduct was established by ^{1}H and ^{13}C NMR spectroscopy. Ynamines react analogously.[64,65] Keto-stabilized phosphorus ylides react first to give cyclic adducts which lose Ph$_3$PO by a Wittig-type elimination. The structure of the latter adducts was proved by independent synthesis.[65] Less reactive alkenes, such as vinyl ethers and vinyl acetates and the electron-poor alkenes methyl acrylate and dimethyl acetylenedicarboxylate, did not give cycload-ducts.[65] Neidlein and Salzmann[61] find that cycloaddition products are also formed with isocyanates, isothiocyanates, and carbodiimides (Scheme 6). The thiatriazolines decompose under the influence of bases.[65] O-Methyl-N-benzyl-N'-tosylsulfonylisourea was obtained when the corresponding thiatriazoline was treated with sodium methoxide in methanol solution.

Kinetic experiments show that the rate of thermal decomposition of 4-benzyl-5-tosylimino-Δ^2-1,2,3,4-thiatriazoline is not influenced by the presence of enamines (the first-order rate constant is 42×10^{-5} s^{-1} in CCl$_4$ at 60°) excluding attack of alkene on the thiatriazoline with simultaneous loss of nitrogen.[65] Instead, this result indicates formation of a discrete intermediate, which apparently is either the thiaziridine (**19**) and/or the 1,3-dipolar species (**20**) [Eq. (24)].

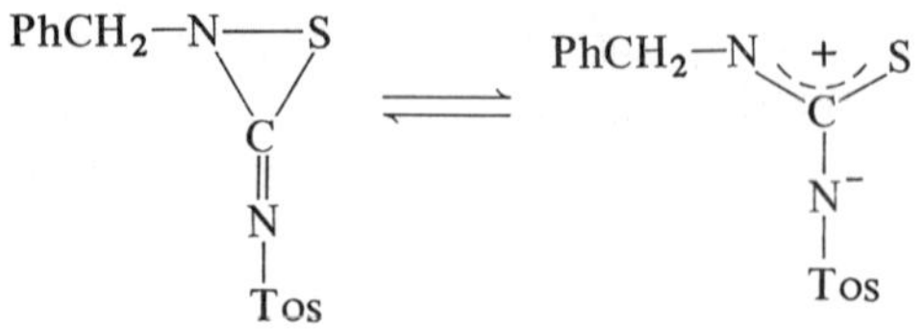

$$\begin{array}{c}\text{PhCH}_2-\text{N}-\text{S}\\ \diagdown\diagup\\ \text{C}\\ \|\\ \text{N}\\ |\\ \text{Tos}\\ (19)\end{array} \quad \rightleftharpoons \quad \begin{array}{c}\text{PhCH}_2-\text{N} \overset{+}{\diagdown}\diagup \text{S}\\ \text{C}\\ |\\ \text{N}^-\\ |\\ \text{Tos}\\ (20)\end{array} \qquad (24)$$

Loss of sulfur from these species would give the carbodiimide, whereas addition of alkenes would give the cycloadducts. The authors infer that this intermediate cannot add concertedly to alkenes in the supra-supra fashion because it would involve a four-electron transition state, but that a stepwise addition would hardly rationalize the stereospecificity observed; this suggests that the thiaziridine (or the 1,3-dipole) participates in an antarafacial reaction.[65] However, according to recent discussions antarafacial addition is seldom observed[68] and stereospecificity need not be lost in a stepwise process.[69]

In a continuation of their previously mentioned work,[29] Neidlein and Salzmann[61] found that 4-substituted 5-sulfonylimino-1,2,3,4-thiatriazolines can also be obtained by the action of diazomethane on 5-

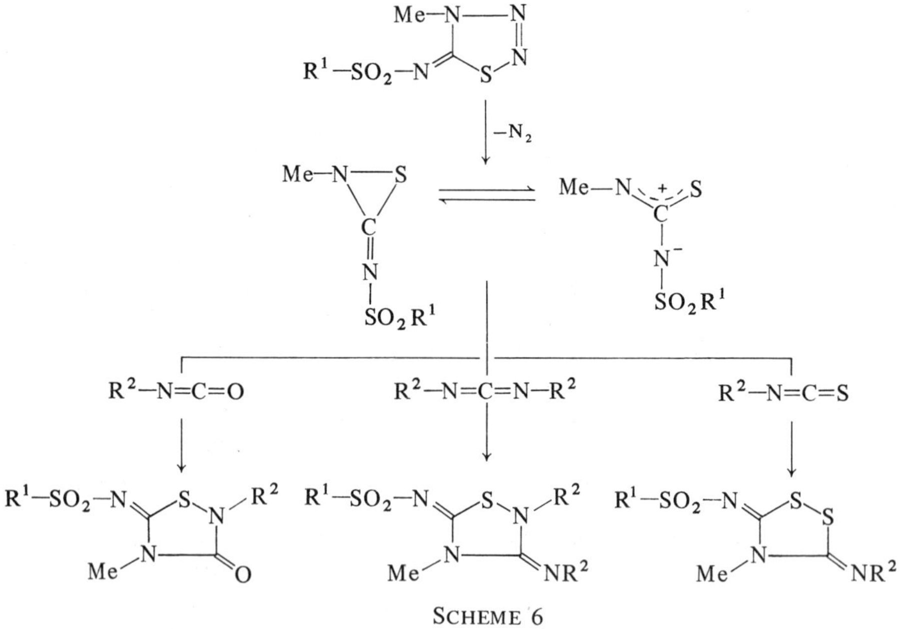

SCHEME 6

[68] P. D. Bartlett, G. M. Cohen, S. P. Elliott, K. Hummel, R. A. Minns, C. M. Sharts, and J. Y. Fukunaga, *J. Amer. Chem. Soc.* **94**, 2899 (1972).
[69] P. D. Bartlett, *Int. Congr. Pure Appl. Chem., 23rd* **4**, 281 (1971).

(sulfonylamino)thiatriazoles, most often in admixture with the isomeric 5-(N-methyl-N-sulfonylamino)thiatriazole [Eq. (25)]. Separation of the mixture was achieved by column chromatography.

$$R-SO_2-NH-\underset{S-N}{\overset{N-N}{\diagdown}} \xrightarrow{CH_2N_2} R-SO_2-N=\underset{S-N}{\overset{Me-N-N}{\diagdown}} + R-SO_2-N(Me)-\underset{S-N}{\overset{N-N}{\diagdown}} \quad (25)$$

These thiatriazolines are analogous to the compounds obtained by L'abbé and co-workers.[64,65] The thermal decomposition intermediates were trapped as shown in Scheme 6, but the presence of the cyclic form has not yet been demonstrated.[61]

B. Alledged 1,5-Dihydro-1,2,3,4-thia(S^{IV})triazoles

S-Alkylated N,N-disubstituted thioamides react with azide ion at room temperature to form what are suggested to be trisubstituted 1,5-dihydro-1,2,3,4-thia(S^{IV})triazoles[70] [Eq. (26)].

$$\underset{R^3R^4N}{\overset{R^1}{\diagdown}}C-S-R^2 + N_3^- \longrightarrow \underset{R^3R^4N}{\overset{R^1}{\diagdown}}\underset{S(R^2)-N}{\overset{N=N}{\diagdown}} \quad (26)$$

An acyclic azide structure was rejected since the compounds show no azide IR absorption. A less likely three-membered ring was also considered. Comparison with other potential dihydro-S^{IV}-thiatriazoles was made. Simple α-azido thioethers show the spectroscopic and chemical properties of azides rather than of dihydro-S^{IV}-thiatriazoles. o-(Methylthio)phenyl azide was prepared but also showed the characteristics of an azide.[70] These results cast doubt on the suggested structure, since the only essential difference in composition is an amino group that is not in a position to stabilize the suggested heterocyclic ring. Structure $R^1(R^3R^4N)C=N-N=N-SR^2$ is an alternative, but further consideration must await X-ray crystallographic analysis.

Notes Added in Proof

Since this manuscript was submitted the crystal and molecular structures of 5-phenylthiatriazole (Section II, A), its 3-oxide (Section II, B, 2),

[70] S. I. Mathew and F. Stansfield, *J. Chem. Soc., Perkin Trans. I*, 540 (1974).

and its 3-ethyl derivatives (Section III, B, 1) have been determined by X-ray methods (T. Ottersen, *Acta Chem. Scand., in press*). Studies on the trapping of intermediates from 4-alkyl-5-sulfonyliminothiatriazolines (Section VIII, A) have also been reported [G. L'abbé, G. Verhelst, C.-C. Yu, and S. Toppet, *J. Org. Chem.* **40**, 1728 (1975)]. In addition, further studies on the reactions of alkyl azides with aryl isothiocyanates (Section VIII, A) have been carried out [G. L'abbé, E. Van Loock, G. Verhelst, and S. Toppet, *J. Heterocycl. Chem.* **12**, 607 (1975)].

ACKNOWLEDGMENTS

The author is indebted to Professors K. A. Jensen and M. G. Ettlinger for valuable discussions and suggestions.

The Nomenclature of Heterocycles

ALAN D. MCNAUGHT

The Chemical Society,
London, England

I. Introduction	176
A. The Function of Nomenclature	177
B. The Genesis of Present-Day Nomenclature Systems	178
C. Conventions	179
1. Terminations for Saturated and Unsaturated Organic Compounds	179
2. Position of Numerals	179
II. Nomenclature of Heterocyclic Skeletons	180
A. Description of Nomenclature Systems	181
1. Replacement Nomenclature ("a" Nomenclature; IUPAC Rule B-4)	181
2. Extended Hantzsch–Widman System for Monocycles (IUPAC Rule B-1)	182
3. Fusion Systems for Polycycles (IUPAC Rules A-21 to A-23 and B-3)	183
4. Bridging Nomenclature for Use with Fused Systems (IUPAC Rules A-34 and B-15)	191
5. Extended von Baeyer System for Bridged Structures (IUPAC Rules A-31, A-32, and B-14)	193
6. Spiro Systems (IUPAC Rules A-41, A-42, B-10, and B-11)	194
7. Ring Assemblies (IUPAC Rules A-51, A-52, A-54, and B-13)	196
B. Application of Nomenclature Systems	197
1. General Aspects	197
2. Specific Problems	205
C. Natural Products	219
1. Alkaloids	219
2. Porphyrins and Corrins	221
3. Steroids and Carbohydrates with Heterocyclic Modifications	221
4. Flavan and Derivatives	223
5. Nucleic Acid Bases	223
D. Other Nomenclature Systems	224
1. Cyclophanes	224
2. Rings Consisting Entirely of Heteroatoms	225
3. Boron and Silicon Heterocycles	227
III. Naming of Derivatives	227
A. General Principles	227
1. Nomenclature Types	227
2. The Principal Characteristic Group	230
3. Numbering	234
4. Ordering of Prefixes	235
5. Heterocycles as Substituents	236
6. Construction of a Substitutive Name	237

  B. Some Specific Topics 239
   1. Ketones and Imines 239
   2. Thio Compounds 240
   3. Dipolar (+, −) Compounds 240
   4. Stereochemistry 242
IV. Miscellaneous Examples 243
  Appendix: IUPAC Rules A-11 to A-56, B-1 to B-15, C-14, and C-15 . 247

I. Introduction

  For the practicing chemist, a working knowledge of present-day chemical nomenclature is indispensable. However, most chemists have neither the time nor the inclination to acquire the detailed knowledge of nomenclature rules that is essential for their correct application. To derive a unique systematic name for a complex chemical structure is an exercise requiring some intellectual agility; the rules to be applied are often highly involved, and their correct interpretation is not always apparent.

  This article is an attempt to provide a guide to the nomenclature of heterocyclic compounds. The various systems available will be described, and problems associated with their use will be discussed. Where appropriate, the advantages and disadvantages of particular systems will be mentioned. In the interests of precision and conciseness, nomenclature rules themselves are not framed in a readily digestible form; they are meant not for reading, but for reference. It is hoped that the following text will be more readily assimilable, and, in conjunction with the original rules (the most relevant of which are provided here as an Appendix), will enable the reader to name the majority of heterocyclic molecules. However, it is beyond the scope of the article to cover all possible heterocyclic structures; in cases where the text proves inadequate, reference to the rules themselves will be necessary.

  Before proceeding to a description of nomenclature systems, it will be useful to consider the function of a vocabular nomenclature in chemistry today. Then will follow a brief outline of the genesis of present-day nomenclature and some notes on conventions. The main body of the text consists of Sections II and III: Section II is devoted to the nomenclature of heterocyclic skeletons and is, as far as possible, a comprehensive survey; Section III deals with the naming of derivatives and is less detailed, only general principles being outlined. Then follow a few miscellaneous examples (Section IV) illustrating the use of the systems described. Reference is made throughout to the substantial extracts from the IUPAC Organic Nomenclature Rules, 1969, given in the Appendix.

A. The Function of Nomenclature

The systematic name of a compound is designed so that one may deduce from it the molecular structure of the compound, as indicated by its graphic formula. In other words, it is essentially a verbal substitute for the graphic formula and, in its most elaborate form, provides precisely the same structural information. However, it is a poor substitute: most chemists if required to indicate the structure of a compound would draw the formula rather than write the name. Similarly, a systematic name is not often readily understood until it has been translated into its graphic equivalent. Inevitably, systematic names tend to be long, and ill-adapted for use in conversation; they are useful only when to employ the graphic form would be impossible or inconvenient.

Nevertheless it is hard to envisage a situation in which the existence of a systematic vocabular nomenclature will not be of value. Such a nomenclature is still indispensable to indexers (although, with the availability of ciphering systems such as the Wiswesser notation,[1] this state of affairs may not be permanent) and to authors of papers and of books. To restrict a writer of a chemical text to using graphic formulas, without their verbal equivalents, would be to hinder severely his ability to communicate (and could make matters difficult for the printer). In writing about chemistry, systematic names should be considered as convenient substitutes for graphical structures; the transfer of information is most efficient when both types of structural designation are employed, the one complementing the other. One should resist the temptation to overwhelm a chemical text with long, cumbersome, systematic names; a phrase such as "the tricyclic enone" or "the hydroxy acid," with reference to a displayed graphic formula, is often all that is necessary.

Trivial names convey little or no structural information. However they are still widely used in the chemical literature, and it seems unlikely that they will ever be eliminated completely from it. Many such names are now very well established (indeed present-day systematic nomenclature is based to a considerable extent on a residue of trivial names), and they are usually brief and well suited to the spoken language. However, the extensive use of trivial nomenclature in the literature places a substantial burden on the memory of the reader, thus impairing an author's ability to communicate, and is not to be encouraged. The introduction of a new trivial name is rarely justified nowadays, and in writing about chemistry it is best to employ only the relatively small number of trivial names generally accepted at the international level.

[1] E. G. Smith, "The Wiswesser Line-Formula Chemical Notation." McGraw-Hill, New York, 1968.

B. The Genesis of Present-Day Nomenclature Systems

Only in recent years has international agreement been reached over a wide area of organic chemical nomenclature. However, the first step in this direction was the establishment of the so-called Geneva Rules[2] in 1892. These rules outlined only a few basic principles and were very limited in scope. They were not revised and extended until 1930, when the International Union of Chemistry published the Liège Rules.[3] Although representing a considerable advance, the Liège Rules were far from comprehensive, and served mainly as a basis for the development of present-day systems.

The principal advances in the systematization of organic nomenclature have come from the International Union of Pure and Applied Chemistry (IUPAC) Commission on the Nomenclature of Organic Chemistry, and from the Chemical Abstracts Service. The IUPAC Definitive Rules for Hydrocarbons and Heterocyclic Systems (1957)[4] and for Characteristic Groups (1965)[5] have been widely accepted by the chemical community, and, in their latest revised form,[6] constitute the standard reference work. These rules are closely related to those developed in parallel by Chemical Abstracts for indexing purposes, and it is fortunate that, as a result of close cooperation between the two bodies, there are few areas of disagreement.

The purpose for which the Chemical Abstracts system was devised is different from that for which the IUPAC Rules were intended. The former system is designed to provide a single, unambiguous, preferred index name for each compound, whereas the IUPAC Rules, devised for use in the chemical literature, frequently give alternatives. Chemical Abstracts names are in inverted form as required to provide a useful alphabetical index, whereas no such restriction applies to IUPAC names. Both systems make use of trivial names, most commonly as a basis for conducting systematic operations, although Chemical Abstracts is at present engaged in a severe pruning of its trivial vocabulary.

---

[2] A Pictet, *Arch. Sci. Phys. Nat.* [3] **27**, 485 (1892).
[3] A. M. Patterson, *J. Amer. Chem. Soc.* **55**, 3905 (1933).
[4] "IUPAC Nomenclature of Organic Chemistry," Definitive Rules for Section A. Hydrocarbons, Section B. Fundamental Heterocyclic Systems. Butterworths, London, 1958.
[5] "IUPAC Nomenclature of Organic Chemistry," Definitive Rules for Section C. Characteristic Groups Containing Carbon, Hydrogen, Oxygen, Nitrogen, Halogen, Sulfur, Selenium and/or Tellurium. Butterworths, London, 1965.
[6] "IUPAC Nomenclature of Organic Chemistry," Definitive Rules for Sections A and B (3rd ed.) and C (2nd ed.), Butterworths, London, 1971.

Sec. I.C] THE NOMENCLATURE OF HETEROCYCLES 179

The need for the general use of a largely systematic nomenclature of the type developed by IUPAC and by Chemical Abstracts is beyond question, otherwise communication among chemists will become increasingly difficult. However, it is unlikely that nomenclature will ever become completely systematic, or that a completely new, entirely systematic nomenclature will ever be acceptable. Nevertheless the future will probably involve increasing systematization and the gradual disappearance of many trivial names. It is to be hoped that before long IUPAC recommendations covering all areas of chemistry will be available, that they will prove acceptable to the chemical community at large, and that their application will become widespread.

C. Conventions

The conventions used in the IUPAC Rules with respect to spelling, position of numerals, italicization, etc., are those of Chemical Abstracts. Current British practice is different in some respects, and it is useful at this point to note two principal differences. British conventions will be used in the present article.

1. *Terminations for Saturated and Unsaturated Organic Compounds*

American practice is to use the terminations "-ane" for saturated and "-ene" for unsaturated compounds in both the hydrocarbon and the heterocyclic series. British usage is essentially the same for hydrocarbons, but for heterocyclic molecules the final "e" is omitted; in fact, a terminal "e" is used in general only for hydrocarbons, bases, and the five-membered ring ending "-ole." This leads in particular to some differences in the spelling of both trivial and systematic names for one-ring heterocycles: e.g., thiophen, oxiren, dioxaphospholan (British); thiophene, oxirene, dioxaphospholane (American).

2. *Position of Numerals*

It is normal British practice to place the numeral referring to the position of a molecular feature (i.e., its locant) immediately before the syllable it qualifies. American usage is more variable (although the same procedure is followed in cases where confusion or ambiguity might result): it is common to find the locant for a multiple bond or for the principal functional group at the beginning of the name (or immediately before the parent name). A few examples from the IUPAC Rules will make the difference in usage clear:

IUPAC
2(3*H*)-Pyrazinone
N-Ethyl-*N*-methyl-8-quinolinecarboxamide
4-Imino-2,5-cyclohexadien-1-one

British
Pyrazin-2(3*H*)-one
N-Ethyl-*N*-methylquinoline-8-carboxamide
4-Iminocyclohexa-2,5-dien-1-one

The British usage is less likely to lead to misunderstanding and is also more consistent. However, exceptions to it are made, especially in the case of contracted trivial names, where its application would result in lack of clarity or of euphony, e.g., 8-quinolyl, 2-pyridone, and 1-naphthoic acid are used rather than quinol-8-yl, pyrid-2-one, and naphth-1-oic acid.

II. Nomenclature of Heterocyclic Skeletons

In 1928, A. M. Patterson, later one of the authors of the Ring Index, wrote: "Any attempt to construct a strictly logical system of names for the large number of parent ring systems now known seems impractical, at least as far as common use is concerned."[7] This sentiment is no less applicable today. For heterocycles in particular, the number of trivial names in current use is large. In the 1969 IUPAC Rules,[6] 63 trivially named heterocyclic skeletons with various degrees of unsaturation are listed, with a further 25 in the 1973 Tentative Rules (Section D).[8] Chemical Abstracts nomenclature rules list 83 such skeletons with maximum unsaturation, many of which are not included by IUPAC. Thus, although systematic operations are often required to derive the name of a heterocyclic skeleton, the parent names to which these operations are applied are frequently trivial.

In this section the various nomenclature systems in common use will be described first, then their application and some particular problems

---

[7] A. M. Patterson, *J. Amer. Chem. Soc.* **50**, 3074 (1928).

[8] "IUPAC Nomenclature of Organic Chemistry," Tentative Rules for Section D. Organic Compounds Containing Elements Which Are Not Exclusively Carbon, Hydrogen, Oxygen, Nitrogen, Halogen, Sulfur, Selenium, and Tellurium, IUPAC Inform. Bull. No. 31, IUPAC, Oxford, 1973.

associated with their use. Finally the naming of natural products and some special nomenclature systems will be considered briefly. (Although the existence of virtually all the skeletons cited as examples is established, the states of hydrogenation illustrated are not necessarily those of any known derivatives.)

A. Description of Nomenclature Systems

The most recent recommendations by IUPAC for naming heterocyclic skeletons include only a small number of basic nomenclature systems, and, as a consequence of its need to provide unique names, the number of systems employed by Chemical Abstracts is even smaller. Of the systems now to be outlined, most are described in detail in the Appendix (to which reference is made where appropriate); the following text will cover only the general principles involved. Many of the systems are modifications of those used for carbocycles.

1. *Replacement Nomenclature ("a" nomenclature, IUPAC Rule B-4)*

A heterocycle can be derived formally from a carbocycle by replacement of one or more carbon atoms with heteroatoms. In a replacement name, this is indicated by use of prefixes terminating in "a," cited in the

TABLE I

Examples of Replacement Prefixes (in Descending Order of Priority)[a]

Element	Valence[b]	Prefix	Element	Valence	Prefix
O	II	Oxa	Sn	IV	Stanna
S	II	Thia	Pb	IV	Plumba
Se	II	Selena	B	III	Bora
Te	II	Tellura	Al	III	Alumina
N	III	Aza	Ga	III	Galla
P	III	Phospha	In	III	Inda
As	III	Arsa	Tl	III	Thalla
Sb	III	Stiba	Be	II	Berylla
Bi	III	Bismutha	Mg	II	Magnesa
Si	IV	Sila	Zn	II	Zinca
Ge	IV	Germa	Cd	II	Cadma
			Hg	II	Mercura

[a] This table is an extended version of that provided in the Appendix as Table I under Rule **B-1**.
[b] The prefixes given indicate replacement by the heteroatom in the prescribed valency state only.

order of their listing in Table I. This type of nomenclature can be applied as an extension of the various systems of carbocyclic nomenclature, as will be indicated later. Examples 1–6 will suffice for the present.

(1) Oxacyclopropane

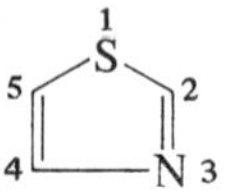

(2) 1-Thia-3-azacyclopenta-2,4-diene

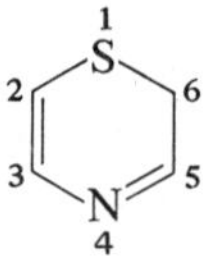

(3) 1-Thia-4-azacyclohexa-2,4-diene (not 1-thia-4-azabenzene)

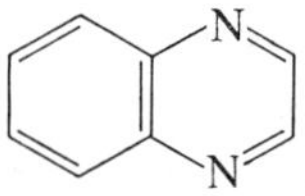

(4) 1,4-Diazanaphthalene

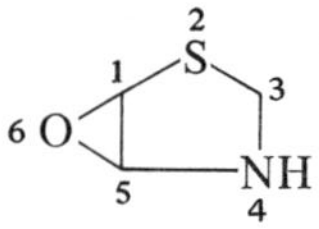

(5) 6-Oxa-2-thia-4-azabicyclo[3.1.0]hexane

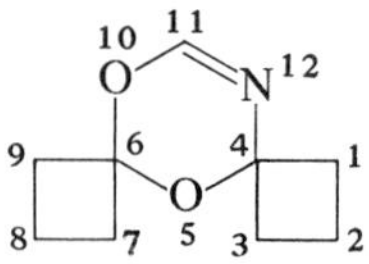

(6) 5,10-Dioxa-12-azadispiro[3.1.3.3]dodec-11-ene

2. *Extended Hantzsch–Widman System for Monocycles (IUPAC Rule B-1)*

This is a method for naming three- to ten-membered monocyclic compounds, of various degrees of unsaturation, containing one or more heteroatoms. It was originally devised for five- and six-membered rings;[9]

[9] A. Hantzsch and J. H. Weber, *Ber.* **20**, 3119 (1887); O. Widman, *J. Prakt. Chem.* [2] **38**, 185 (1888).

the extended version was adopted in the Ring Index[10] and in the IUPAC 1957 Rules.[4] Names are derived by combining the appropriate replacement prefix(es) (oxa-, aza-, etc., see Section II, A, 1) with a suffix denoting ring size and whether saturated or unsaturated (see Table II in IUPAC Rule B-1). Suffixes corresponding to partially saturated four- and five-membered rings are also available (Rule B-1.2). Examples 7–11 illustrate the use of the system for compounds having commonly used trivial names, though it is most frequently used for compounds lacking a trivial designation.

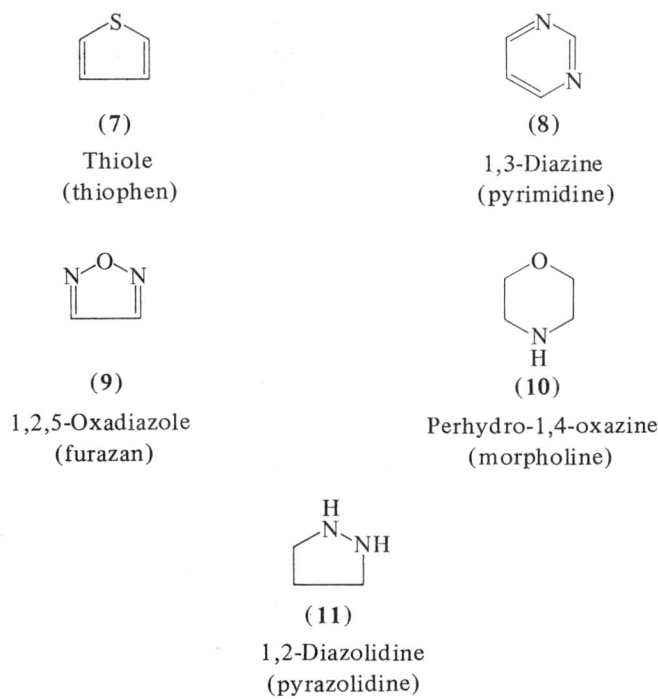

(7)
Thiole
(thiophen)

(8)
1,3-Diazine
(pyrimidine)

(9)
1,2,5-Oxadiazole
(furazan)

(10)
Perhydro-1,4-oxazine
(morpholine)

(11)
1,2-Diazolidine
(pyrazolidine)

3. *Fusion Systems for Polycycles (IUPAC Rules A-21 to A-23 and B-3)*

For derivation of a fusion name, the molecule is considered as a combination of two or more cyclic units by a "fusion" (or fusions) such that each component has at least one bond in common with another. All the components should normally have maximum unsaturation (i.e., contain the maximum number of noncumulative double bonds); names based on

[10] A. M. Patterson and L. T. Capell, "The Ring Index," 2nd ed., Amer. Chem. Soc. 1960.

wholly or partially saturated components present problems that will be discussed later. Names of components can be trivial or systematic.

Two types of fusion nomenclature are available; of these the older Stelzner system[11] is not recommended for use by IUPAC (except as in Rule B-23.5) or by Chemical Abstracts. Nevertheless, it will be described briefly here since it still appears commonly in the literature. The principles of both systems are best understood in application to polycarbocycles.

a. *Chemical Abstracts/IUPAC System for Carbocycles (Rules A-21 to A-23).* The components are trivially named except for monocycles other than benzene (see later). The largest possible unit with a trivial name is chosen as base component, and its peripheral bonds, starting with the 1,2-bond, are consecutively assigned italic letters in alphabetical order, by following the direction of the numbering system: thus in **12** the 1,2-bond is assigned letter *a*, the 2,3-bond letter *b*, the 3,3a-bond letter *c*, etc.

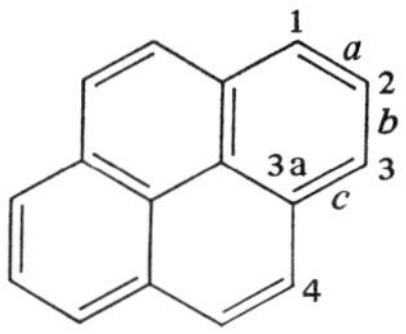

(12)

A component fused to the base component (referred to as a primary component) is cited as a prefix terminating usually in "o," with the locants of the bond engaged in fusion specified in square brackets in the order corresponding to the direction of lettering of the base component, as in example **13**. A (secondary) component fused to a primary component is cited as a further prefix, with the use of primes to indicate the

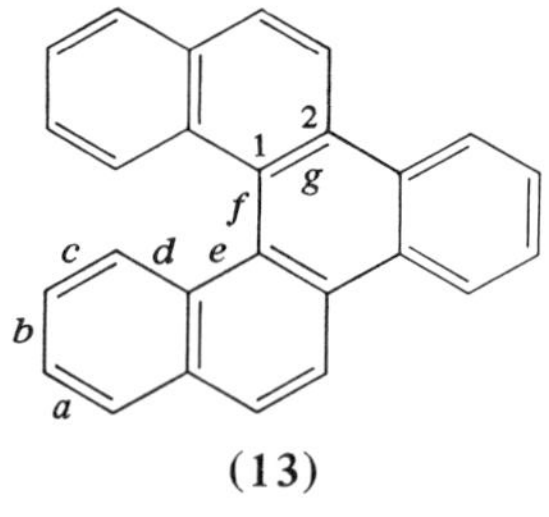

(13)

Naphtho[1,2-*g*]chrysene

[11] R. Stelzner, "Literatur-Register der organischen Chemie," Vol. III, 1914–1915, pp. 22–24. Springer-Verlag, Berlin, 1931.

bond engaged in fusion. The name thus derived is that of the system containing the maximum number of noncumulative double bonds, which is assigned a new numbering system not related to those of the components. Thus, **14a** is named as indicated and numbered as in **14b**.

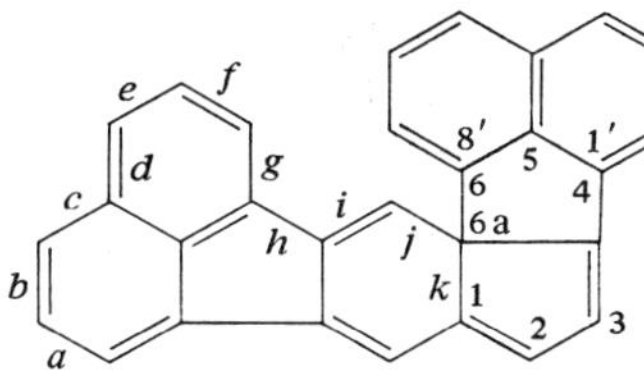

(14a) 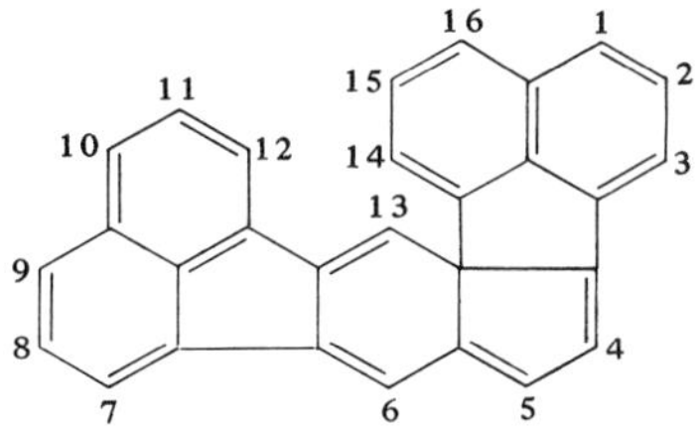   (14b)

Naphtho[1′,8′-4,5,6]pentaleno[1,6a-k]fluoranthene

For monocyclic components other than benzene the nomenclature is based on that of the corresponding cycloalkanes. Prefixes of the type cyclopenta-, cyclohepta-, etc., are used, and for names of monocyclic base components the termination "-ene" is employed, this termination indicating maximum unsaturation in the *whole* system (not just a single double bond in the cycloalkene unit). Thus **15** and **16** are named as shown.

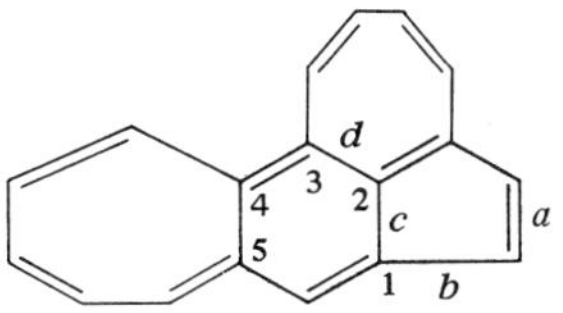

(15)

Cyclohepta[4,5]benz[1,2,3-cd]azulene

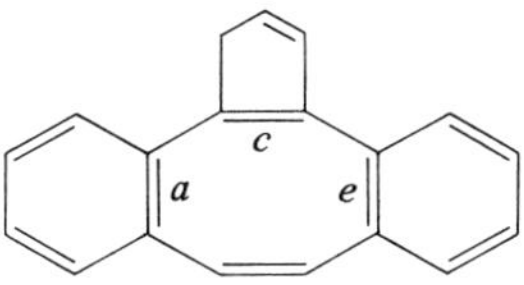

(16)

1H-Dibenzo[a,e]cyclopenta[c]cyclo-octene

b. *Stelzner System*[11] *for Carbocycles*. This system differs from the IUPAC/Chemical Abstracts system in three principal respects: (i) the name derived refers to a molecule in which the bond distribution in the components is retained (i.e., the skeleton named is not necessarily that containing the maximum number of noncumulative double bonds); (ii) the bonds engaged in fusion are indicated by numbers in parentheses (no letters are used), those of the base component being unprimed, those of primary components carrying a single prime, etc.; (iii) the whole molecule retains the numbering of the individual components. These differences are well exemplified by the fusion of two indene units: structure **17** is named indeno[1,2-*a*]indene by the IUPAC system, whereas the structure corresponding to the Stelzner name indeno(2′,3′:2,3)indene is **18**. The IUPAC name for the latter would be 9,10-dihydroindeno[1,2-*a*]indene.

(17)

(18)

(IUPAC numbering illustrated)

The disadvantages of the Stelzner system, as applied to carbocycles, are not very serious. The original system did not indicate how to deal with cases of fusion involving two adjacent bonds in each component, but is capable of logical extension to cover such cases. However, the IUPAC/Chemical Abstracts system is simpler in that it *always* names the form of the molecule with maximum unsaturation and yields a uniform numbering system for the skeleton; it is also more able to deal with complex cases.

In applying the two systems to heterocyclic compounds, the disadvantages of the Stelzner system become more apparent, as will become clear from the following (see below, Section c, ii).

c. *Application of Fusion Nomenclature to Heterocycles.* Methods of applying the fusion systems in the heterocyclic series are of two types.

i. *By use of replacement prefixes.* Fused heterocycles can be named by applying replacement nomenclature (see Section II, A, 1) to the corresponding carbocycles. Two procedures exist for doing this: the Stelzner method[12] (not to be confused with the Stelzner fusion system;[11] see above) and the Chemical Abstracts method. Both were described in the IUPAC 1957 Rules, but the Stelzner method was discarded in the 1969 revised version.

[12] R. Stelzner, "Literatur-Register der organischen Chemie," Vol. V, pp. ix–xv, 1926.

Sec. II.A]   THE NOMENCLATURE OF HETEROCYCLES   187

CHEMICAL ABSTRACTS METHOD (IUPAC RULE B-4.2). This is a straightforward application of replacement principles to the Chemical Abstracts fusion name of the corresponding carbocycle. The resulting name is that of the heterocycle containing the maximum number of noncumulative double bonds (which may be different from the number of such bonds in the carbocycle). The numbering system of the carbocycle is retained. For examples, see **19** and **20**.

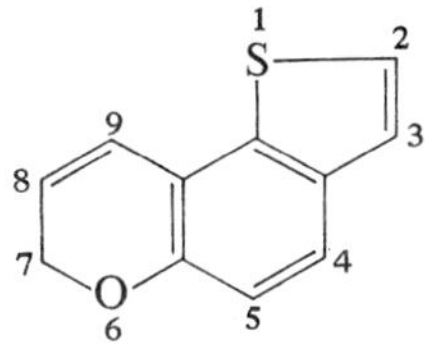

(19)

7*H*-6-Oxa-1-thiacyclopenta[*a*]naphthalene

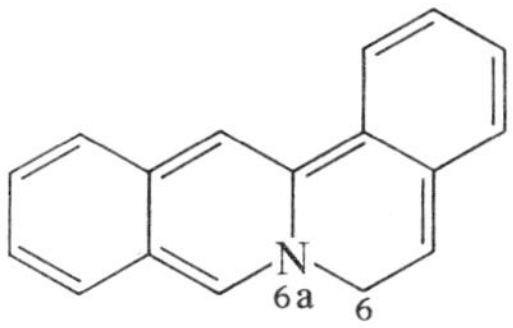

(20)

6*H*-6a-Azabenz[*a*]anthracene

STELZNER METHOD. This is a general method of applying replacement nomenclature to ring systems; it differs from the Chemical Abstracts procedure in that replacement principles are applied in all cases to the name of the hydrocarbon with the same bond distribution in the rings as the heterocycle to be named. This leads to no difference for monocycles (see examples **1–3**), but in the case of fused skeletons the "parent" hydrocarbon of a fully unsaturated heterocycle is frequently a partially hydrogenated molecule. The parent hydrocarbon name for application of either replacement method can be trivial or a name derived by fusion principles (as above). The examples **21–24** of both

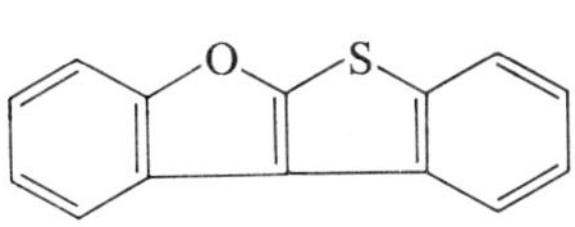

(21)

(S) 9,10-Dihydro-9-oxa-10-thiaindeno[1,2-*a*]indene
(CA) 9-Oxa-10-thiaindeno[1,2-*a*]indene
(SF) 1-Oxa-1'-thiaindeno(2',3':2,3)indene

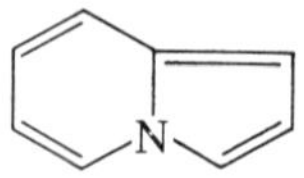

(22)

(S) 3a-Aza-3a*H*-indene
(CA) 3a-Azaindene

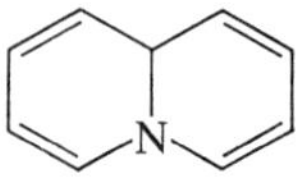

(23)

(S) 4a,8a-Dihydro-4a-azanaphthalene
(CA) 8a*H*-4a-Azanaphthalene

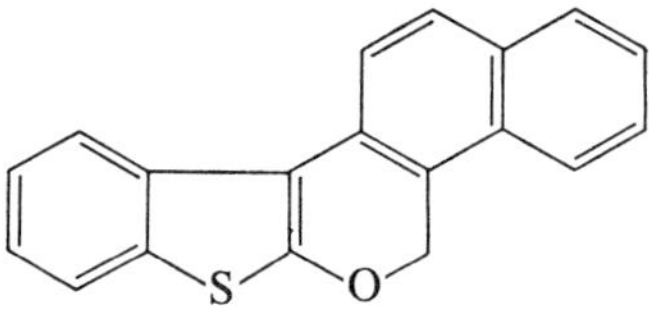

(24)

(S) 5,6-Dihydro-6-oxa-7-thiaindeno[1,2-*a*]phenanthrene
(CA) 6-Oxa-7-thiaindeno[1,2-*a*]phenanthrene
(SF) 3,4-Dihydro-3-oxa-1′-thiaindeno(3′,2′:1,2)phenanthrene

Stelzner (S) and Chemical Abstracts (CA) types of replacement name for the same molecules will make the difference clear.

As will be apparent, the (S) names for **21** and **24** are obtained by applying Stelzner replacement principles to Chemical Abstracts fusion names for the parent carbocycles. However, the Stelzner principles were originally intended to be applied to Stelzner fusion (SF) names, as shown in **21** and **24**.

Thus in many cases Stelzner replacement names require "hydro-" prefixes where the heterocyclic skeleton is in fact maximally unsaturated. This is a notable disadvantage, which often causes confusion, and the method is not recommended.

ii. *By use of heterocyclic components*

IUPAC/CHEMICAL ABSTRACTS SYSTEM (IUPAC RULE B-3). The fusion operations described in Section II, A, 3, a are applied to trivial and/or Hantzsch–Widman names of heterocyclic (and, if present, carbocyclic) components; see examples **25–27**. In addition a unit consisting of a benzene ring fused to a one-ring heterocycle can be considered as a single component. If this one-ring heterocycle contains 3–10 members it

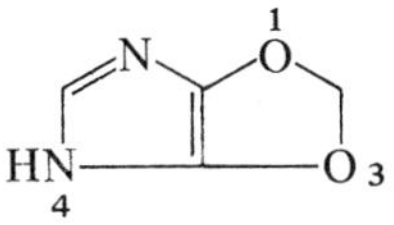

Sec. II.A]   THE NOMENCLATURE OF HETEROCYCLES   189

(25)
4*H*-1,3-Dioxolo[4,5-*d*]imidazole

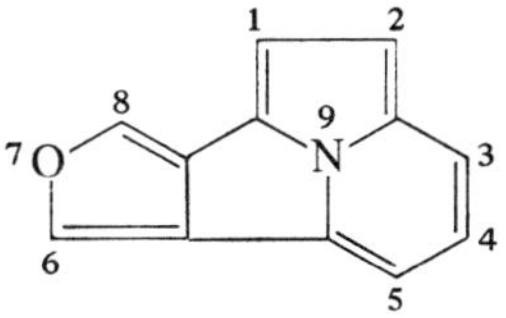

(26)
Furo[3,4-*a*]pyrrolo[2,1,5-*cd*]indolizine

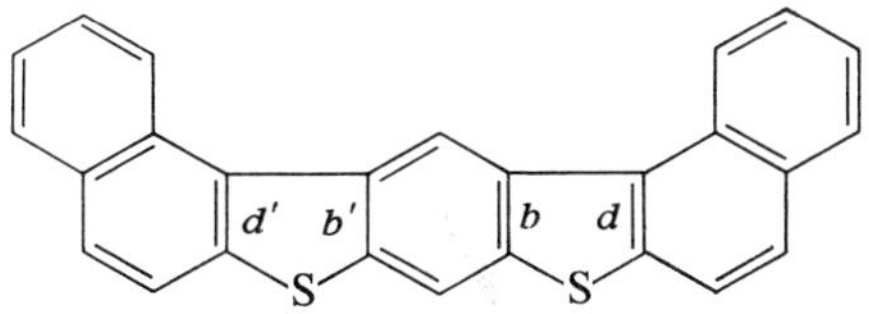

(27)
Dinaphtho[1,2-*d*:1′,2′-*d*′]benzo[1,2-*b*:5,4-*b*′]dithiophen

is given a trivial or a Hantzsch–Widman name, and the whole component is named as a "benzoheterocycle," with, if necessary, the locants of the heteroatoms (not the locants of the fusion bond) cited at the beginning, as in examples **28** and **29**. If the one-ring heterocycle contains

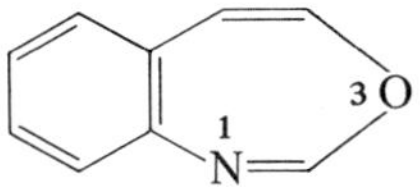

(28)
3,1-Benzoxazepine

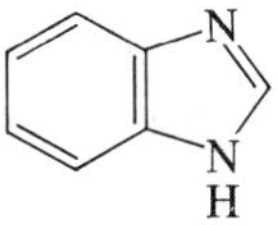

(29)
Benzimidazole

more than 10 members, it is given a replacement name with the termination "-in" (or "-ine" if nitrogenous) to indicate maximum unsaturation, as in **30** and **31**. This extension to rings of more than 10 members, which

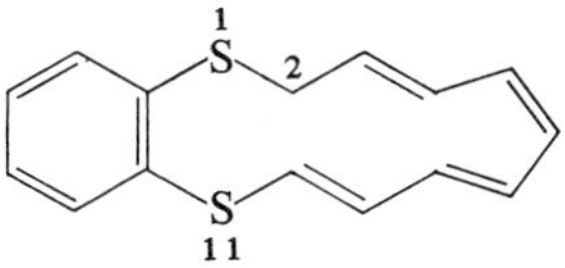

(30)
2*H*-1,11-Benzodithiacyclotridecin

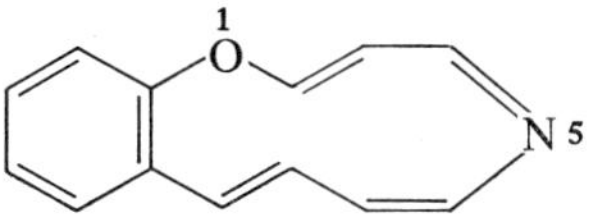

(31)
1,5-Benzoxazacycloundecine

can be applied generally in fusion nomenclature, is not described by IUPAC, but it is used by Chemical Abstracts and is very convenient. The IUPAC Rules do not indicate how to deal with rings of this size in fusion names. For examples, see **32** and **33**.

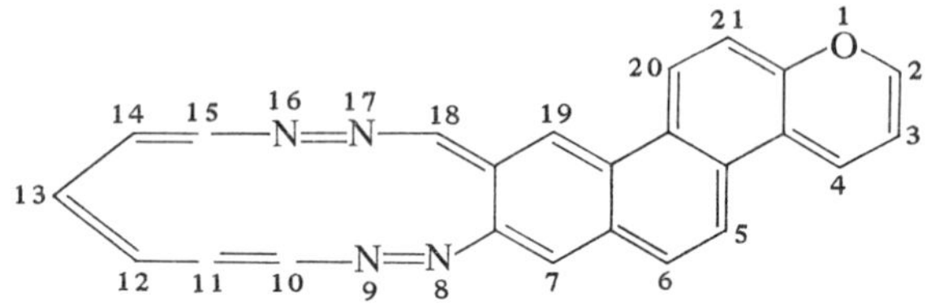

(32)

Pyrano[2′,3′:7,8]phenanthro[2,3-c][1,2,6,7]tetra-azacyclotridecine

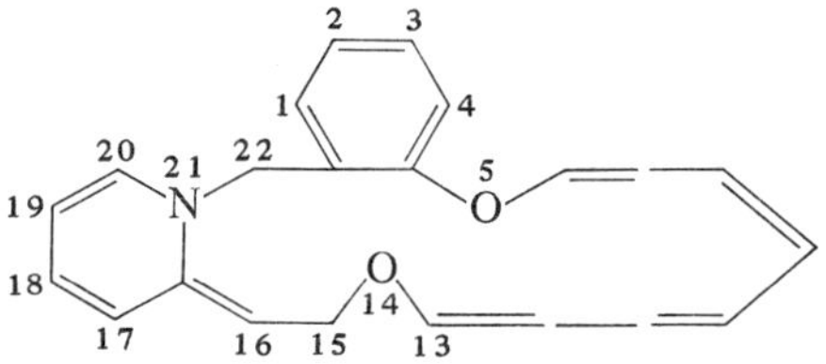

(33)

15H,22H-Pyrido[2,1-m][1,10,14]benzodioxazacycloheptadecine

STELZNER FUSION SYSTEM. Stelzner fusion principles can be applied to heterocyclic components. However, strict application of the system is not always possible, and problems associated with deciding upon the degree of hydrogenation of the parent fused molecule often arise. For example, **34** and **35** are named thieno[3,2-a]indolizine and thieno-[2,3-b]indole by the IUPAC system. The Stelzner name for **35** would be thieno(2′,3′:2,3)indole, but **34** cannot be similarly named, since in no

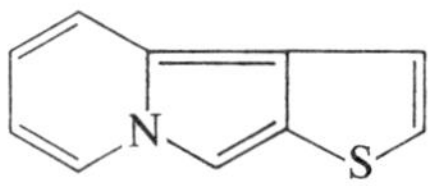

(34) 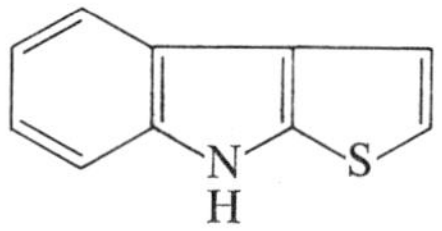   (35)

Kekulé form does the thiophen ring contain two double bonds. To refer to the compound as 2′,3′-dihydrothieno(3′,2′:1,2)indolizine is obviously unsatisfactory. Thus although for relatively simple molecules the Stelzner system is adequate, it would require modification to cover the more complex cases. Since the IUPAC system is capable of straightforward application to a much wider range of compounds, there seems

Sec. II.A] THE NOMENCLATURE OF HETEROCYCLES 191

to be no good reason for retaining the Stelzner system, which will therefore not be discussed further.

4. Bridging Nomenclature for Use with Fused Systems (IUPAC Rules A-34 and B-15)

Two methods are available for naming bridged cyclic skeletons. One (the von Baeyer system; see the following section, 5) is particularly suitable for naming predominantly saturated skeletons; the other, to be described now, is appropriate for predominantly unsaturated skeletons and involves citing the bridge as a prefix to a fusion name.

There are two convenient ways of citing such bridges:

Method a. The bridge is considered as a normal divalent radical substituent, e.g., ethylene ($-CH_2-CH_2-$), *o*-phenylene (*o*-C_6H_4), retaining its own numbering system (but with primes and starting at the lower-numbered bridgehead), e.g., **36a** and **37a**.

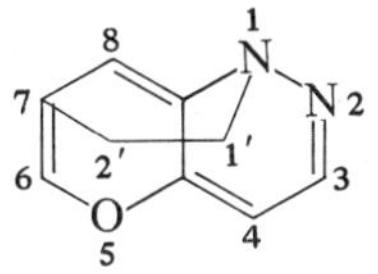

(36a)

1,7-Ethylene-1*H*-pyrano[3,2-*c*]pyridazine

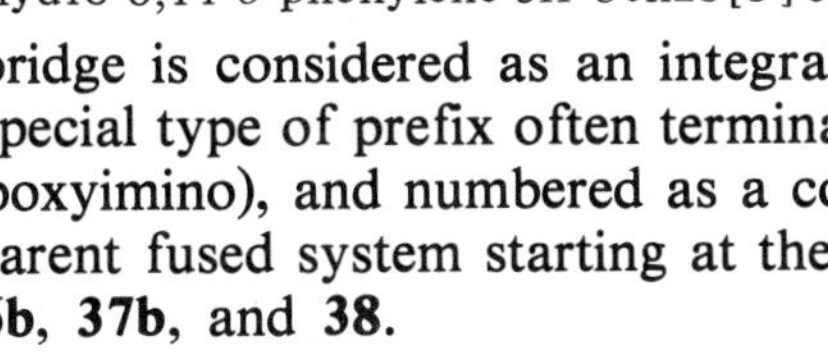

(37a)

6,11-Dihydro-6,11-*o*-phenylene-5*H*-benzo[*b*]carbazole

Method b. The bridge is considered as an integral part of the ring system, cited as a special type of prefix often terminating in "-o" (e.g., ethano, benzeno, epoxyimino), and numbered as a continuation of the numbering of the parent fused system starting at the higher-numbered bridgehead, e.g., **36b**, **37b**, and **38**.

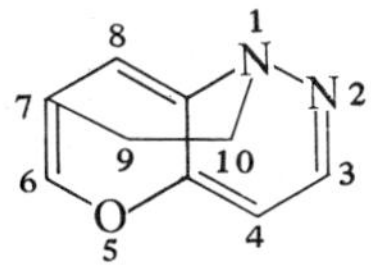

(36b)

1,7-Ethano-1*H*-pyrano[3,2-*c*]pyridazine

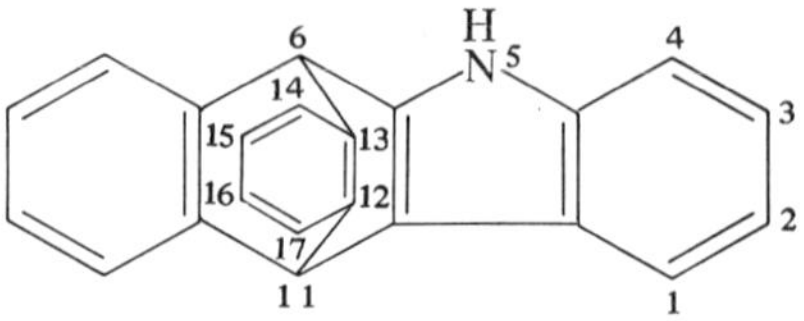

(37b)

6,11-Dihydro-*o*-benzeno-5*H*-benzo[*b*]carbazole

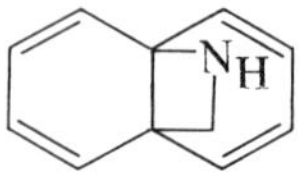

(38)

8a,4a-(Iminomethano)naphthalene

Method a is in several ways the easier to apply (it uses well known radical prefixes and leads to no problems in deciding on the degree of hydrogenation of the ring system), and was given as an alternative to method b in the 1957 IUPAC Rules. However, it was removed from the 1969 version and is not used by Chemical Abstracts. Method b has the advantage that the polycyclic system is considered as a whole; its use is more general than that of method a, and it seems advisable for method a to be abandoned.

Chemical Abstracts has developed the use of method b to quite a high level of sophistication. Trivalent (e.g., metheno) and tetravalent units and many ring systems can be cited as bridging groups, e.g., **39–41**.

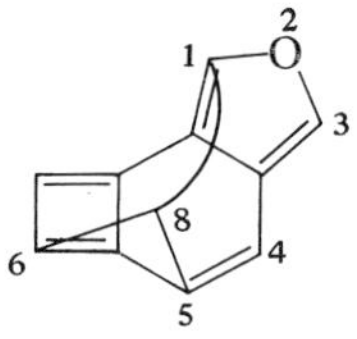

(39)

1,5,6-Methenocyclobut[*e*]isobenzofuran

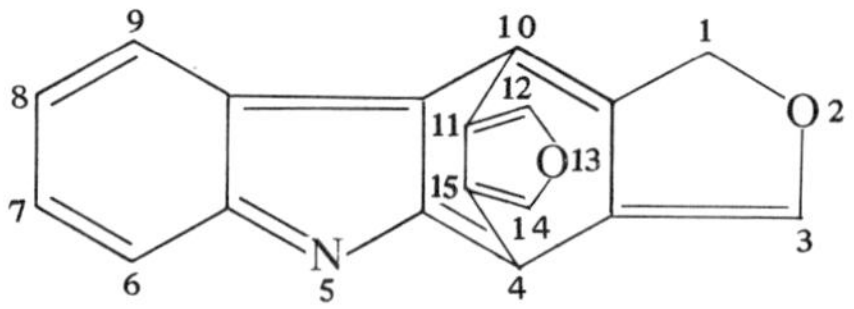

(40)

4,10[3',4']-Furano-1*H*-furo[3,4-*b*]carbazole

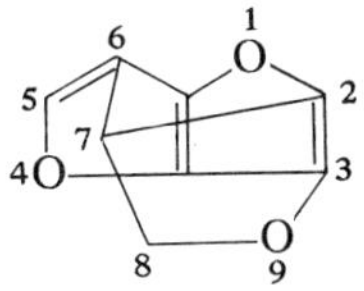

Sec. II.A] THE NOMENCLATURE OF HETEROCYCLES 193

(41)

3,2,6-(Epoxyethanylylidene)furo[3,2-b]furan

Bridging nomenclature should not be used when addition of the bridge is equivalent to adding a further fused ring, e.g., **42** is named 1,2,3,4-tetrahydrobenzo[a]phenazine, not 1,2-butanophenazine.

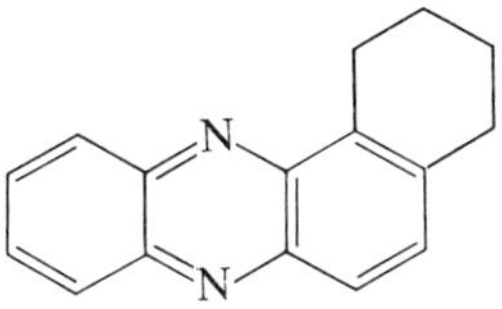

(42)

5. Extended von Baeyer System for Bridged Structures (IUPAC Rules A-31, A-32, and B-14)

The original von Baeyer system[13] provided a means of naming saturated bicyclic bridged hydrocarbons and has been extended to cover skeletons containing any number of rings (this number being defined as the minimum number of bond scissions required to convert the skeleton into a monocyclic hydrocarbon). The name obtained in the bicyclic case consists of the prefix bicyclo-, followed in square brackets by the number of carbon atoms separating the bridgeheads by the three possible routes, in descending numerical order, followed in turn by the name of the alkane containing the same number of carbon atoms as the whole bicyclic skeleton, e.g. **43**.

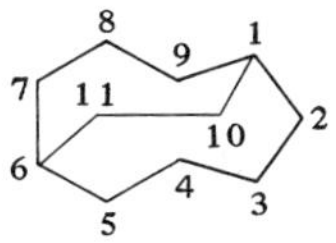

(43)

Bicyclo[4.3.2]undecane

For skeletons containing three or more rings (tricyclo-, tetracyclo-, etc.), superscripts are used to specify the positions of the additional bridges, e.g., **44**.

[13] A. von Baeyer, *Ber.* **33**, 3771 (1900).

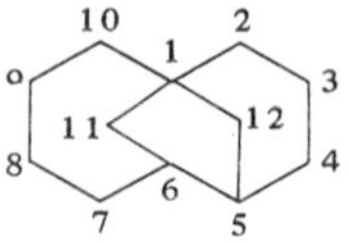

(44)

Tricyclo[4.4.1.1^{1,5}]dodecane

Names of heterocycles of this type are readily derived by straightforward application of replacement principles (see Section II, A, 1), e.g., **45** and **46**.

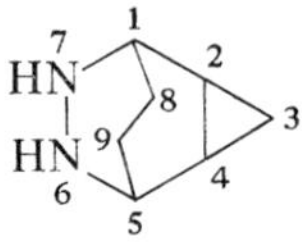

(45)

6,7-Diazatricyclo[3.2.2.0^{2,4}]nonane

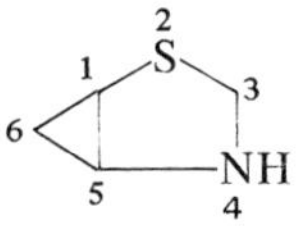

(46)

2-Thia-4-azabicyclo[3.1.0]hexane

6. Spiro Systems (IUPAC Rules A-41, A-42, B-10, and B-11)

Two systems for naming spiro compounds are available. Both are described, without a statement of preference, in the IUPAC 1969 Rules; however only one (method a) is used by Chemical Abstracts.

Method a. Two types of name are used.

i. For saturated hydrocarbons containing a single spiro linkage, names are of the type "spiro[x.y]alkane," where x and y are the numbers, in ascending order, of carbon atoms other than the spiro atom in each ring, e.g., **47**. Heterocyclic analogues are named by replacement principles (see Section II, A, 1), e.g., **48**.

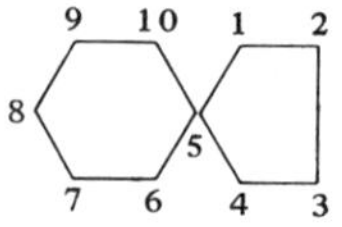

(47) (48)

Spiro[4.5]decane   5-Oxa-8-thiaspiro[3.4]octane

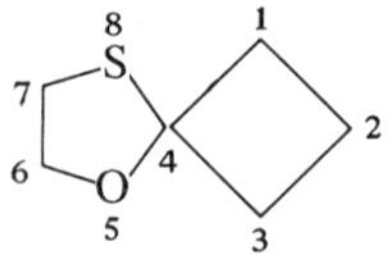

Sec. II.A]   THE NOMENCLATURE OF HETEROCYCLES   195

ii. When one or both components of the spiro compound are fused or bridged molecules, the names of the individual components are cited (in square brackets and in alphabetical order) with the prefix "spiro" and separated by the locants of the spiro linkage. The components retain their original numberings (with primes for the component cited second), in contrast to the saturated hydrocarbon case i, where the skeleton is numbered as a single unit. The components can be carbocyclic or heterocyclic, e.g., **49** and **50**.

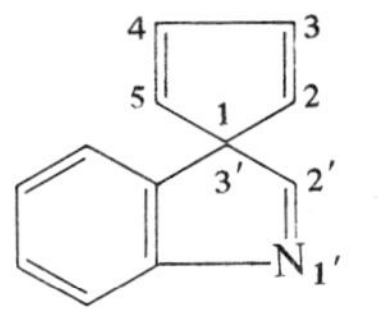

(49)

Spiro[cyclopenta-2,4-diene-1,3'-[3H]indole]

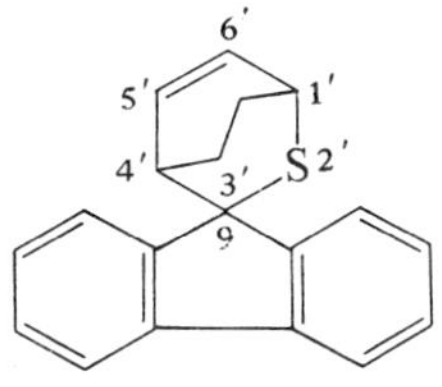

(50)

Spiro[fluorene-9,3'-[2]thiabicyclo[2.2.2]oct[5]ene]

Heterocyclic components can be given replacement names; e.g., **49** would then be named spiro-[3H-1-azaindene-3,1'-cyclopenta[2,4]diene]. Both types of method a spiro nomenclature are readily extensible to molecules containing more than one spiro linkage, e.g., **51** and **52**.

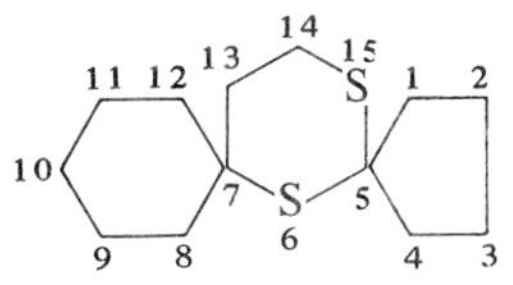

(51)

6,15-Dithiadispiro[4.1.5.3]pentadecane

(52)

Trispiro[cyclopropane-1,2' : 2,2" : 3,2'"-tris[1,3]benzodioxole]

Method b. In this method only one type of name is used, in contrast to method a. For a molecule containing one spiro linkage the names of the components (with the larger component cited first) are separated by the infix "spiro," with appropriate locants before and after, those of the second component being primed. As in method a, ii, the components retain their original numbering in the complete skeleton, and the components may be carbocyclic or heterocyclic. Names derived in this way are 1,3-oxathiolan-2-spirocyclobutane (or 1-oxa-3-thiacyclopentane-2-spirocyclobutane) (**48**), 3*H*-indole-3-spiro-1'-cyclopenta-2',4'-diene (**49**), and cyclohexanespiro-4'-[1,3]dithian-2'-spirocyclopentane (**51**).

Each of methods a and b has its own advantages. For example, method a immediately indicates the presence of a spiro linkage by citing "spiro" as a prefix, and yields a uniform numbering system for the simpler saturated molecules; method b applies a single type of name to all spiro compounds and always cites the larger component first. However, although both types of name appear extensively in the literature, there is no good reason for continuing to use both methods, and since method a is the one employed by Chemical Abstracts, it is to be preferred.

7. *Ring Assemblies (IUPAC Rules A-51, A-52, A-54, and B-13)*

A special nomenclature exists for assemblies of identical cyclic units linked by single or double bonds in such a way that no new rings are created. Names are derived by combining the appropriate multiplying prefix (bi-, ter-, quater-, etc.) and either (a) the name of the cyclic unit or (b) that of the derived radical, with locants, primed where necessary, before the name indicating the positions of linkages, e.g., **53** and **54**.

Chemical Abstracts uses method a for most cases except polyphenyls and bi(cycloalkyl)s. Method b names are the more readily applied to molecules containing double-bond internuclear linkages, but on the

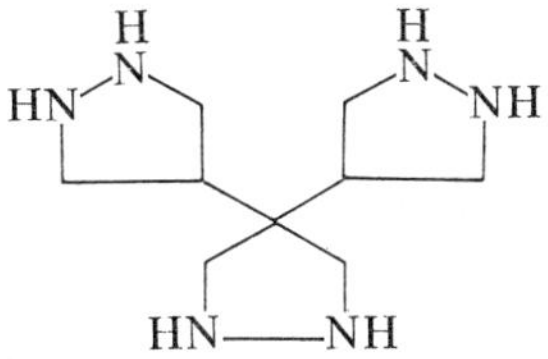

(53)

(a) 4,4′:4′,4″-Terpyrazolidine (or 3,3′,3″,4,4′,4″-hexa-aza-1,1′:1′1″-tercyclopentane)
(b) 4,4′ : 4′,4″-Terpyrazolidinyl

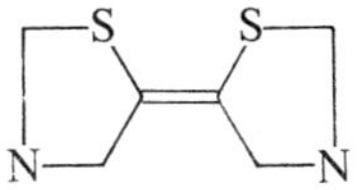

(54)

(a) $\Delta^{5,5'}$-Bithiazolidine
(b) 5,5′-Bithiazolidinylidene

whole it seems better to employ method a, thus reserving the -yl termination for radical substituents.

B. Application of Nomenclature Systems

As will be apparent, the foregoing descriptions of the systems available give only a broad outline of the operations involved. For each system, derivation of a unique name and numbering for a particular skeleton requires use of the full text of the appropriate rules provided in the Appendix. Construction of the name is governed by lists of priorities, the application of which is illustrated in many of the examples. However some general aspects of the use of the systems will be considered here, in particular their applicability in various contexts, and a number of specific problems.

1. *General Aspects*

a. *Trivial Names.* The use of many of the systems described in the preceding section involves conducting systematic operations with trivial names. Both IUPAC and Chemical Abstracts have attempted to limit the number of such names which can be treated in this way, and their lists of allowed trivial names show a large measure of agreement. Most of the retained trivial names are those of molecules having maximum unsaturation; use of trivially named partially or completely saturated units in systematic operations is desirable in only a few specific cases. Those trivial names that are retained by IUPAC for wholly or partially unsaturated carbocycles are listed in Rules A-21.2 and A-23.1, and those for heterocycles are in Rule B-2. (An extended version of the list in Rule B-2 is provided in the 1973 Section D Tentative Rules.[8]) Of the nomenclature systems 1–7 (above), only systems 2 (Hantzsch–Widman), 5 (von Baeyer), and 6a, i (spiro for simple saturated skeletons) are essentially independent of trivial nomenclature; all the others rely to a considerable extent on trivial names as a basis for operations.

b. *Applicability of Replacement Nomenclature.* Since, in principle, the name of *any* heterocycle can be derived from that of the corresponding carbocycle by use of replacement principles, the replacement system is extremely flexible. However, it has not hitherto been widely applied in the literature, preference having been given frequently to names based on trivial heterocycle names. For fused skeletons in particular the most extensively used names are those formed by fusion operations from Hantzsch–Widman names and the trivial names listed in IUPAC Rule B-2, e.g., **25–27**; for these the replacement names would be 1,3-dioxa-4,6-diazapentalene (**25**), 7-oxa-8c-azapentaleno[1,2,3-*cd*]indene (**26**), and 7,9-dithiadinaphtho[1,2-*a*:2′,1′-*h*]-*s*-indacene (**27**). The principal reason for this is probably the fact that both the Ring Index[10] and Chemical Abstracts prefer this procedure; in fact Chemical Abstracts uses replacement names for fused skeletons only when the trivially based system cannot be applied (for example, when two or more components are fused to each other as well as to the base component, e.g., **55**, which

(55)

6*H*-1,7-Dioxacyclopent[*cd*]indene

cannot be named as a cyclopentafuropyran). The existence of two ways of applying replacement nomenclature to fused skeletons (Chemical Abstracts and Stelzner methods) may also have contributed to the unpopularity of replacement names, since a knowledge of which method was being used was often necessary for correct interpretation of a name.

The main advantages of fusion names based on heterocyclic components would appear to be the clear indication given by such names of the natures of the individual rings involved (e.g. pyridopyrimidine vs.

triazaindene) and convenience for indexing (the system gives a wider range of index headings, and these headings are more informative, e.g., a search under "pyrido" will yield many skeletons containing pyridine rings). However, it is becoming increasingly clear that these advantages do not compensate sufficiently for the limited applicability of such names to complex skeletons, and the virtues of the replacement system (with the Stelzner method of application abandoned, as now recommended by IUPAC) make it distinctly more attractive. Admittedly, the replacement system as now described by IUPAC will require some elaboration to make it universally applicable and to yield a unique name in every case. However, it is essentially simple and easy to use. For example, for fused skeletons the only trivial names it is necessary to know are those of the 35 carbocycles listed by IUPAC (Rule A-21.2). Indeed for monocycles too the replacement system is entirely adequate, and it would be possible, if desired, to abandon the Hantzsch–Widman system completely.

As long as Chemical Abstracts continues to limit severely the use of replacement nomenclature for heterocycles, it would not be helpful to recommend that replacement principles be used universally in preference to other systems. However, a recent publication[14] by the American Chemical Society does indeed advocate the replacement system as the preferred way of naming all heterocycles, and this may well be a pointer toward a general change in attitude. It is to be hoped that authors and editors alike will appreciate the advantages of replacement principles and will, by increasing their use in the literature, encourage such a trend. In the remainder of this article, replacement names will be given to examples as alternatives wherever appropriate.

In using replacement nomenclature it is imperative to apply replacement prefixes *only* to a carbocyclic parent name, *never* to a heterocyclic

(56)

$1H$-Triazolo[4,5-d]pyrimidine   } correct
$1H$-1,2,3,4,6-Penta-azaindene

$1H$-2,3,4,6-Tetra-azaindole
$1H$-1,3,4,6-Tetra-azaisoindole
$1H$-3,4,6-Triazaindazole } incorrect
8-Azapurine (purine is numbered nonsystematically)

[14] "Nomenclature of Organic Compounds" (J. H. Fletcher, O. C. Dermer, and R. B. Fox, eds.), Amer. Chem. Soc. Advan. in Chem. Ser., No. 126, 1974.

parent. Thus names of the type "5-azaindole" are inadmissible: to allow such usage would lead to the possibility of a multiplicity of replacement names for molecules containing several heteroatoms (see **56**).

c. *Monocycles.* The naming of one-ring heterocycles presents relatively few problems. Three types of name are available: (i) trivial, (ii) Hantzsch–Widman, and (iii) replacement. Replacement names are applicable to all cases, whereas the Hantzsch–Widman system covers only rings with 3–10 members. Only a few trivially named molecules (14 unsaturated, 9 partially and fully saturated) are listed as allowed by IUPAC, and alternative Hantzsch–Widman names can be derived for all these (see **7–11**).

For replacement names the parent carbocycles, apart from benzene, are all named as cycloalkanes (modified to -ene, -diene, etc., as necessary) with the heteroatom replacing $>$CH$_2$, $=$CH, or $=$C$-$ as appropriate to its valency. Replacement names derived from benzene are not used unless the three formal double bonds are retained; otherwise names of the type oxacyclohexa-2,4-diene are appropriate [see **1–3** and **57–60**; replacement names for **7–11** are thiacyclopenta-2,4-diene (**7**), 1,3-diazabenzene (**8**), 1-oxa-2,5-diazacyclopenta-2,4-diene (**9**), 1-oxa-4-azacyclohexane (**10**), and 1,2-diazacyclopent-3-ene (**11**)].

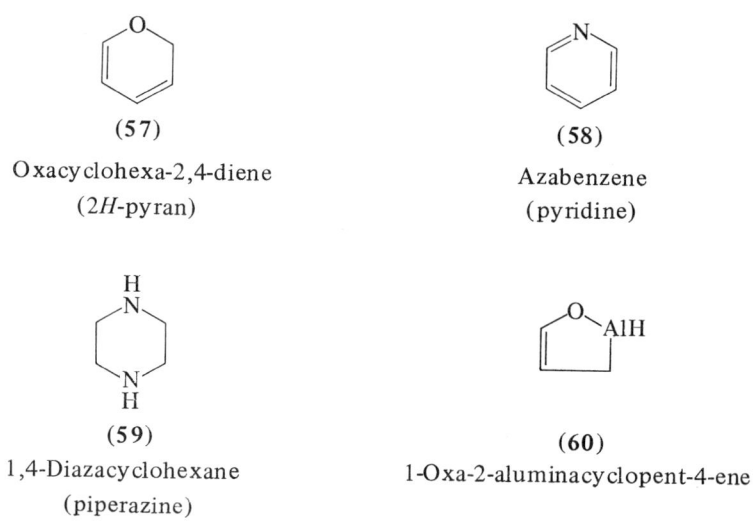

(**57**)
Oxacyclohexa-2,4-diene
(2*H*-pyran)

(**58**)
Azabenzene
(pyridine)

(**59**)
1,4-Diazacyclohexane
(piperazine)

(**60**)
1-Oxa-2-aluminacyclopent-4-ene

The Hantzsch–Widman system provides names for both fully saturated and fully unsaturated molecules. Partially saturated molecules are named by attaching appropriate "hydro" prefixes to the names of the fully saturated rings; however the special suffixes applicable to four- and five-membered rings may also be used (this is not now recommended by Chemical Abstracts). Examples of partially saturated rings are 2,3-dihydro-1,2-oxaluminole (or Δ⁴-1,2-oxaluminolen) (**60**), and **61** and **62**.

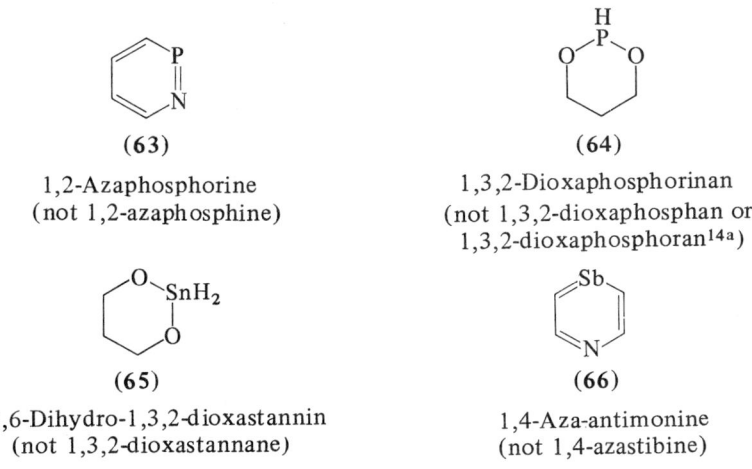

(**61**)

1,2-Dihydro-1,3-diazete
(or 1,3-diazetine)

(**62**)

3,6,7,8-Tetrahydro-2H-1,4-oxazocine

Some complications arise with certain heteroatoms when strict application of the Hantzsch–Widman system could lead to confusion; in these cases prefixes or suffixes are suitably modified, e.g., **63–66**.

(**63**)

1,2-Azaphosphorine
(not 1,2-azaphosphine)

(**64**)

1,3,2-Dioxaphosphorinan
(not 1,3,2-dioxaphosphan or
1,3,2-dioxaphosphoran[14a])

(**65**)

5,6-Dihydro-1,3,2-dioxastannin
(not 1,3,2-dioxastannane)

(**66**)

1,4-Aza-antimonine
(not 1,4-azastibine)

Replacement names: 1-aza-2-phosphabenzene (**63**)
1,3-dioxa-2-phosphacyclohexane (**64**)
1,3-dioxa-2-stannacyclohexane (**65**)
1-aza-4-stibabenzene (**66**)

No such complications arise with replacement nomenclature.

d. *Polycycles.* It is desirable for the parent name of a carbocyclic or heterocyclic skeleton containing more than one ring to be either (I) that of a fully unsaturated molecule (which can be modified by appropriate "hydro" prefixes in naming a partially saturated molecule) or (II) that of a fully saturated molecule (which can be modified by changing the ter-

---

[14a] The American version (as in the IUPAC Rules) carries a terminal "e." The confusion with phosphorane (PH₅) is not so apparent when, as here, the terminal "e" is omitted (British convention, see Section I, C, 1).

mination "-ane" to "-ene," "-diene," etc.). Systems that do not meet one or other of these qualifications (e.g., the Stelzner fusion and replacement systems) are in general unsatisfactory since their use frequently gives rise to problems in defining precisely what the degree of hydrogenation of the parent is. Type I names are afforded by fusion systems (other than Stelzner), and type II names by the von Baeyer system and the spiro system 6a, i for simple saturated skeletons.

It is also desirable (for the sake of convenience) for the skeleton to be numbered as a whole, rather than for its components to retain their individual numberings. This again argues against the use of Stelzner nomenclature, and also against the bridging nomenclature method a and the spiro nomenclature method b as opposed to a, i.

Systems that do not meet all the above specifications, but the use of which is acceptable because of their convenience or the lack of suitable alternatives are the bridging nomenclature, system 4, method b (in which saturated bridging groups are permitted), the spiro method, 6a, ii (in which the components may be fully unsaturated *or* saturated and retain their original numberings), and system 7 for assemblies of identical cyclic components.

Thus, in the light of the foregoing arguments (and of others mentioned when the individual nomenclature systems were described), the following systems are recommended here for use in naming polycyclic skeletons:

3, fusion nomenclature (excluding systems of the Stelzner type)
4, bridging nomenclature method b
5, von Baeyer system
6, spiro nomenclature methods a, i and ii
7, ring assembly nomenclature method a

Replacement principles can be applied satisfactorily to all these systems in order to name heterocycles. Indeed, this is the only way in which the von Baeyer system and the spiro method 6a, i (i.e., the systems of type II; see above) can be used in the heterocyclic series. However, as explained in Section II, B, 1, b, replacement nomenclature has not previously been applied extensively in other areas, although it has been used to a limited extent for fused skeletons. It is to be hoped that its use will increase.

e. *Which System to Use.* The nomenclature systems recommended in the preceding section are essentially those employed by Chemical Abstracts. However, the need in an index for assigning to every molecule a unique name has made it necessary for Chemical Abstracts to elaborate their nomenclature rules to a degree somewhat beyond the needs of authors. In other words, although it is essential for a name used by an author to define a single molecule clearly and unambiguously, it is not so crucial that the structure should have only one "correct" name. What is important is that an author should be able to use the nomenclature systems at his disposal with sufficient flexibility, so that they are an aid to communication, not a hindrance. Rigid application of Chemical Abstracts names in a paper can easily obscure the sense and make the argument more involved than it need be. For example, two structures differing in only a minor aspect may even have Chemical Abstracts index names derived from two different nomenclature systems (e.g., **67a** and **67b**), yet it is often important from an author's point of view that they should be named in a similar way.

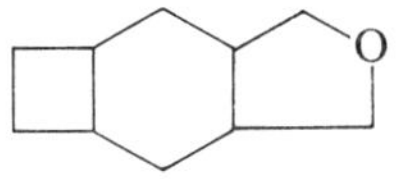

(67a)

Decahydrocyclobut[*f*]isobenzofuran

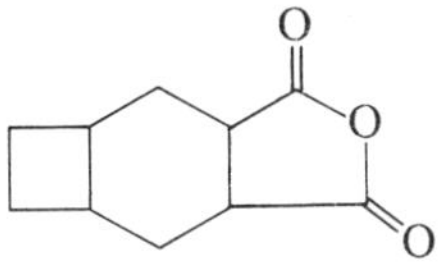

(67b)

Octahydrobicyclo[4.2.0]octane-3,4-dicarboxylic anhydride

However, it would be unwise to take this attitude too far: although some freedom of choice among alternative nomenclature systems is desirable, it is seldom wise to apply an individual system other than in a completely rigorous way, and it is not at all advisable to "mix" systems; i.e. to use names such as 5,6-benzo-2-azabicyclo-

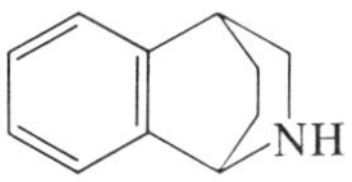

(68)

[2.2.2]octene (for **68**) rather than 1,2,3,4-tetrahydro-1,4-ethanoisoquinoline, 1,2,3,4-tetrahydro-2-aza-1,4-ethanonaphthalene, or even 9-azatricyclo[6.2.2.0^{2,7}]dodeca-2(7),3,5-triene.

As will be apparent, the problems involved in deciding how to name a multiring heterocyclic skeleton are considerably greater than those for monocycles. However, with regard to the systems recommended above, it is possible to give some general advice as to when and how these systems should be used.

The application of replacement nomenclature has already been discussed (Section II, B, 1, b) and will not be dealt with further here (although replacement names will be given where appropriate). Also it is unnecessary to discuss the use of the trivial names allowed by IUPAC; they may be applied whenever it appears appropriate. Those cases in which ring assembly nomenclature should be used are generally obvious and require no comment. Problems do arise, however, in deciding whether to use fusion nomenclature or von Baeyer nomenclature.

The choice here is essentially between a fully unsaturated parent (type I, see Section II, B, 1, d) and a saturated parent (type II), and should be made with regard to simplicity, brevity, clarity, and suitability in context. (As already noted, choice of an index name by Chemical Abstracts in cases of this type is governed by an elaborate system of rules.) Application of these criteria (apart from the last, which cannot readily be exemplified) is best illustrated by a set of examples (**69–74**; of course the reader may not agree with the indicated preferences). It must be emphasized that partially saturated skeletons should not be used as parents. In particular the partially and fully saturated units listed by IUPAC in Rule B-2.12 should never be used in fusion names, since any saturation is automatically removed by operation of fusion principles.

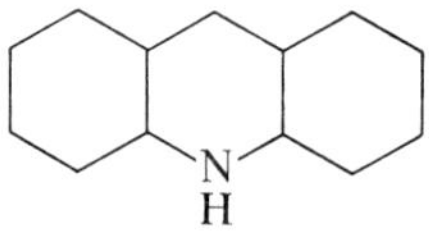

(69)

Tetradecahydroacridine or tetradecahydro-9-aza-anthracene, rather than 2-azatricyclo[8.4.0^{3,8}]tetradecane

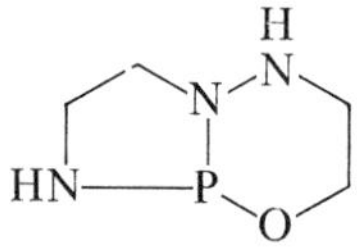

(70)

5-Oxa-1,2,7-triaza-6-phosphabicyclo[4.3.0]nonane
rather than tetrahydro-7-oxa-1,3a,4-triaza-7a-phosphaindene
or tetrahydro[1,3,2]diazaphospholo[2,1-b][1,3,4,2]oxadiazaphosphorine

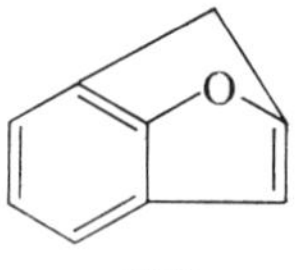

(71)

1-Oxa-2,7-methanoindene or 2,7-methanobenzofuran
rather than 10-oxatricyclo[3.3.1.1^{3,9}]deca-1(9),3,5,7-tetraene

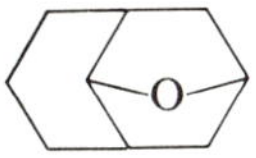

(72)

10-Oxatricyclo[3.3.1.1^{3,9}]decane
rather than octahydro-1-oxa-2,7-methanoindene
or octahydro-2,7-methanobenzofuran

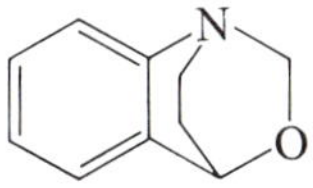

(73)

2H,4H-3-Oxa-1-aza-1,4-ethanonaphthalene
or 2H,4H-1,4-ethano-3,1-benzoxazine
rather than 9-oxa-1-azatricyclo[6.2.2.0^{2,7}]dodeca-2,4,6-triene
(*not* 5,6-benzo-3-oxa-1-azabicyclo[2.2.2]hexene)

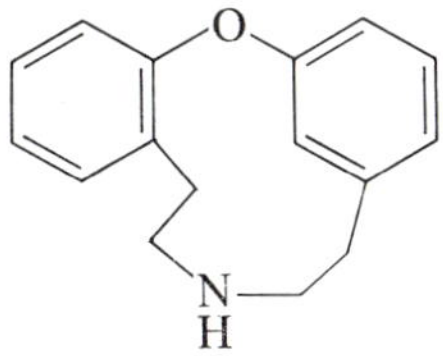

(74)

11H-12,13,14,15-Tetrahydro-5-oxa-13-aza-6,10-methenobenzocyclotridecene
or 7H-8,9,10,11-tetrahydro-2,6-metheno-1,9-benzoxazacyclotridecine
rather than 2-oxa-11-azatricyclo[12.3.1.0^{3,8}]octadeca-1(17),3(8),
4,6,14(18),15-hexaene (and preferably *not* a dibenzo compound, since
this would require an extension of the fusion nomenclature system).
The fully saturated form of **74** would be best named as 2-oxa-11-
azatricyclo[12.3.1.0^{3,8}]octadecane.

With regard to spiro nomenclature, method 6a, i is applicable unless one at least of the components is a fused or bridged unit; then method 6a, ii should be used. In using method 6a, ii it is not necessary to ensure that all units are either fully saturated or fully unsaturated; a mixture of both types is frequently convenient, e.g., **50** and **52** (however, see Section II, B, 2, a, ii).

2. *Specific Problems*

a. *Hydrogenation*

i. *Fused skeletons*. Confusion frequently arises over the designation of the extent of saturation (or unsaturation) of a fused polycyclic skeleton. However, the likelihood of error is minimized if it is remembered that (for the systems recommended) the parent name is always that of the skeleton containing the maximum number of noncumulative double bonds. The only exception to this arises in the use of bridging nomenclature (system 4), where the state of saturation of the bridge is indicated in the bridging prefix (e.g., etheno or ethano). In fact there is a further, associated, problem in the use of bridging nomenclature (not referred to specifically by IUPAC): insertion of the maximum number of noncumulative double bonds is considered to take place *after* addition of the bridge. Thus **75** is named 9,10-diaza-4a,8a-ethenoanthracene rather than the 4a,8a-dihydro derivative. (Similarly, **38** is not a 4a,8a-dihydro derivative.)

(75)

The convenience of always deriving a name from a fully unsaturated parent is the main reason for abandoning the Stelzner nomenclature systems for fused skeletons. However, it should be emphasized that, as a consequence of the rejection of Stelzner nomenclature, all fusion names terminating in "-cycloalkene" imply full unsaturation (i.e., the "ene" here does not indicate only one double bond), e.g., **76** is 5-azabenzocyclo-octene, and **77** is the 5,6,7,8,9,10-hexahydro derivative.

(76) (77)

There is often more than one way of inserting the maximum number of double bonds into a cyclic skeleton. In order to specify the parent precisely in such cases, it is necessary to denote the location of at least one hydrogen atom (in a $-CH_2-$ or $-NH-$ group, etc.). This is done by citation of "indicated hydrogen" (IUPAC Rule A-21.6) before the parent name (see examples **19**, **20**, **25**, **30**, etc.). In heterocyclic skeletons, which are often tautomeric, designation of "indicated hydrogen" limits the name to a particular tautomer. When there is a choice, "indicated hydrogen" is always given the lowest possible locant, even in partially saturated molecules (see also the later section on numbering), e.g., **78** and **79**.

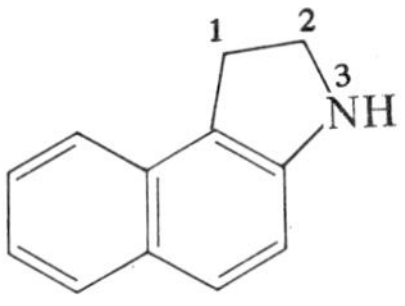

(78)

2,3-Dihydro-1*H*-benz[*e*]indole or 2,3-dihydro-1*H*-3-azacyclopenta[*a*]naphthalene (*not* 1,2-dihydro-3*H*-benz[*e*]indole)

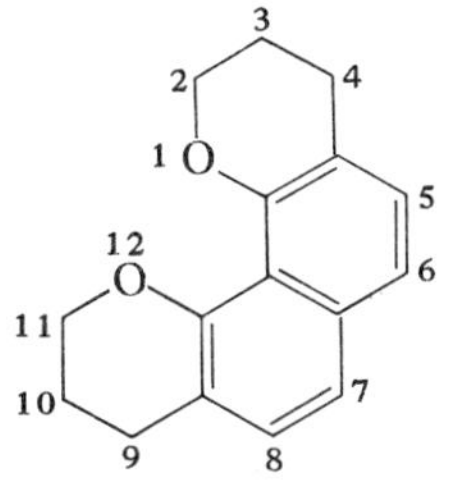

(79)

3,4,10,11-Tetrahydro-2*H*,9*H*-naphtho[1,2-*b*:8,7-*b'*]dipyran, or 3,4,10,11-tetrahydro-2*H*,9*H*-1,12-dioxabenzo[*c*]phenanthrene

In replacement names, the position of "indicated hydrogen" is determined *after* performing the replacement operation(s) and inserting the maximum number of double bonds; its location does not necessarily bear any relation to that of "indicated hydrogen" in the parent hydrocarbon, cf. **80** and **81**.[14b]

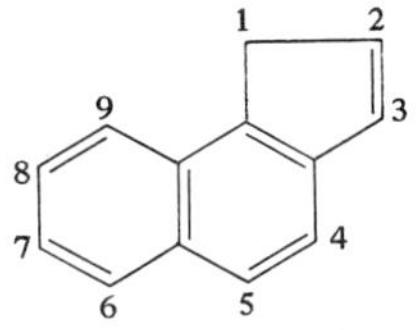

(80)   (81)

1*H*-Cyclopenta[*a*]naphthalene   5*H*-4-Oxa-9b-azacyclopenta[*a*]naphthalene

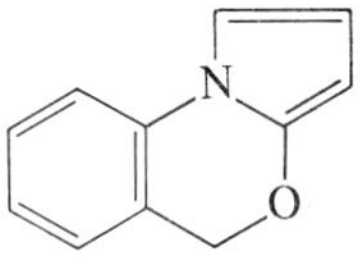

ii. *Spiro systems.* In using method 6a, ii for naming spiro compounds, it is necessary to ensure that the names of the components are such that no double bonds are attached to the point of spiro union. When a component has a maximally unsaturated parent, the correct procedure may not be apparent. There is no problem when the spiro

---

[14b] Although this procedure for replacement names is recommended in the Chemical Abstracts Ninth Collective Index Nomenclature Manual, the Index Guide now employs a more complicated system.

atom carries "indicated hydrogen" (see preceding section). However, when this is not the case the spiro atom must be considered as involved in a double bond which has been hydrogenated. The obvious way to do this is by means of a "dihydro" prefix, and indeed this is permissible. However, Chemical Abstracts uses a system whereby a further hydrogen atom is "indicated" (that on the other ring atom involved in the formally hydrogenated double bond; cf. examples **82** and **83**). The

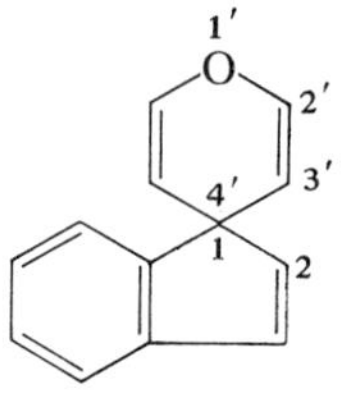

(82)

Spiro[indene-1,4'-[4H]pyran]
or spiro[indene-1,4'-oxacyclohexa[2,5]diene]

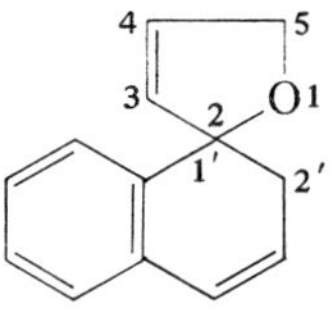

(83)

Spiro[furan-2(5H),1'(2'H)-naphthalene] rather than a 1',2,2',5-tetrahydro derivative; or spiro[naphthalene-1(2H),2'-oxacyclopent[3]ene] rather than a 1,2-dihydro derivative

IUPAC Rules do not describe this system, but do employ it in some examples of radical names (Rules B-12), and a similar principle can be used in naming ketones, imines, bivalent radicals, and some cations (see later). It is to be preferred on the grounds that it is less cumbersome than

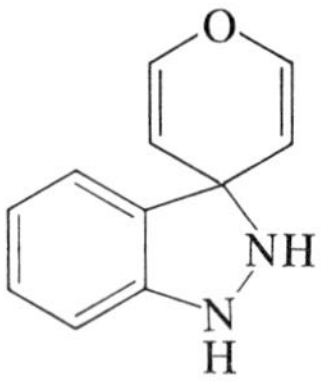

(84)

1,2-Dihydrospiro-[3H-1,2-diazaindene-3,4'-oxacylohexa[2,5]diene],
or 1,2-dihydrospiro-[3H-indazole-3,4'-[4H]pyran];
not spiro-[1H-1,2-diazaindene-3(2H),4'-oxacyclohexa[2,5]diene]

Sec. II.B]   THE NOMENCLATURE OF HETEROCYCLES   209

employing "hydro" prefixes. However, it must be remembered that in such cases the state of hydrogenation of the parent system is determined by introducing the maximum number of double bonds *after* formation of the spiro linkage, e.g., **84**.

iii. *Use of Δ to indicate unsaturation.* It is often convenient to denote the position of a double bond in a partially saturated skeleton by means of the prefix Δ with a superscript locant. This usage is common with the partially saturated four- and five-membered ring Hantzsch–Widman names [e.g., Δ^2-oxazoline (**85**)] and with polyhydro fused skeletons [e.g., Δ^7-octahydro-1-azanaphthalene (**86**)]. However, particularly in certain areas of natural product chemistry, Δ has been used as an "operator," i.e., to indicate the presence of unsaturation not otherwise specified in the name. This latter usage is to be avoided [i.e., **86** should not be named Δ^7-perhydro- (or decahydro-)1-azanaphthalene].

(85) 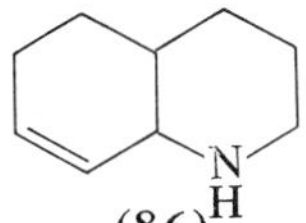   (86)

b. *Numbering.* It is important to assign the correct numbering system to a compound so that its chemistry can be discussed without ambiguity by referring to the locants of the various skeletal positions. However, the numbering of a skeleton is generally a consequence of the way in which the skeleton is named. Different nomenclature systems usually give rise to different numberings, as illustrated by examples **87–89**. It is therefore advisable not to vary nomenclature systems when

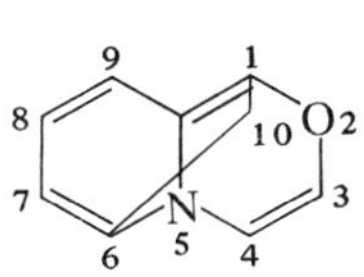

(87)

1,6-Methanopyrido[2,1-*c*][1,4]oxazine

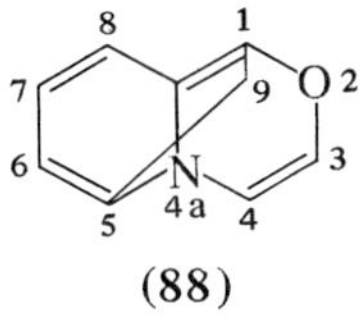

(88)

2-Oxa-4a-aza-1,5-methanonaphthalene

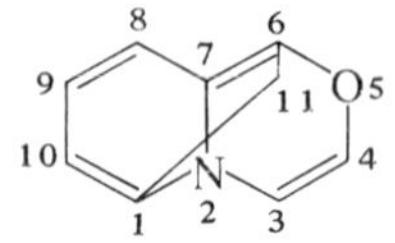

(89)

5-Oxa-2-azatricyclo[4.4.1.0^{2,7}]undeca-1(10),3,6,8-tetraene

discussing a series of compounds (even if this means "bending" the rules). Those systems that number the skeleton as a whole (e.g., Chemical Abstracts fusion system) are more convenient than those that do not (e.g., Stelzner fusion system).

For each of the nomenclature systems recommended here, correct numberings are obtained by application of elaborate series of priorities. These are set out clearly in the IUPAC Rules (Appendix), and no useful purpose would be served by a further detailed description. Their use is illustrated in many of the examples given here. However, the following points deserve some emphasis or further explanation.

i. *Lowest number.* It is necessary to be quite clear as to the meaning of "lowest number" when applied to two or more sets of locants. When the locants in each set are arranged in ascending numerical order, that set is "lowest" which contains the lowest number at the first point of difference. The "lowest" set is *not* that with the lowest arithmetical sum. For example, **90** is 1,7,14-triazabenzo[*b*]triphenylene (sum of locants = 22), not the 2,8,9-triaza compound (sum = 19).

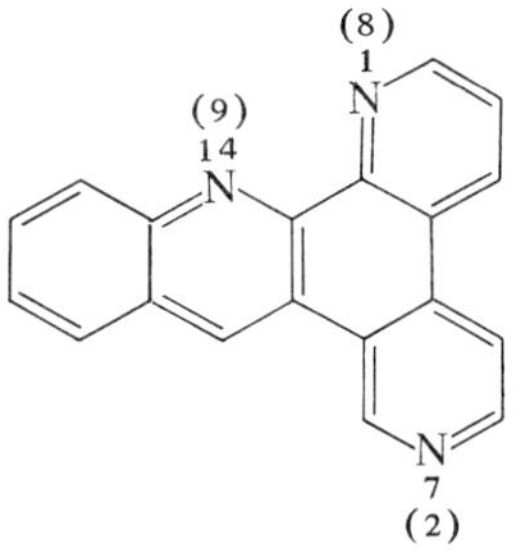

(90)

ii. *Nonsystematic numberings.* The nonsystematic numberings for anthracene, phenanthrene, acridine, carbazole, purine, and xanthen are so well established in the literature that they have been retained by IUPAC (see Rules A-22.5 and B-2.11). Heterocycles derived by replacement principles from anthracene and phenanthrene retain the carbocycle numbering. When these six skeletons are used as base components in fusion names, the procedure for lettering the bonds is *not* modified to follow the nonsystematic locants, and the complete skeleton is numbered

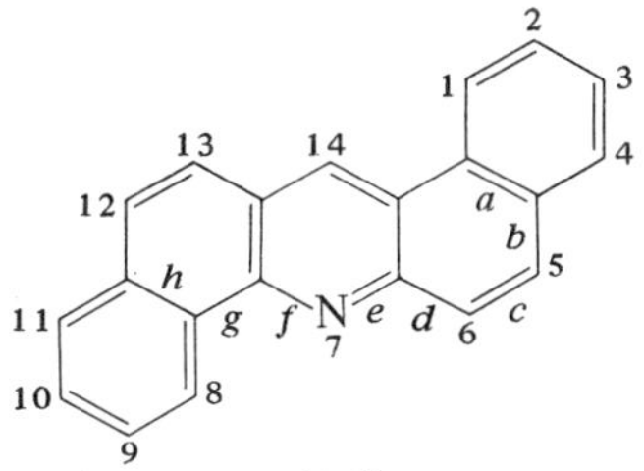

(91)

Dibenz[*a,h*]acridine or 7-azadibenz[*a,h*]anthracene

systematically, e.g. **91**. Cyclopenta[*a*]phenanthrene is also numbered nonsystematically, by analogy with the steroid nucleus. Replacement derivatives retain the steroid numbering, e.g., **92**.

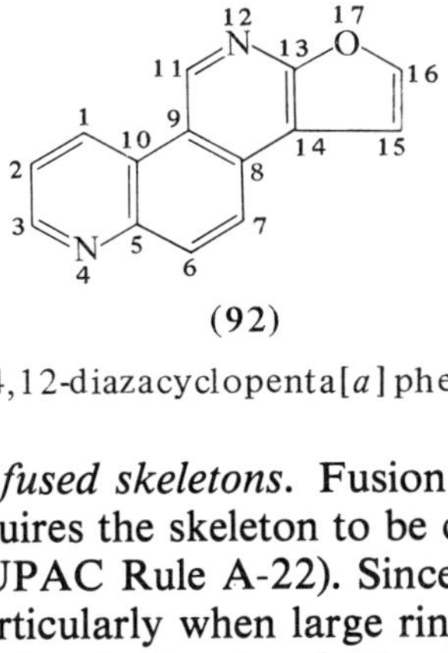

(92)

17-Oxa-4,12-diazacyclopenta[*a*]phenanthrene

iii. *Orientation of fused skeletons.* Fusion nomenclature, in contrast to other systems, requires the skeleton to be correctly oriented before it can be numbered (IUPAC Rule A-22). Since errors in interpreting this rule are common, particularly when large rings or rings containing odd numbers of atoms are involved, a few further (carbocyclic) examples are given (**93–96**). Examples **94–96** illustrate the ways in which

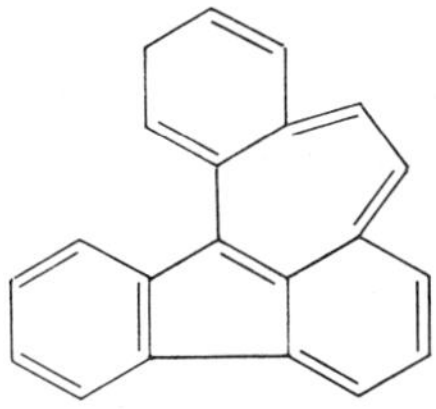

correct

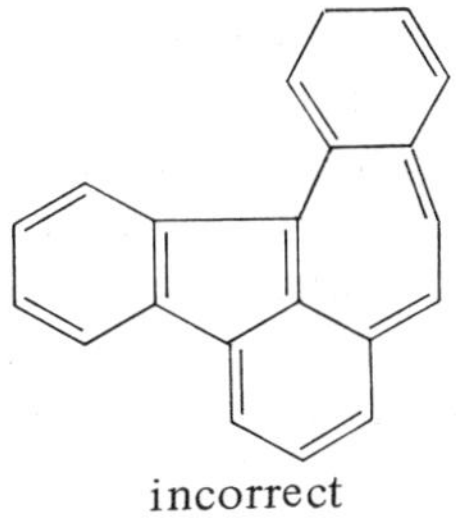

incorrect

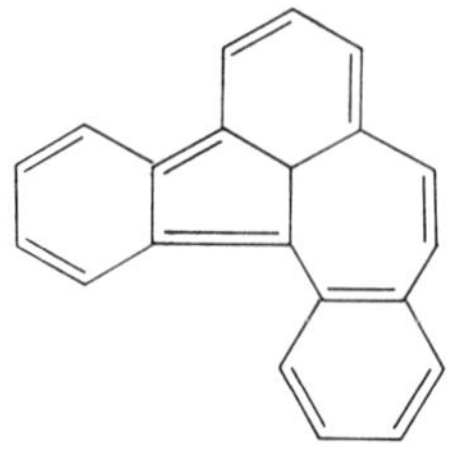

incorrect 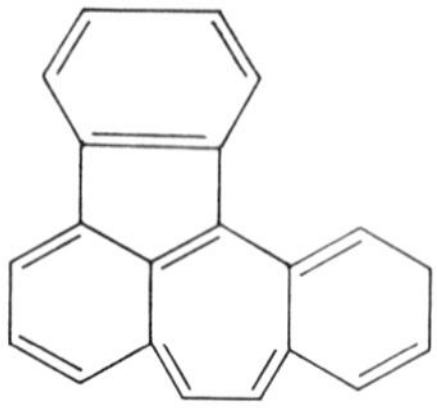   incorrect

(93)

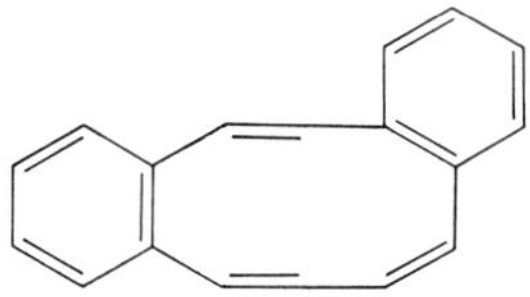

correct 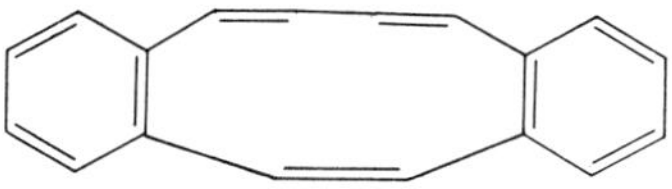   incorrect

(94)

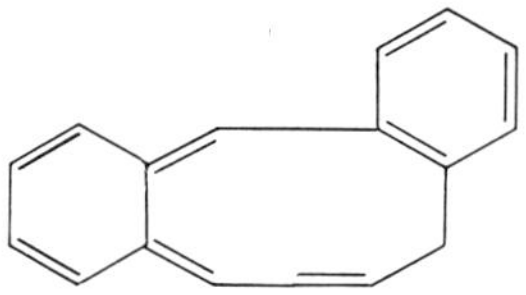

correct 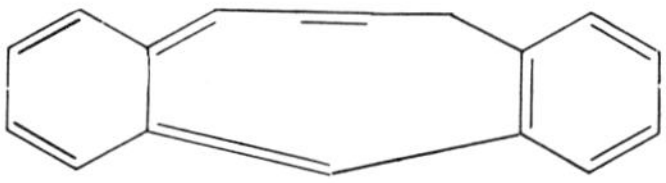   incorrect

(95)

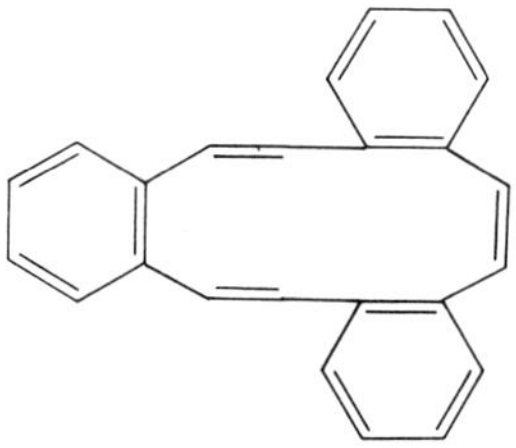

correct 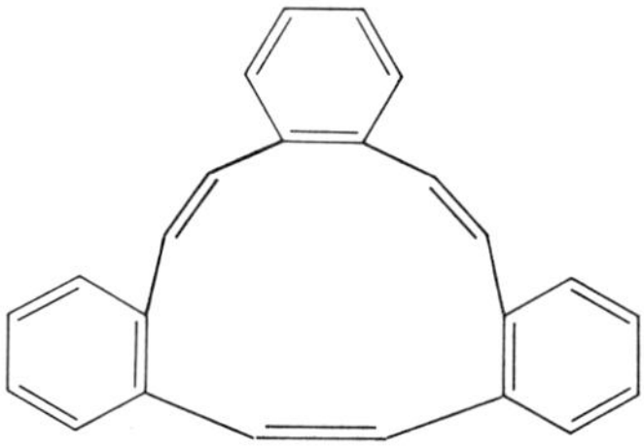   incorrect

(96)

macrocycles should be drawn for the purpose of orientation, i.e., with only two bonds vertical (at the ends of the polygon) and the numbers of atoms in the two chains linking these bonds equal or differing by one. (The incorrect orientation of **96**, which in fact yields the same numbering as the correct one, is the orientation illustrated in the Chemical Abstracts Index Guide; however, this method of representation is misleading in its implication that one can draw the skeleton with three rings in a horizontal row.)

Sec. II.B] THE NOMENCLATURE OF HETEROCYCLES 213

iv. *Priorities for heteroatoms.* It should be noted that there is a difference in order of priorities between the rules for numbering monocycles and those for polycyclic skeletons. In the monocycle case (Rule B-1.53), the locant 1 is given to a heteroatom as high as possible in the table of replacement prefixes, and the numbering is then that which gives overall lowest locants to all the heteroatoms. In the polycyclic case this order is reversed (Rule B-3.4): the overall numbering of the heteroatoms is considered before their individual priorities. For example, **97** (thiopyrano[3,4-*b*][1,4]oxazine or 4-oxa-6-thia-1-azanaphthalene) is numbered starting with N (**97a**) rather than O (**97b**), whereas in 1,4-oxazine itself the reverse applies.

(97)

(a) correct   (b) incorrect

v. *Replacement in fused systems.* The numbering of a fused skeleton named by replacement principles is frequently different from that of the same skeleton named by fusion of heterocyclic units (with trivial or Hantzsch–Widman names). This is because in the latter system heteroatoms at a ring junction are given individual numbers, whereas a ring junction carbon atom in a carbocycle has the same locant as an adjacent, nonjunction carbon atom but with a suffix "a" or "b," etc. Thus care must be taken in matching the numbering system to the nomenclature system used (compare the numbering systems for **98** named as in **a** and **b**).

(98)

(a) Imidazo[2,1-*b*]thiazole   (b) 1-Thia-3a,6-diazapentalene

If application of the numbering rules in any system still leaves a choice, this may be settled by consideration of substituents (see Section III, Naming of Derivatives).

c. *Charged Skeletons.* The naming of heterocyclic skeletons carrying a positive or a negative charge is at present in a confused state. The IUPAC Rules provide a rudimentary treatment [see Rules B-6 and B-10, and also Rules C-82 to 84 (not included in the Appendix)], but will require elaboration to cover complex cases. Chemical Abstracts uses a slightly different system. Both IUPAC and Chemical Abstracts have the matter under consideration.

i. *Cations.* A cation can be considered to be derived formally from a neutral molecule either (I) by addition of H^+ or (II) by removal of H^-. IUPAC recommends use of the terminations "-ium" for case I and "-ylium" for case II, e.g., 9a*H*-quinolizinium (**99**) and quinolizinylium (**100**). Chemical Abstracts uses almost the same principles, but with one

(**99**)   (**100**)

important exception. When, in case I, for a fused skeleton, the additional free valency of the charged heteroatom is used in forming a ring bond, the Chemical Abstracts name of the skeleton is based on the least hydrogenated form of the *cation* (not the neutral molecule); thus the Chemical Abstracts name for **100** is quinolizinium and for **99** is 5,9a-dihydroquinolizinium. Hence the problem in naming such species is at what stage to insert the maximum number of double bonds in order to define the hydrogenation state of the parent skeleton. Neither of the foregoing systems is really satisfactory: it would be preferable not to use the exceptional treatment employed by Chemical Abstracts (since this leads to direct conflict with the IUPAC method), but the IUPAC procedure (i.e., use of a case II name for the "exceptional" situation) is not always very clear to envisage, requiring apparent formal removal of H^- from a heteroatom in an unusually high valency state (e.g., from **101**). Clarification could be provided by indicating the unusual valency

(**101**)

(see the following section); thus **101** could be named as $5\lambda^5$-quinolizine and the cation (**100**) derived by loss of H^- as $5\lambda^5$-quinolizin-5-ylium. (Chemical Abstracts and IUPAC have a system of this type under consideration.) In fact all cations could be named by the use of case II (-ylium) names with specification of unusual valency [e.g., $1\lambda^5$-indol-1-ylium (**102**)], but it would probably be better to restrict the use of this

(**102**)

system to skeletons in which the heteroatom carries a ring bond in excess of the usual number.

A further associated problem arises with case II (-ylium) names when it is necessary to consider loss of H⁻ as taking place formally from a dihydro form of a skeleton (i.e., when the heteroatom carrying the charge has no hydrogen attached in the neutral molecule, but cannot form an extra ring bond in the charged species), e.g., **103**. This can be

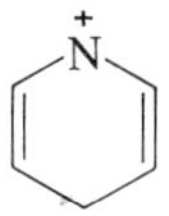

(103) 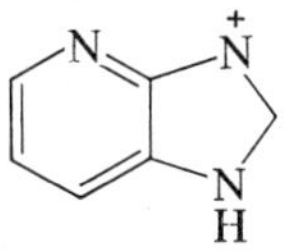   (104)

overcome either by adding "-ylium" to the dihydro name, or by use of "indicated hydrogen", e.g., **103** could then be named 1,4-dihydropyridinylium or pyridin-1(4*H*)-ylium. The latter system is less cumbersome and perhaps to be preferred, but it should be remembered, if it is to be employed, that maximum unsaturation is introduced *after* formation of the cation; e.g., **104** is 1,2-dihydroimidazo[4,5-*b*]pyridin-3-ylium, not 1*H*-imidazo[4,5-*b*]pyridin-3(2*H*)-ylium.

IUPAC Rules B-6 and B-10 give examples of heterocyclic cations named by replacement principles. The normal replacement prefixes (oxa-, aza-, etc., indicating replacement of —C— by —O— or >N—) are modified to oxonia-, azonia-, etc. (indicating replacement of —C— by >Ö— or —N⁺—). The application is straightforward provided that one assumes, for fused skeletons, that the parent skeleton is obtained by adding maximum unsaturation *after* the replacement has been carried out; e.g. **100** is thus named 4a-azonianaphthalene, and **105** is named 9-thionia-10-aza-anthracene. However, this method can only be used for

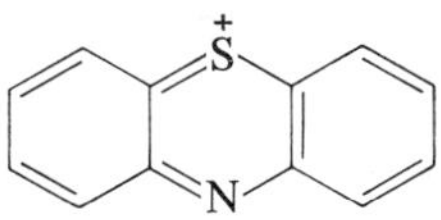

(105)

case I cations, or in case II where the heteroatom carries an extra ring bond in the charged species; it cannot be applied to cations such as **103** (i.e., "azonia" cannot be used to mean replacement of —C— by —N⁺—).

With this limitation in mind, one should consider the possibility of applying "-ium" and "-ylium" suffixes to ordinary replacement names (instead of using "-onia-" prefixes); this sort of treatment could be very useful in certain cases (see, for example, Section iii), and may prove to be the method of choice for naming cations if the use of replacement nomenclature becomes more widely accepted. By this method, **100** could be named 4aλ^5-4a-azanaphthalen-4a-ylium; **99**, 8a*H*-4a-azanaphthalen-4a-ium; and **103**, azabenzen-1(4*H*)-ylium or azacyclohexa-2,5-dien-1-ylium.

ii. *Anions*. For almost all heterocycles it is only necessary to consider anions formally derived by loss of H$^+$ from a neutral skeleton; this is indicated by the termination "ide." The IUPAC Rules list no heterocyclic examples, but application of this system is straightforward. Use of "indicated hydrogen" is necessary in certain cases. For examples, see **106** and **107**. There are no recommended replacement prefixes for

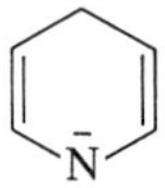

(106)   Pyridin-1(4*H*)-ide, or azacyclohexa-2,5-dien-1-ide

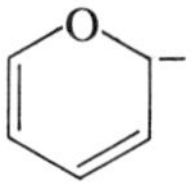

(107)   Pyran-2-ide, or oxacyclohexa-3,5-dien-2-ide

anions equivalent to the "-onia" series for cations. (However, the IUPAC Section D Tentative Rules[8] recommend the use of "borata" for replacement of —C— by —B$^-$—, in a section devoted specifically to the nomenclature of boron compounds.)

iii. *Dipolar* (+, −) *Skeletons*. The naming of heterocycles containing both positively and negatively charged ring atoms is not dealt with specifically in the IUPAC Rules. Chemical Abstracts employs an exceedingly cumbersome system for such skeletons, involving formal addition and removal of water, leading to names of the type 1*H*-naphtho[1,8-*de*]triazinium hydroxide inner salt (for **108**). Although

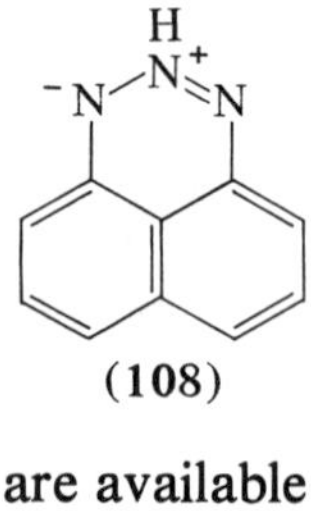

(108)

cationic replacement prefixes are available (see Section i), corresponding prefixes for negatively charged heteroatoms have not been devised (though no doubt terms of the type "oxida-," "azida-," etc., could be introduced). However, it would appear reasonable to name dipolar skeletons of this type by use of appropriate "-ium" or "-ylium" and "-ide" suffixes, as in the examples 109–111 (108 would thus be named

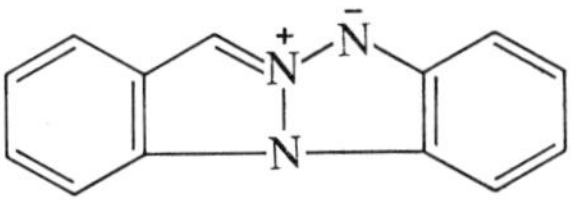

(109)

5H-6λ^5-Indazolo[1,2-a]benzotriazol-6-ylium-5-ide,
or 10H-4b,9aλ^5,10-triazaindeno[1,2-a]inden-9a-ylium-10-ide

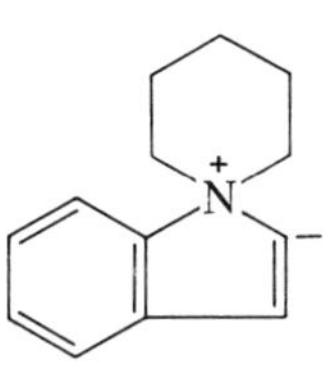

(110)

Spiro[indole-1,1'-piperidinium]-2-ide,
or spiro[azacyclohexane-1,1'-[1]azaindenium]-2'-ide

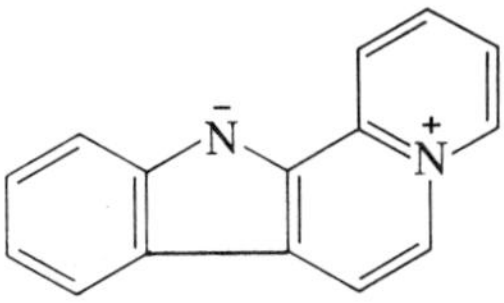

(111)

12H-5λ^5-Indolo[2,3-a]quinolizin-5-ylium-12-ide,
or 11H-4aλ^5,11-diazabenzo[a]fluoren-4a-ylium-11-ide

1H-naphtho[1,8-de]triazin-2-ium-1-ide or 1H-1,2,3-triazaphenalen-2-ium-1-ide). Alternatively one could treat the formally zwitterionic systems as neutral skeletons containing atoms of unusual valency (see Section d below); thus 108 and 111 could be represented as 112 and 113, and named as 1,2$\lambda^5\sigma^3$,3-triazaphenalene and 4a$\lambda^5\sigma^3$,11-diazabenzo[a]fluorene.

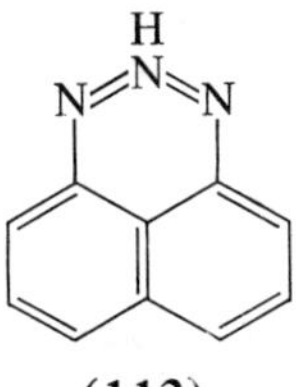

(112) 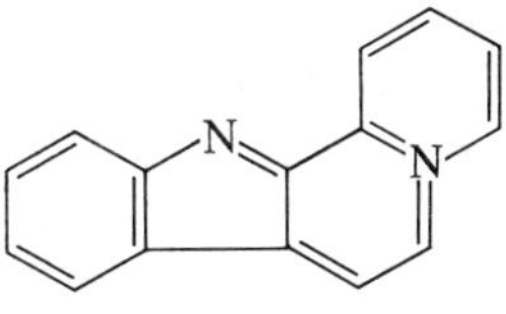   (113)

d. *Unusual Valencies.* There is as yet no accepted, universally applicable system of indicating unusual valency states for heteroatoms. Systems at present in use by Chemical Abstracts include (i) use of hydro prefixes, e.g., 1,1-dihydrothiophen (for **114**); (ii) use of indicated hydrogen, e.g., 1*H*-1,2,6-thiadiazine (for **115**); (iii) use of dehydro

(114) (115)

(116) (117)

(118) (119)

prefixes, e.g., 3,3-didehydro-2,4-dioxa-3-plumbabicyclo[3.3.1]nonane (for **116**); (iv) use of additive nomenclature (see later), e.g., pyridine*N*-oxide; and (v) indication of the unusual valency as a superscript roman numeral affixed to the element symbol, e.g., 1,6,6a-trithia(*6a-S*^{IV})-cyclopenta[*cd*]pentalene (for **117**). This last method is probably the most widely applicable, but can give rise to ambiguity in certain cases, e.g., thiophen-S^{IV} could be **114** or **118**.

In the Tentative Section D Rules,[8] IUPAC has proposed the use of the symbols λ^n, indicating the "connecting number n" of the heteroatom (where n is the sum of all classical valence bonds and/or units of charge), and σ^m, indicating the "sigma number m" (i.e., the number of sigma bonds attached to the heteroatom). Use of λ and/or σ can define uniquely a skeleton containing a heteroatom of unusual valency, but the IUPAC proposals require some elaboration in order for this to be so. Some examples of the use of the system are: 1*H*-1λ^4-thiophen (for **114**), 1λ^4,2,6-thiadiazine (for **115**), 2,4-dioxa-3λ^2-plumbabicyclo[3.3.1]nonane (for **116**), 1,6,6aλ^4-trithiacyclopenta[*cd*]pentalene (for **117**), 1$\lambda^4\sigma^2$-thiophen (for **118**), 1λ^5-pyridin-1-one (for pyridine *N*-oxide), and 2$\lambda^4\sigma^2$,1,3-benzothiadiazole (or 2$\lambda^4\sigma^2$-thia-1,3-diazaindene) (for **119**).

c. *Arynes.* Parent names corresponding to maximally unsaturated skeletons may be modified by "dehydro" prefixes to indicate the presence of triple, rather than double bonds, e.g., **120** and **121**. Although

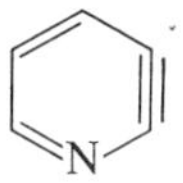

(120)

2,3-Didehydropyridine or 2,3-didehydroazabenzene
or azacyclohexa-1,3-dien-5-yne

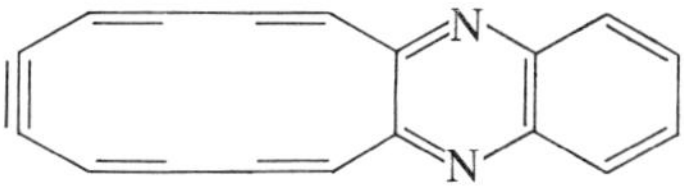

(121)

10,11-Didehydrocyclododeca [b] quinoxaline,
or 10,11-didehydro-5,16-diazacyclododeca[b] naphthalene

trivial designations such as benzyne and pyridyne are well established, this usage should not be extended to more complex molecules. It should be noted that "dehydro" implies loss of *one* hydrogen atom, not two (errors of this type are common in the literature).

C. Natural Poducts

Current nomenclature for heterocyclic natural products is to a large extent trivially based; however, some degree of semisystematization has been achieved. The following notes will indicate briefly the nature of the systems available and their sources, where appropriate.

1. *Alkaloids*

Although all alkaloids can be named by the principles already outlined in this article, the cumbersome nature of such names for complex ring systems makes it desirable to use trivial parent names for some large heterocyclic skeletons. It is preferable for such trivial names to refer to skeletons with no substituents (or very few), and it is often convenient for them to carry inherent stereochemical implications. The most extensive source of these names is the Chemical Abstracts Index Guide (or the Ninth Collective Index Nomenclature Manual), but the names given here do not correspond, in many cases, to those in common use, and IUPAC recommendations, when they appear, may well differ in some respects. Some of the principal skeletons listed by Chemical Abstracts are illustrated (**122–130**).

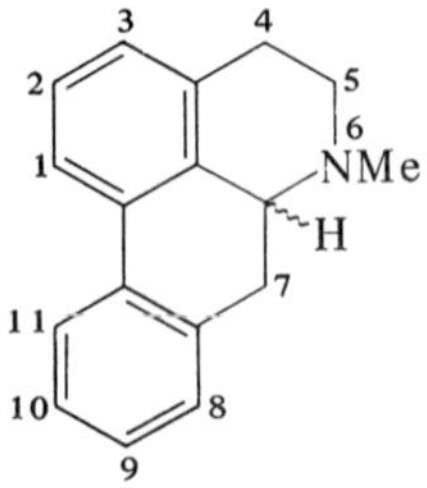

(122) Aporphine

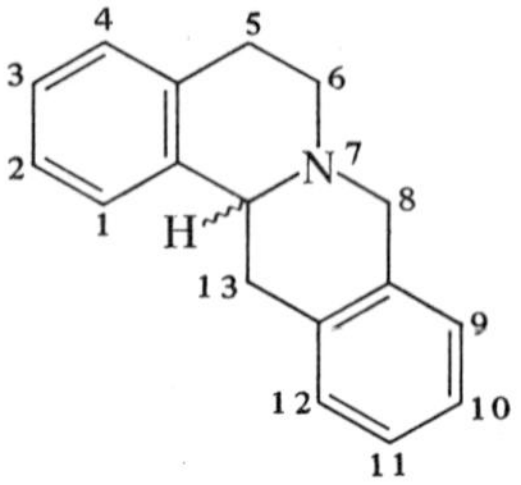

(123) Berbine

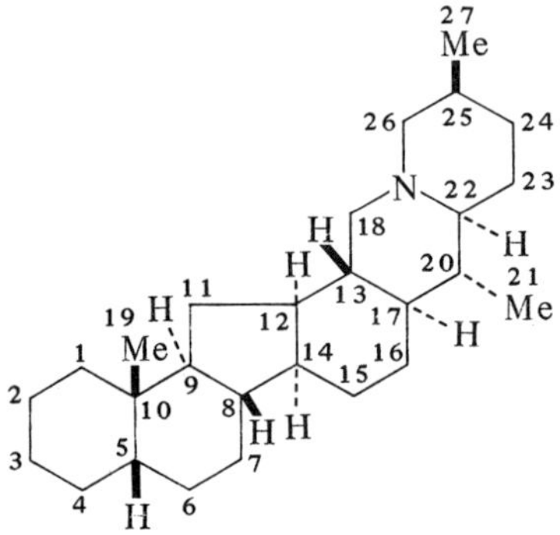

(124) Cevane[14c]

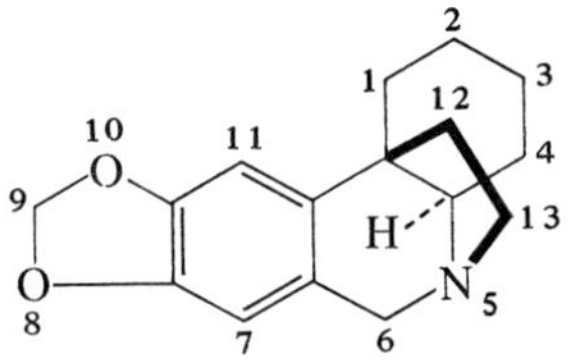

(125) Crinan

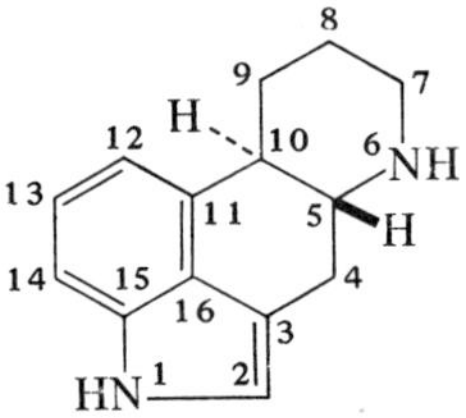

(126) Ergoline

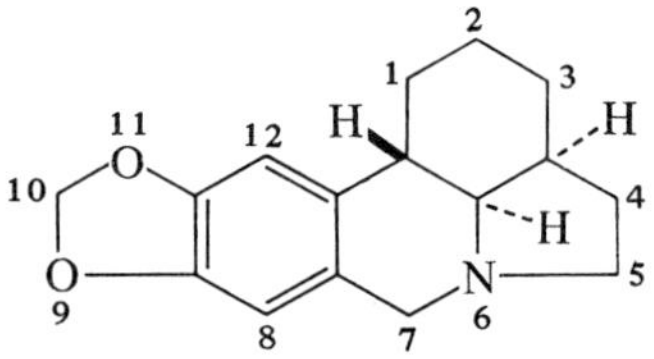

(127) Lycoran

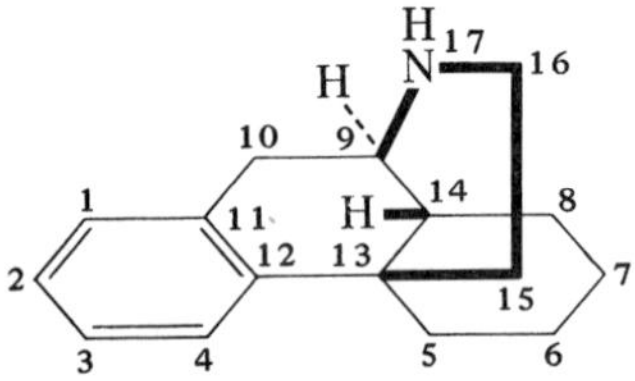

(128) Morphinan

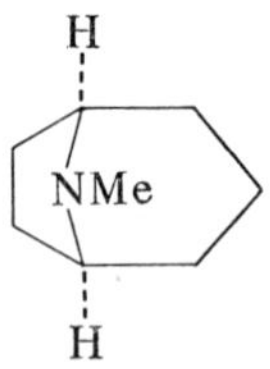

(129) Tropane

[14c] The IUPAC Rules for Nomenclature of Steroids name this as (22S,25S)-5α-cevanine.

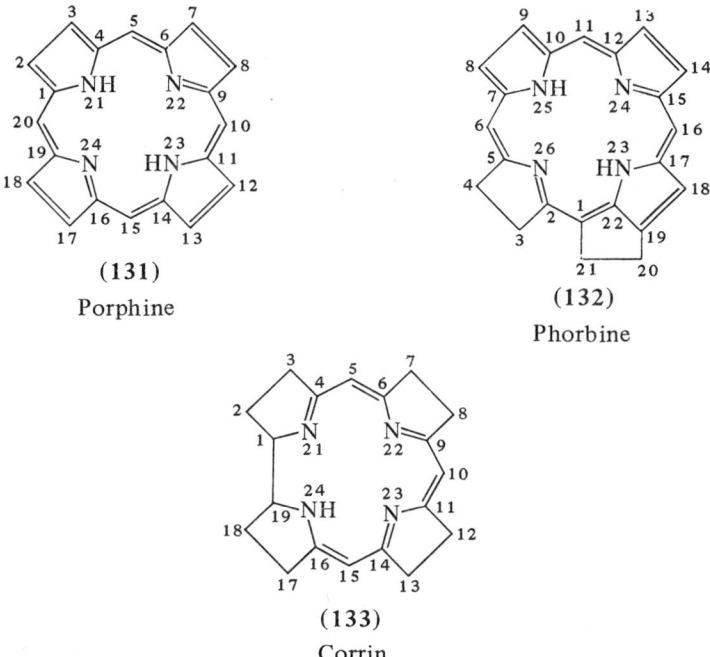

(130) Yohimban

2. Porphyrins and Corrins

The nomenclature of this group is at present under discussion by IUPAC. As for alkaloids, trivial names for large fundamental cyclic skeletons are used. Those employed in Chemical Abstracts are illustrated (131–133).

(131) Porphine

(132) Phorbine

(133) Corrin

3. Steroids and Carbohydrates with Heterocyclic Modifications

The IUPAC Definitive Rules for Nomenclature of Steroids[15] contain proposals for naming steroid alkaloids and steroids containing fused

---

[15] "IUPAC Nomenclature of Organic Chemistry and IUPAC-IUB Biochemical Nomenclature," Definitive Rules for Nomenclature of Steroids, *J. Pure Appl. Chem.* **31**, 283 (1972).

heterocycles. Some examples are illustrated (**134–137**). Similar principles to those used for naming **137** are under consideration for naming analogous carbohydrate structures (e.g., **138**). Replacement principles can be applied conveniently to naming steroids containing skeletal heteroatoms, e.g., **139**.

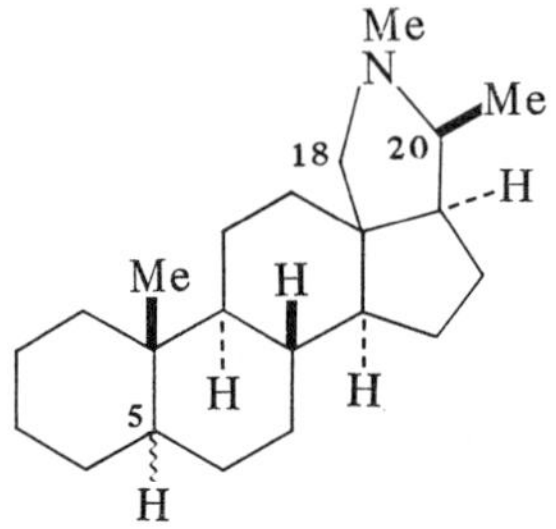

(134) Conanine

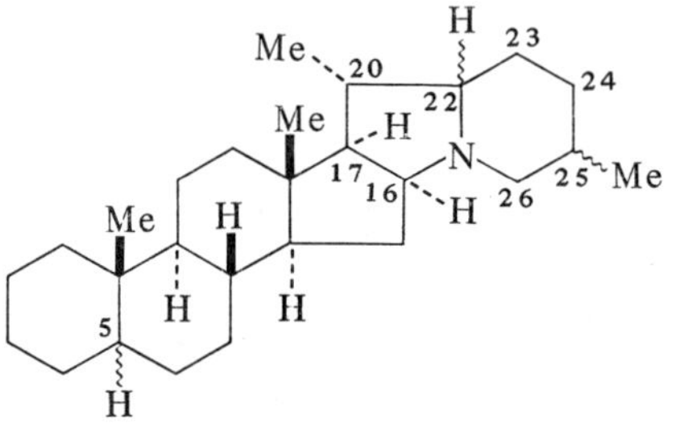

(135) Solanidanine

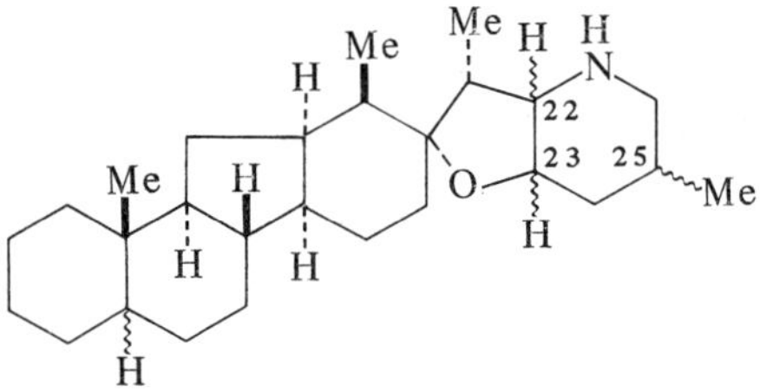

(136) Jervanine

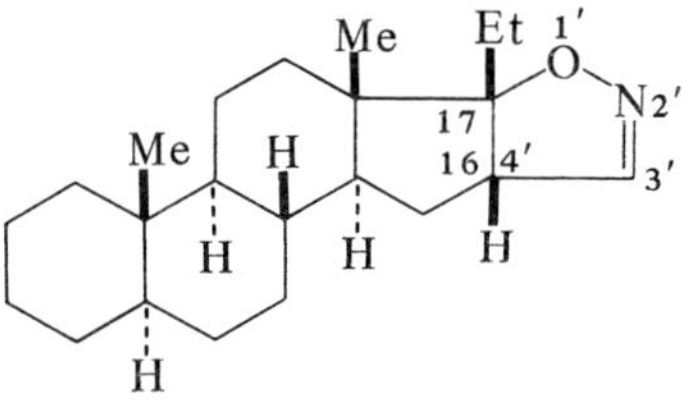

(137) $4'\beta H$-5α-Pregnano[16,17-d]isoxazole

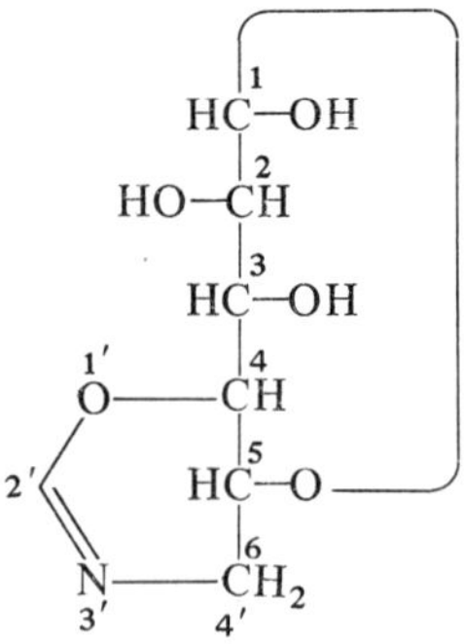

(138)

4,6-Dideoxy-$5',6'$-dihydro-α-D-idopyranoso[6,5,4-de][1,3]oxazine

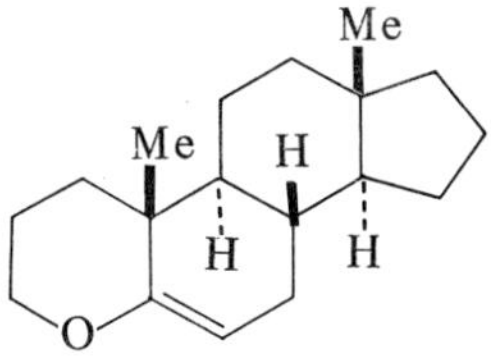

(139)

4-Oxa-androst-5-ene

4. *Flavan and Derivatives*

2-Phenylchroman (3,4-tetrahydro-2-phenyl-2*H*-1-oxanaphthalene, **140**), the parent of a large group of natural products, is commonly named flavan, and the species **141–143** are referred to as flaven, flavone, and flavylium, respectively (although these modifications are not strictly systematic).

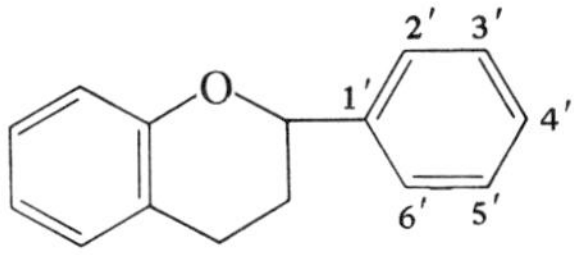

(140) 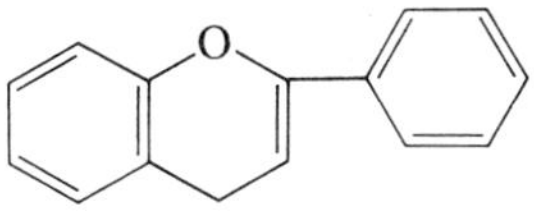   (141)

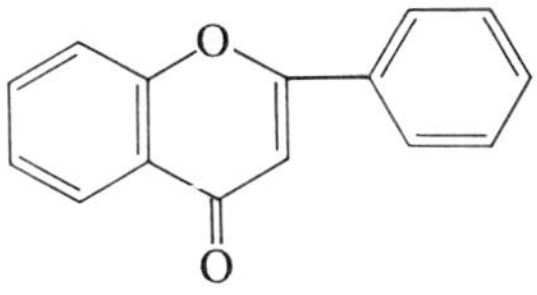

(142) 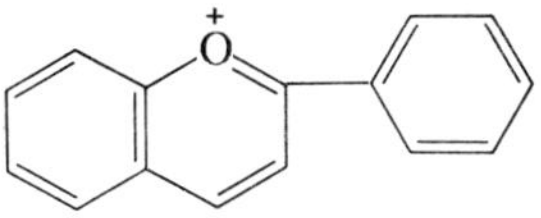   (143)

5. *Nucleic Acid Bases*

Accepted trivial names are given to the *substituted* purines and pyrimidines **144–148**, which are the principal component bases of nucleic acids.

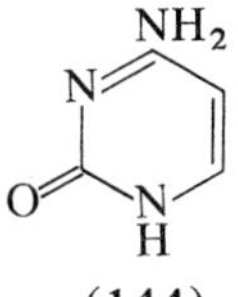

(144)

Cytosine [4-aminopyrimidin-2(1*H*)-one or 4-amino-1,3-diazacyclohexa-3,5-dien-2-one]

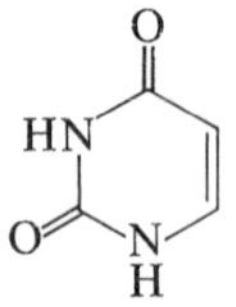

(145) Uracil

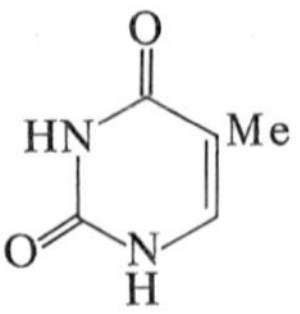

(146) Thymine

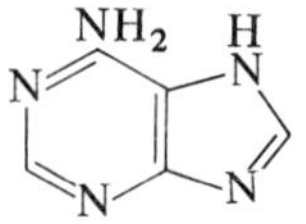

(147) Adenine

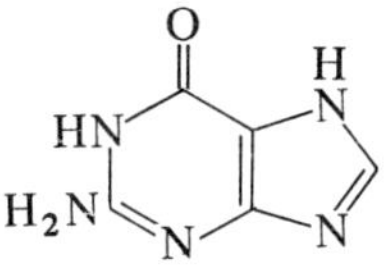

(148)
Guanine (2-amino-1,7-dihydropurin-6-one,
or 5-amino-1,6-dihydro-1,3,4,6-tetra-azainden-7-one)

D. OTHER NOMENCLATURE SYSTEMS

1. *Cyclophanes*

A specialized system of nomenclature has been developed, principally by Smith,[16] for naming skeletons consisting of aromatic residues linked in various ways by saturated bridges. Those skeletons containing benzene residues only are termed cyclophanes, and names for heteroaromatic analogues are based on the name of the heterocycle with the termination "-ophane." The numbers of atoms in the bridges are indicated in square brackets, and the orientation of substitution on the aromatic residue(s) is shown in parentheses. A few simple examples are given (**149–151**), with alternative names.

The system has recently been elaborated to an extraordinary degree by various authors,[17] yielding in many cases names almost as complex

---

[16] B. H. Smith, "Bridged Aromatic Compounds." Academic Press, New York, 1964.
[17] F. Voegtle and P. Neumann, *Tetrahedron Lett.*, 5329 (1969); *Tetrahedron* **27**, 5847 (1970); K. Hirayama, *Tetrahedron Lett.*, 2109 (1972); Th. Kauffmann, *Tetrahedron* **28**, 5183 (1972).

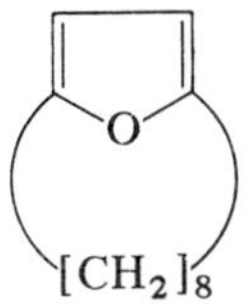

(149)

[8](3,5)Furanophane
{13-Oxabicyclo[8.2.1]trideca-1(12),10-diene
or 1,14-epoxycyclododeca-1,3-diene}

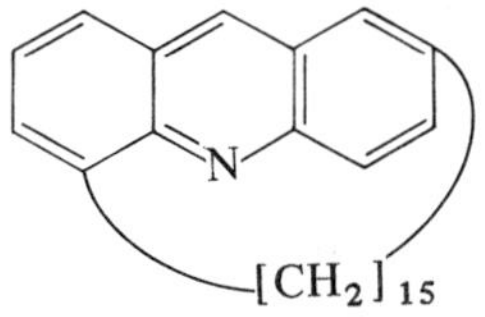

(150)

[15](2,5)Acridinophane
{26-Azatetracyclo[15.7.5.0^{19,27}.0^{21,25}]nonacosa-
1(24),17,19(27),20,22,25,28-heptaene}

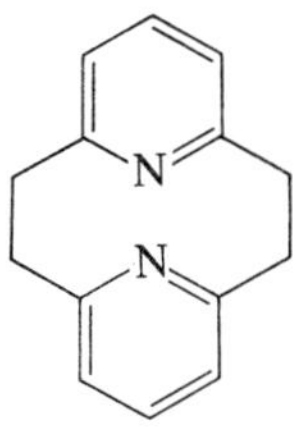

(151)

[2.2](2,6)Pyridinophane
(1,5:12,8-dinitrilotetradeca-1,3,9,11-tetraene)

as those derived by more conventional methods. Whether or not it fulfils a real need, rather than providing only an interesting intellectual exercise, remains to be seen.

2. *Rings Consisting Entirely of Heteroatoms*

Methods for naming homogeneous rings and rings of repeating units consisting entirely of heteroatoms are outlined in the IUPAC 1973 Tentative Rules (Section D).[8] Such molecules can, of course, be named by strict replacement principles, and most organic chemists will doubtless wish to use replacement names for them. (Hantzsch–Widman names could also be used.) However, this practice may seem unreasonable in application to compounds containing no carbon. Some examples of these alternative names are given here (a), along with their replacement equivalents (b) (see **152–156**).

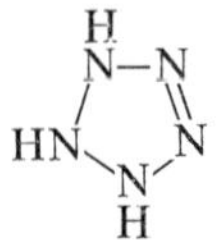

(152)

(a) Cyclopenta-azene
(b) Penta-azacyclopentene

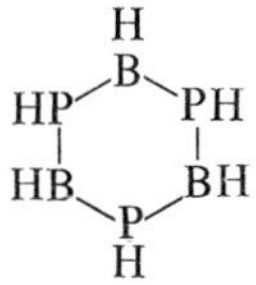

(153)

(a) Cyclotriboraphosphane
(b) 1,3,5-Triphospha-2,4,6-triboracyclohexane

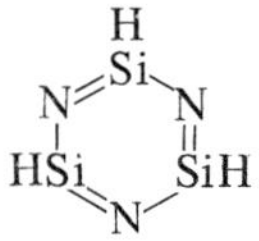

(154)

(a) Cyclotri(silazene)
(b) 1,3,5-Triaza-2,4,6-trisilabenzene

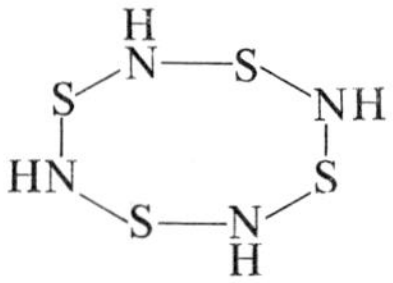

(155)

(a) Cyclotetra-azathiane
(b) 1,3,5,7-Tetrathia-2,4,6,8-tetra-azacyclo-octane

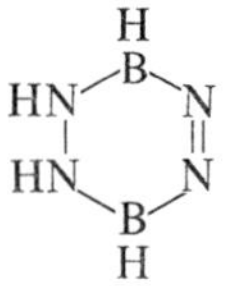

(156)

(a) Cyclodi(boradiazan)-1-ene
(b) 1,2,3,4,5-Tetra-aza-3,6-diboracyclohex-1-ene

3. Boron and Silicon Heterocycles

The naming of organoboron and organosilicon compounds receives special treatment by Chemical Abstracts and by IUPAC, and a discussion of the systems used is beyond the scope of this article. The reader is referred to the IUPAC Section D Tentative Rules[8] for a comprehensive exposition.

III. Naming of Derivatives

A comprehensive treatment of the nomenclature of substituted heterocycles will not be given here. This section will deal only with the general principles involved and with a few specific topics. For a detailed exposition, the reader should consult the IUPAC Organic Nomenclature Rules, Section C.[6]

As already pointed out, the nomenclature rules employed by Chemical Abstracts yield a unique index name for every structure, whereas the IUPAC Rules allow some latitude. This latitude is considerable with regard to the naming of substituted structures, and it is usually possible for an author to find in the rules a procedure suitable for any specific purpose.

A. General Principles

1. Nomenclature Types

a. *Substitutive.* This, the most widely applicable type of nomenclature, involves adding a substituent prefix or suffix to a parent name. Prefixes are of the radical type (often terminating in "-yl"), implying the presence of one or more free valencies. A suffix indicates the principal characteristic group in the compound (see Section III, A, 2) and also implies the presence of a free valency or valencies (whereby the parent skeleton is attached). Thus the overall process involves formal replacement of H atoms by other atoms or groups. For example, **157** would thus be named 4-chloro-3-nitroquinoline-2-carbaldehyde (or 4-chloro-3-nitro-1-azanaphthalene-2-carbaldehyde) and **158** would be 6-hydroxyimino-7,7-dimethyl-2-azabicyclo[2.2.1]heptan-5-one.

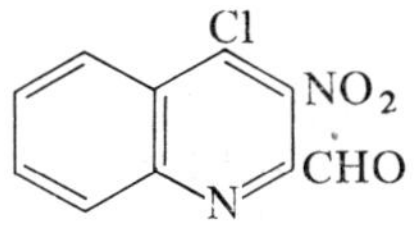

(157) 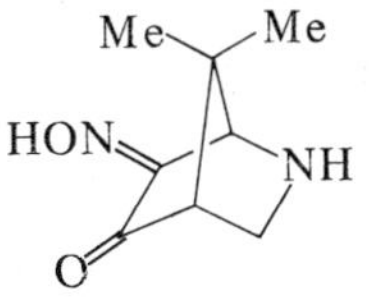   (158)

Since other types of nomenclature cannot be applied to as wide a range of structures, the Sections III, A, 2–6 will deal only with principles relating to the use of substitutive nomenclature.

b. *Conjunctive*. Conjunctive nomenclature involves the joining of two "parent" names with formal loss of the same number of H atoms from each component, as in examples **159** and **160**. [Substitutive names

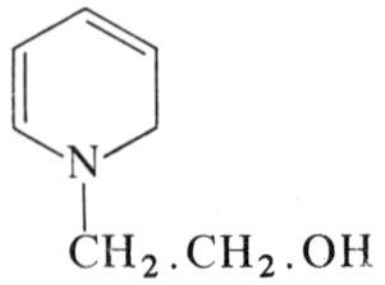

(159)

Pyridine-1(2*H*)-ethanol,
or azacyclohexa-2,4-diene-1-ethanol

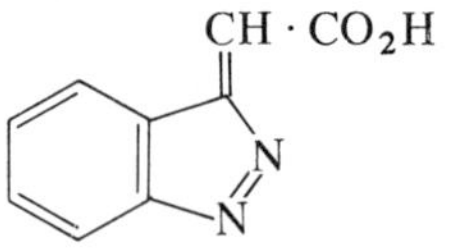

(160)

3*H*-Indazole-$\Delta^{3,\alpha}$-acetic acid,
or 3*H*-1,2-diazaindene-$\Delta^{3,\alpha}$-acetic acid

for these are 2-(azacyclohexa-2,4-dien-1-yl)ethanol **(159)** and 3*H*-1,2-diazainden-3-ylideneacetic acid **(160)**.] This type of nomenclature is used extensively by Chemical Abstracts because it frequently provides a more suitable index entry than the substitutive system [e.g., **161** is "quinoline-2-acetic acid" (listed under quinoline) rather than "acetic acid, 2-quinolyl" (listed under acetic acid)].

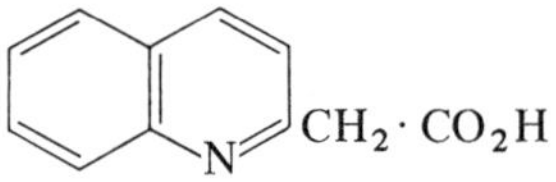

(161)

c. *Radicofunctional*. In this type of name the functional class name of the compound (e.g., alcohol, ketone, sulphoxide, etc.) is cited as a separate word after the name(s) of the radical(s) attached to the functional group. For examples, see **162** and **163**. Substitutive nomenclature is in general to be preferred [e.g., 1-(3-pyridyl)propan-1-one **(162)** and 2-oxainden-1-ol **(163)**].

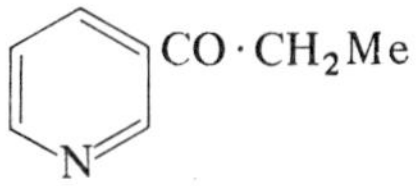

Sec. III.A]   THE NOMENCLATURE OF HETEROCYCLES   229

(162)
Ethyl 3-pyridyl ketone

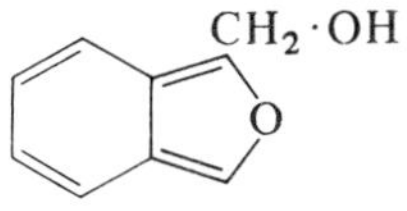

(163)
Isobenzofuran-1-ylmethyl alcohol, or 2-oxainden-1-ylmethyl alcohol

d. *Additive.* In an additive name an atom or group is formally added to a parent structure, usually by placing the name of the added unit, in ionic form, as a separate word after the name of the parent, as in examples **164** and **165**. This type of nomenclature is rarely of use nowadays.

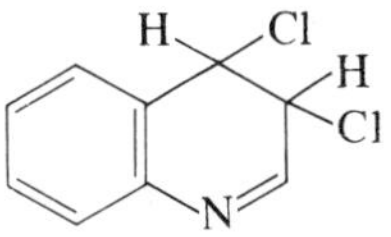

(164)
Quinoline 3,4-dichloride
(3,4-dichloro-3,4-dihydroquinoline)

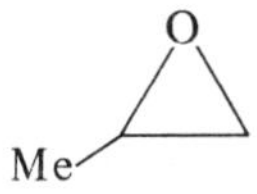

(165)
Propene oxide
(2-methyloxiran or 2-methyl-oxacyclopropane)

However, a few additive procedures are still employed for specific purposes. Thus the prefix "hydro," although normally treated as a substituent prefix, does in fact represent *addition* of a hydrogen atom. Also, in forming complex radical prefixes for substitutive nomenclature, an additive operation is often necessary [e.g., methoxy-(MeO—) + carbonyl ($>$C=O) = methoxycarbonyl (MeO$_2$C—)].

Additive names are still used in some cases to express unusual heteroatom valencies (e.g., pyridine 1-oxide; see Section II, B, 2, d).

e. *Subtractive.* Prefixes indicating removal of atoms or groups can be useful in certain circumstances. They are often applied to trivial or semitrivial names of natural products, when these refer to a substituted skeleton, to indicate removal of one or more substituents. For examples, see **166–168**. The prefix "dehydro" is particularly useful in naming arynes (see Section II, B, 2, e).

(166)
De-*N*-methyltropane or nortropane

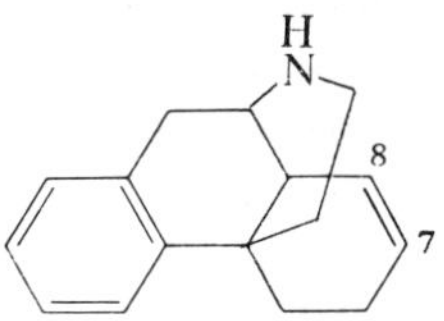

(167)
7,8-Didehydromorphinan

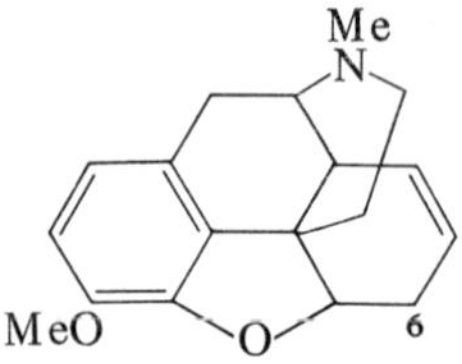

(168)

6-Deoxycodeine (replacement of 6-OH by 6-H)

2. The Principal Characteristic Group

An understanding of the concept of the "principal characteristic group" is vital for the correct application of substitutive nomenclature; this will therefore be dealt with in some detail.

Substituents can be divided for nomenclature purposes into two classes: those that are always cited as prefixes (Table II) and those that can be cited as either prefixes or suffixes (Table III). If any of the latter

TABLE II

COMMON GROUPS CITED ONLY AS PREFIXES

Group	Prefix
Halogen (—X)	e.g., Chloro, fluoro
—XO	e.g., Chlorosyl
—XO$_2$	e.g., Iodyl
—XO$_3$	e.g., Perchloryl
—IX$_2$	e.g., Dichloroiodo (additive)[a]
H (added)[a]	Hydro
=N$_2$	Diazo
—N$_3$	Azido
—NO	Nitroso
—NO$_2$	Nitro
=N(O)OH	aci-Nitro
—OR	R-oxy (additive)[a]
—SR (and —SeR, —TeR)	e.g., R-thio

[a] See Section III, A, 1, d.

are present, the name must include a suffix. However, *only one* kind of group can be cited as suffix, and this is chosen according to the order of the list in Table III, those nearest the head of the list being the most

Sec. III.A]   THE NOMENCLATURE OF HETEROCYCLES

preferred. The group thus chosen is known as the principal characteristic group (or principal group). Table III is not a fully comprehensive list, but it covers most cases commonly encountered and is

TABLE III[a]

COMMON GROUPS THAT CAN BE CITED AS EITHER PREFIXES OR SUFFIXES

Groups	Prefix	Suffix
Anionic centres	-ido, -ato	-ide, -ate
Cationic centres	-io, -ia	-ium
$-CO_2H$[b]	Carboxy	-carboxylic acid
$-(C)O_2H$[b]	—	-oic acid
$-SO_3H$[b]	Sulpho	-sulphonic acid
$-SO_2H$[b]	Sulphino	-sulphinic acid
$-SOH$[b]	Sulpheno	-sulphenic acid
$-CO_2M$[b](salts)	—	M -carboxylate
$-(C)O_2M$[b]	—	M -oate
$-CO_2R$[b](esters)	R-oxycarbonyl	R -carboxylate
$-(C)O_2R$[b]	—	R -oate
$-COX$[b]	Halogenoformyl	-carbonyl halide
$-(C)OX$[b]	—	-oyl halide
$-CONH_2$[b]	Carbamoyl	-carboxamide
$-(C)ONH_2$[b]	—	-amide
$HN=\overset{\mid}{C}-NH_2$	Amidino	-carboxamidine
$HN=(\overset{\mid}{C})-NH_2$	—	-amidine
$-CN$	Cyano	-carbonitrile
$-(C)N$	Nitrilo	-nitrile
$-CHO$[b]	Formyl	-carbaldehyde
$-(C)HO$[b]	Oxo	-al
$>(C)=O$	Oxo	-one
(R)-OH	} Hydroxy	-ol
(Ar)-OH		
$-SH$	Mercapto	-thiol
$-NH_2$	Amino	-amine
$=NH$	Imino	-imine

[a] This table includes only groups containing C, H, O, N, halogen, S, Se, and/or Te. Suggestions for extension to other elements are given in the 1969 Section D IUPAC Tentative Rules.[8]
[b] Followed in order by the corresponding thio, seleno, and telluro derivatives.

capable of logical extension. When more than one of the same principal group is present, all these groups are cited as a suffix if possible. For examples, see **169–173**. (These compounds can, of course, all be named as 1-oxa-anthracene derivatives.)

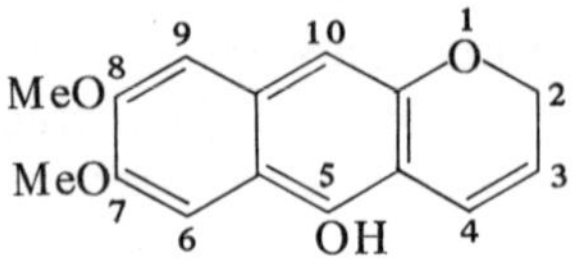

(169)

7,8-Dimethoxy-2H-naphtho[2,3-b]pyran-5-ol

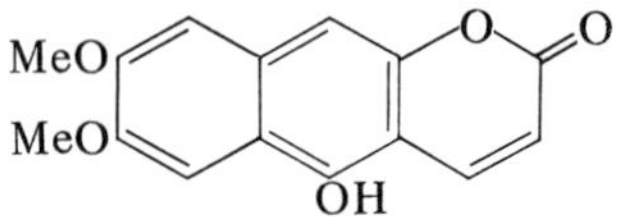

(170)

5-Hydroxy-7,8-dimethoxynaphtho[2,3-b]pyran-2-one

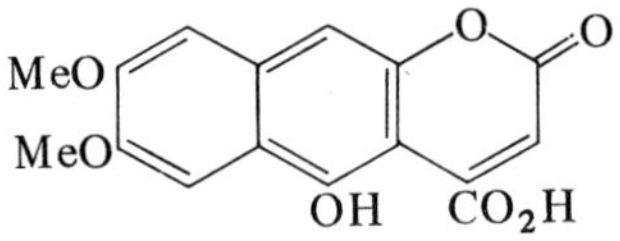

(171)

5-Hydroxy-7,8-dimethoxy-2-oxonaphtho[2,3-b]pyran-4-carboxylic acid

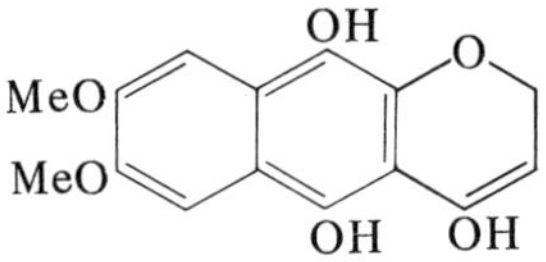

(172)

2,8-Dimethoxy-2H-naphtho[2,3-b]pyran-4,5,10-triol

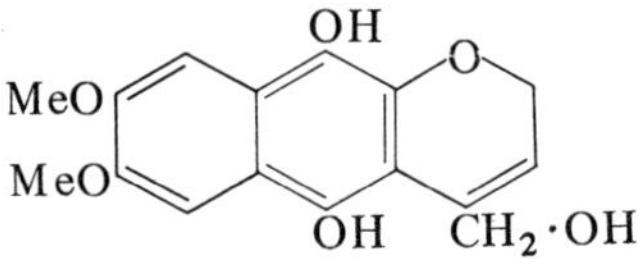

(173)

4-Hydroxymethyl-7,8-dimethoxy-2H-naphtho[2,3-b]pyran-5,10-diol

A correct choice and citation of the principal group is essential because of its effect on the numbering of the skeleton (see the following section). After "indicated hydrogen," the point of attachment of the principal group has the highest priority for assignment of lowest number; thus if there is any choice in numbering the unsubstituted skeleton, a decision is often controlled by the location of the principal group (see examples **174** and **175**).

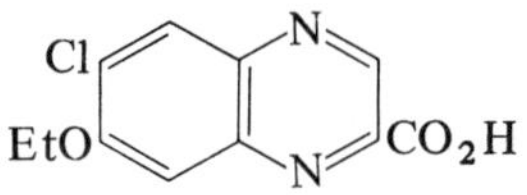

Sec. III.A] THE NOMENCLATURE OF HETEROCYCLES 233

(174)

6-Chloro-7-ethoxy-1,4-diazanaphthalene-2-carboxylic acid

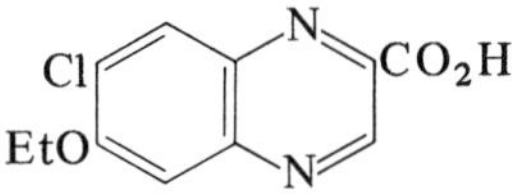

(175)

7-Chloro-6-ethoxy-1,4-diazanaphthalene-2-carboxylic acid

In complex structures the location of the principal group may also decide which portion of the molecule should be considered as the parent for carrying out substitutive operations. For example: (a) if the principal group occurs in a chain carrying a cyclic substituent, the chain is the

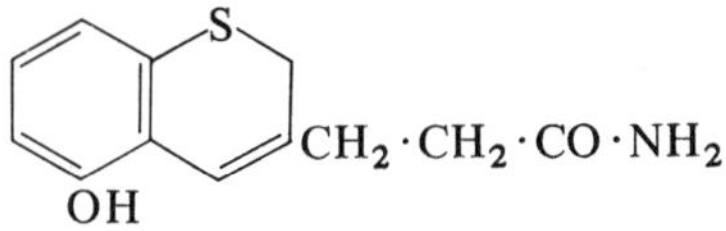

(176)

3-(4-Hydroxy-2H-1-thianaphthalen-3-yl)propionamide

nomenclature parent (e.g., 176); (b) if the group occurs in the cyclic substituent and not in the chain, the ring is the nomenclature parent (e.g., 177); (c) if the compound contains several rings, only one of which

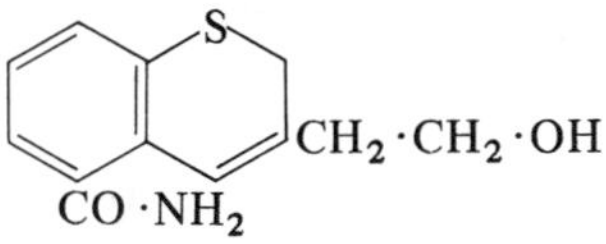

(177)

3-(2-Hydroxyethyl)-2H-1-thianaphthalene-5-carboxamide

carries the principal group, that ring is the parent (e.g., 178); (d) if the principal group occurs in more than one ring, the nomenclature parent

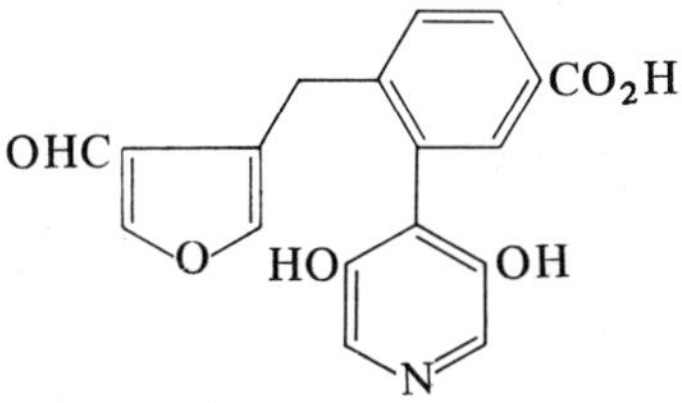

(178)

3-(3,5-Dihydroxy-4-pyridyl)-4-(4-formyl-3-furylmethyl)benzoic acid

is, in order of preference, (i) the ring containing the most principal groups, (ii) the senior ring system as outlined in IUPAC Rule C-14 (see Appendix) (e.g., **179–181**).

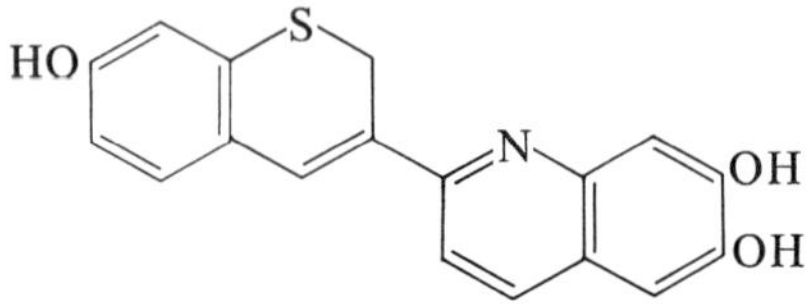

(179)

2-(7-Hydroxy-2*H*-1-thianaphthalen-3-yl)-1-azanaphthalene-6,7-diol

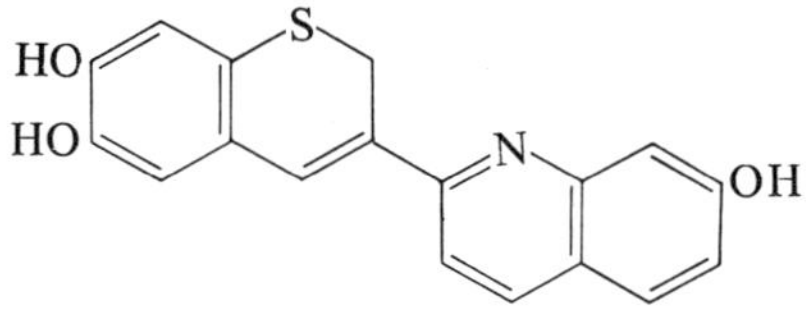

(180)

3-(7-Hydroxy-1-azanaphthalen-2-yl)-2*H*-1-thianaphthalene-6,7-diol

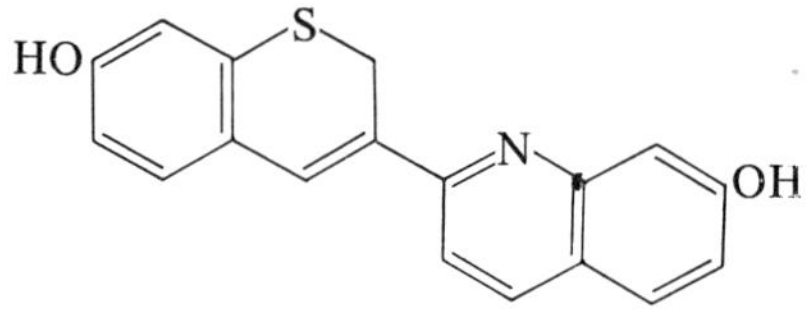

(181)

2-(7-Hydroxy-2*H*-1-thianaphthalen-3-yl)-1-azanaphthalen-7-ol

It may not always be advantageous to apply this system rigorously, but one should have good reasons for deviating from it. (For example, in comparing a number of compounds with widely different substitution patterns, it would be useful to use the same parent name for the whole series.)

3. *Numbering*

The rules for numbering heterocyclic skeletons have been mentioned already (see Section II, B, 2, b). However, these rules may not be sufficient to assign a unique numbering when substituents are present. Further rules are provided by IUPAC to deal with such cases (Rule C-15; see Appendix). These require no comment except to note (a) the high priority for lowest number given first to "indicated hydrogen" and then to the principal group, and (b) the fact that when the cyclic unit is named

Sec. III.A] THE NOMENCLATURE OF HETEROCYCLES 235

as a radical prefix (the principal group being elsewhere in the molecule), the point of radical attachment is next in line after "indicated hydrogen" for assignment of lowest number (see examples **182** and **183**).

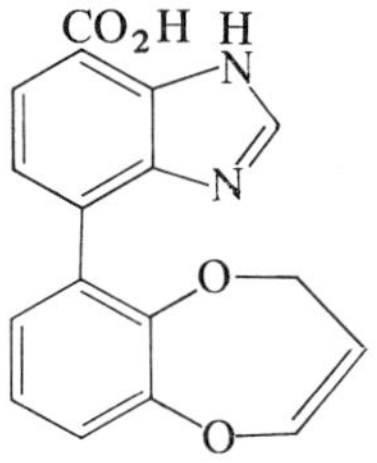

(182)

4-(2*H*-1,5-Benzodioxepin-9-yl)-1*H*-benzimidazole-7-carboxylic acid

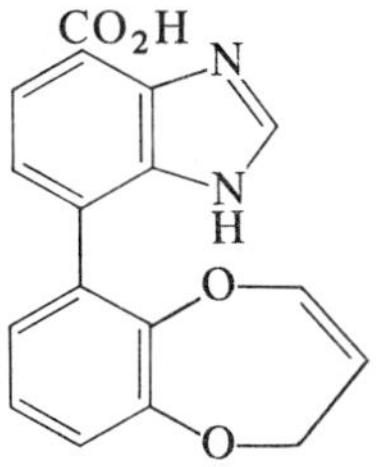

(183)

7-(2*H*-1,5-Benzodioxepin-6-yl)-1*H*-benzimidazole-4-carboxylic acid

4. *Ordering of Prefixes*

Prefixes are of two kinds, detachable and nondetachable. Nondetachable prefixes, such as bicyclo, spiro, benzo, aza, ethano, are integral parts of the parent name and should not be separated from it. Their use has been described already. Detachable prefixes almost invariably refer to substituents and appear in *alphabetical* order before the parent name. "Hydro" and subtractive prefixes (e.g., demethyl) have sometimes been treated as nondetachable, appearing immediately before any other nondetachable prefixes, but it is preferable to treat them as detachable except when they do not have associated locants. This exception arises in particular when the name of a fully hydrogenated cyclic molecule is derived from a fully unsaturated parent name, e.g., 2-iodoperhydro-1-azanaphthalene (or 2-iododecahydro-1-azanaphthalene), 2-nitrotetrahydrofuran.

The use of alphabetical order for detachable prefixes is of some importance even when problems of indexing are ignored, since in some cases the order of prefixes may determine the numbering system [see for example IUPAC Rule C-15.11(e) (Appendix)]. However, it requires some care in application. The following examples are illustrative (the critical characters are in bold type):

> 3-chloro-2,5-dimethylpyridin-4-ol
> 2-chloro-3-dimethylaminopyridin-4-ol
> 2-chloro-3,5-bisdimethylaminopyridin-4-ol
> 2-(4-ethyl-3,5-dinitrophenyl)pyridine
> 2-methyl-3-methylaminopyridin-4-ol (short prefixes before long)
> 2-(1-aminoethyl)-3-(2-aminoethyl)pyridin-4-ol
> 2-(1-chloroethyl)-3-(2-chloromethyl)pyridine

Even in cases where a parent heterocyclic skeleton is not numbered as a whole (e.g., spiro method 6a, ii and ring assembly nomenclature) it is important to realize that substituent prefixes (including hydro) should appear before the whole parent name, *not* before the individual component names (see example **184**).

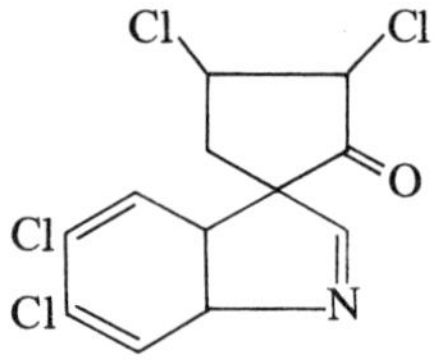

(184)

3′,4′,5,6-Tetrachloro-3a,7a-dihydrospiro-[3H-1-azaindene-3,1′-cyclopentan]-2′-one

5. *Heterocycles as Substituents*

Heterocyclic substituent prefixes generally terminate in "-yl" for single-bond and "-ylidene" for double-bond links. Radical prefixes signifying free valencies of more than one ring atom terminate in "-diyl," triy.," etc. (the alternative "-ylene" to "-diyl" is of no value and could well be discarded).

The only problem of any significance arises with an "-ylidene" type of radical formed from a maximally unsaturated parent, when formal saturation of one double bond is required before the radical can be derived. [Similar problems arise with spiro compounds (Section II, B, 2, a, ii), cations (Section II, B, 2, c, i) and ketones (Section III, B, 1).] In such cases the use of "indicated hydrogen" rather than a "dihydro" prefix is recommended (see example **185**) [with the proviso that maximum

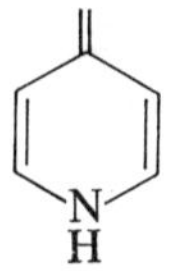

(185)
4(1H)-Pyridylidene

unsaturation is formally introduced *after* the two radical bonds (e.g., 1,2-dihydro-3H-indol-3-ylidene is the correct name for **186**, not 1H-indol-3(2H)-ylidene].

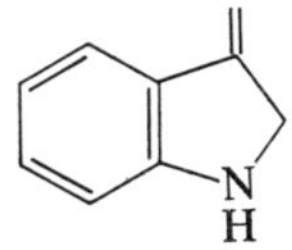

(186)

If the numbering of a ring system is not already fixed, a radical position normally takes precedence over all others, except a position carrying "indicated hydrogen," for lowest number. However, when the "indicated hydrogen" is cited only as a result of formation of a bivalent radical, this order is reversed (e.g., **187**).

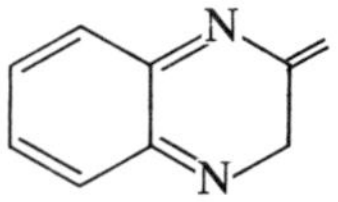

(187)
Quinoxalin-2(3H)-ylidene

When two different unsubstituted cyclic skeletons are bonded to each other it is necessary to name one as a substituent of the other. A decision as to which skeleton should be regarded as the parent is governed by seniority rules (IUPAC Rule C-14; see Appendix).

A number of contracted radical names derived from trivial parent names are in common use, e.g., 2-furyl, 8-quinolyl, 2-benzo[b]thienyl (see IUPAC Rule B-2, tables).

6. *Construction of a Substitutive Name*

It is useful at this point to summarize briefly the procedure for deriving a substitutive name. The following operations should be carried out in the order given:

   a. Choose the principal group (see Section 2 above)

b. Choose and name the parent skeleton (ring or chain; this must contain the principal group at least once; see Section 2 again)

c. Number the parent skeleton (see Section 3)

d. Cite all substituents other than the principal group alphabetically as prefixes (see Section 4) and the latter as suffix.

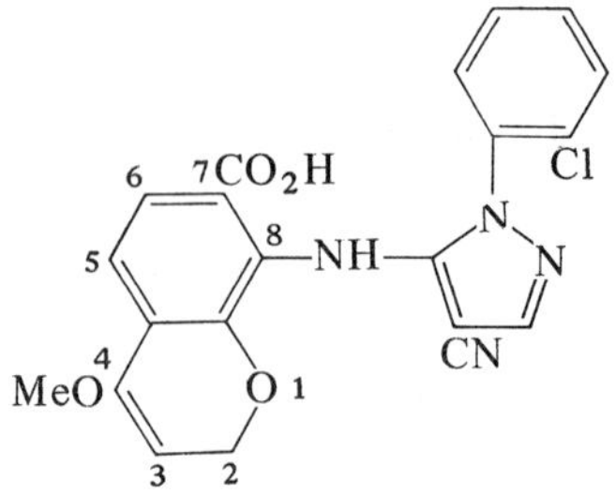

(188)

Principal group: CO_2H
Parent skeleton: $2H$-chromen (or 1-oxanaphthalene)
Name: 8-[1-(2-Chlorophenyl)-4-cyanoimidazol-5-ylamino]-4-methoxy-$2H$-chromen-7-carboxylic acid

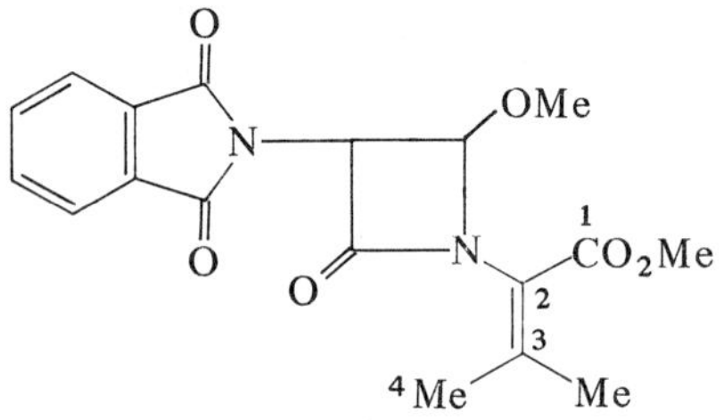

(189)

Principal group: CO_2Me
Parent skeleton: but-2-ene
Name: Methyl 2-(2-methoxy-4-oxo-3-phthalimidoazetidin-1-yl)-3-methylbut-2-enoate

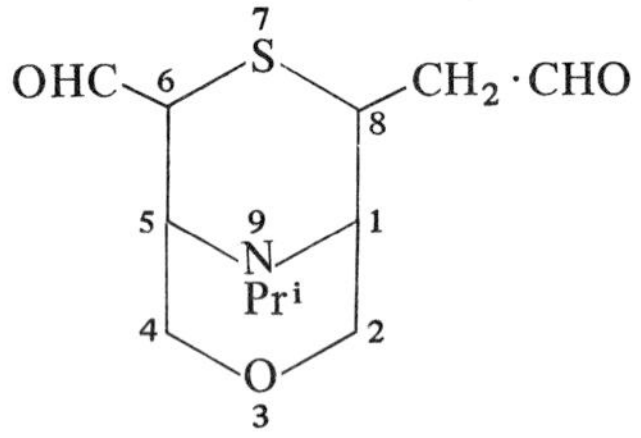

(190)

Principal group: CHO
Parent skeleton: oxa-thia-azabicyclo[3.3.1]nonane
Name: 8-(Formylmethyl)-9-isopropyl-3-oxa-7-thia-9-azabicyclo[3.3.1]nonane-6-carbaldehyde

Sec. III.B]   THE NOMENCLATURE OF HETEROCYCLES   239

These directions are somewhat simplified, but will apply to the majority of structures encountered. An extensive list of substituent prefixes is given at the end of the 1969 IUPAC Organic Nomenclature Rules, Section C.[6] For some examples, see **188–190**.

B. Some Specific Topics

1. *Ketones and Imines*

When the principal group attached to a fully unsaturated parent skeleton is a ketone or an imine, the same problem arises as that noted for spiro compounds (Section II, B, 2, a, ii), cations (Section II, B, 2, c, i), and bivalent radicals (Section III, A, 5), and the same solution is applied. Formal hydrogenation of one double bond is necessary before the group can be introduced, and this is shown by "indicated hydrogen" (see example **191**). However, as before, when this procedure is applied,

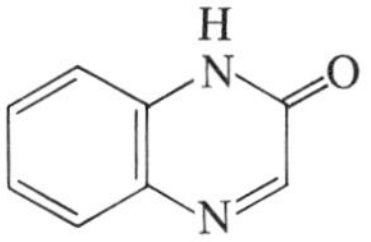

(191)

Quinoxalin-2(1*H*)-one

the state of hydrogenation of the parent skeleton is considered to be that obtained by introducing maximum unsaturation *after* inserting the ketone or imine group [i.e., quinoxaline-2,3-dione, not 2,3-dihydroquinoxaline-2,3-dione (**192**); 1,2-dihydroindol-3-one, not 1*H*-indol-3(2*H*)-one (**193**)]. Strict application of this rule (IUPAC Rule C-315) is essential in order to avoid confusion. The problem arises less

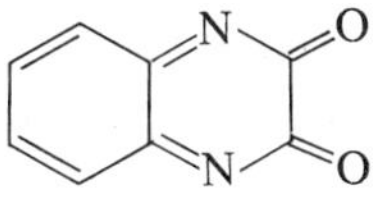

(192) 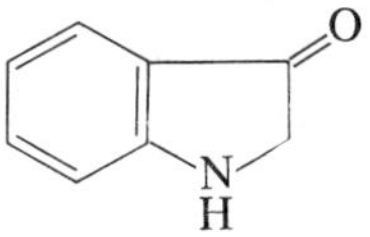   (193)

often with imines, since the priority for citation of "-imine" as a suffix is very low (Table II), and use of "oxo-" and "imino-" as prefixes presents no difficulty.

The following are examples of contracted names for heterocyclic ketones allowed by IUPAC: 2-pyridone, 4-quinolone, 4-oxazolone, 5-pyrazolone, 9-acridone, piperidone, 2-pyrrolidone, 4-thiazolidone.

2. Thio Compounds

The prefix "thio-" (not to be confused with the *carbon* replacement prefix thia-) signifies replacement of an oxygen by a sulphur atom and is widely used in both trivial and systematic names, e.g., thiopyran, pyridine-2-thiocarbaldehyde. Except insofar as a trivial component name may contain a thio prefix (e.g., thioxanthen), "thio" should never be used in naming heterocyclic skeletons: it is not necessary and the likelihood of confusion with "thia" is great. Its principal usefulness is in naming sulphur analogues of oxygen-containing substituents, e.g., -thiocarboxylate (—CS·OR), -thiosulphinate (—SO·SR), thioformyl- (HCS—), etc. The analogous oxygen replacement prefixes "seleno-" and "telluro-" may be used similarly. It is pertinent to note the difficulties in interpreting the trivial contraction "thiopyrone"; is this **194** or **195**?

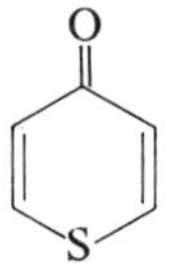

(194) 1-Thio-4-pyrone

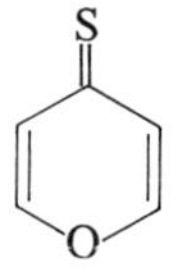

(195) Pyran-4-thione

3. Dipolar (+, —) Compounds

The naming of skeletons in which both a positive and a negative charge can be formally located on ring atoms has already been discussed (Section II, B, 2, c, iii). When one of the charges is located on a substituent, the name can be formed in one of two ways: either (a) the cationic group can be considered as a substituent of the anionic group, and cited as a *prefix* formed by changing the "-ium" ending into "-io", or (b) the anionic group can be considered as a substituent of the cationic group and cited as a *suffix* (see IUPAC Rule C-87, not included in the Appendix). Examples **196–198** illustrate both methods.

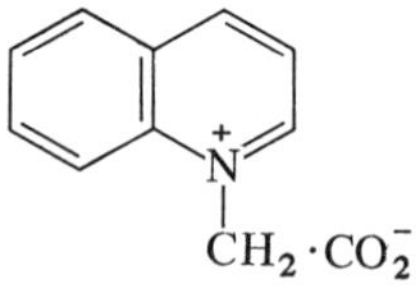

(196)

(a) (1-Quinolinio)acetate
[or (1-azanaphthalenio)acetate]

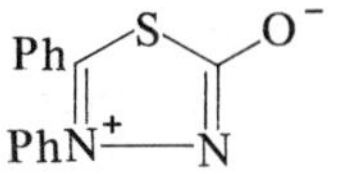

$$\text{(197)}$$

Ph—S—O⁻, PhN⁺—N (4,5-diphenyl-1,3,4-thiadiazol structure)

(b) 4,5-Diphenyl-1,3,4-thiadiazol-4-ium-2-olate
(or 4,5-diphenyl-1-thia-3,4-diazacyclopenta-2,4-dien-4-ium-2-olate)

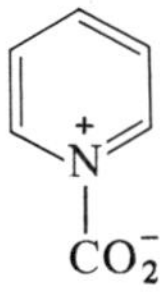

(198) Pyridinium with CO_2^- substituent on N

(a) (1-Pyridinio)formate
(b) Pyridinium-1-carboxylate

This type of nomenclature, although simple and straightforward, is not used by Chemical Abstracts, which prefers the less elegant "inner salt" method already described (e.g., **197** would thus be named 5-hydroxy-2,3-diphenyl-1,3,4-thiadiazolium hydroxide, inner salt). (Use of the prefix "anhydro-" instead of the words "inner salt" is quite widespread, but represents only a marginal improvement.) Chemical Abstracts also uses a further system when the compound can be considered as an ylide (i.e., when the charges can be formally located on adjacent atoms, with the negative charge on carbon). In such cases the cation and anion names are cited as separate words, with the latter terminating in "ylide"; thus, **199** would be named pyridinium dicarboxymethylide rather than dicarboxy-(1-pyridinio)methanide. This is unsatisfactory if one accepts the IUPAC recommendation that the

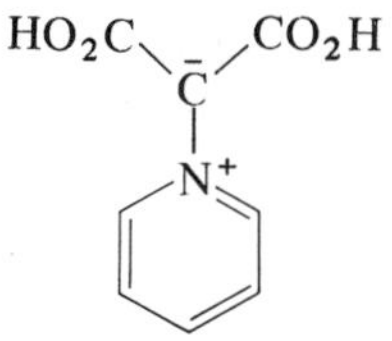

(199) Pyridinium N-C(CO₂H)₂ ylide structure

"-ylide" termination is to be used to indicate a discrete radical anion (Rule C-84.4). [However the termination "-idyl" would seem to be more suitable for radical anions, by analogy with the ending "-iumyl" for radical cations (Rule C-83.3).]

It must be remembered when dealing with compounds of this nature that a systematic name will usually describe only one "resonance" form of a structure. (Indeed this statement is generally applicable; for example, the systematic name of any polyunsaturated compound usually refers to a structure in which the double bonds have been localized.) In a case where it is not desired to represent the molecule in any one particular charge-localized form, systematic nomenclature cannot readily be used. The sydnones (and sydnone imines and related compounds) are a case in point. Chemical Abstracts depicts these as in example **200**, and

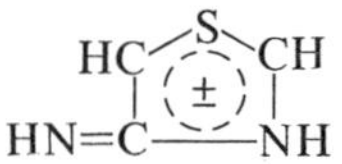

(200)

4-Iminothiazolidine mesoionic didehydro derivative

deals with complex cases by naming the dihydro derivative of the ketone (or imine) form of molecule and adding the words "mesoionic didehydro derivative," as shown. This method is cumbersome, and it is preferable both to draw and to name either the charge-localized form, e.g., **201a**, or at least the partially localized structure (**201b**) (omitting in the name the locant for the positive charge).

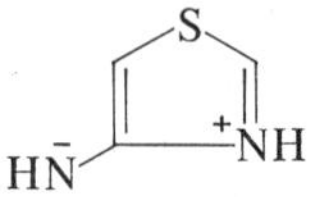

(201a) (201b)

Thiazol-3-ium-4-aminide   Thiazolium-4-aminide

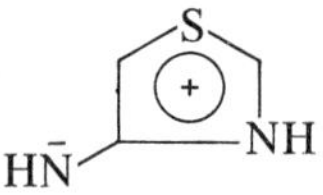

N-Oxides and their analogues (e.g., *N*-imides) can be considered as a special type of dipolar structure. They are usually dealt with by use of simple additive nomenclature (e.g., pyridine *N*-oxide), but alternatives are possible. For example, other methods of indicating unusual valency could be applied, e.g., $1\lambda^5$-pyridin-1-one (see Section II, B, 2, d).

4. *Stereochemistry*

Methods of designating geometrical (*ZE* nomenclature) and chiral isomers (*RS* nomenclature) are given in the IUPAC Rules for Nomenclature of Organic Chemistry, Section E,[18] and will not be discussed here.

---

[18] "IUPAC Nomenclature of Organic Chemistry," Definitive Rules for Section E. Fundamental Stereochemistry, *Pure Appl. Chem.* **45**, 11 (1976).

IV. Miscellaneous Examples

The following examples are taken from recent issues of the *Journal of the Chemical Society*.

Example:

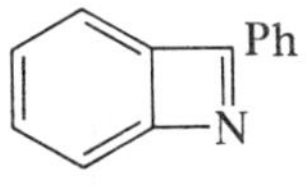

(202)

2-Phenylbenzazete

Chemical Abstracts would name this as 8-phenyl-7-azabicyclo[4.2.0]-octa-1,3,5,7-tetraene, which is unnecessarily complicated. Use of a replacement modification of a fusion name presents difficulties: although the corresponding carbocycle should, in accord with IUPAC Rules, be named cyclobutabenzene, it is almost exclusively referred to in the literature as benzocyclobutene. Thus the name 2-phenyl-1-azacyclobutabenzene would be strictly correct, but 2-phenyl-1-azabenzocyclobutene is more in line with current usage. The fusion name based on "azete" avoids this problem.

Example:

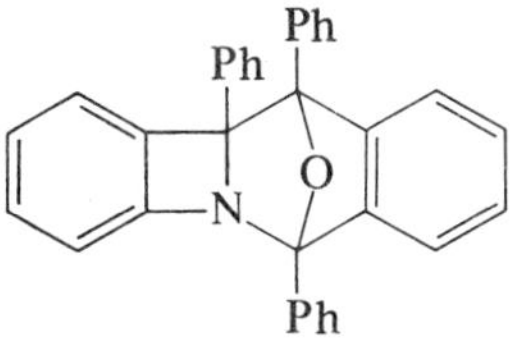

(203)

10,10a-Dihydro-5,10,10a-triphenyl-5,10-epoxy-5*H*-4b-azabenzo[3,4]-cyclobuta[1,2-*b*]naphthalene, or 11,11a-dihydro-6,11,11a-triphenyl-6,11-epoxy-6*H*-[1]benzazeto[1,2-*b*]isoquinoline

A name based on oxa-azapentacycloheptadecahexaene would be much more complex and less appropriate. The difference in numbering between the above alternatives arises from the assignment of an individual locant, 5, to the N atom in the latter.

Example:

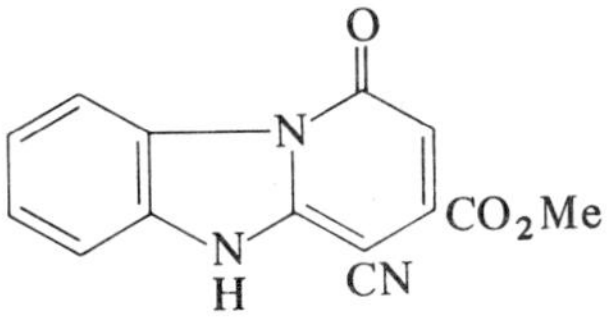

(204)

Methyl 4-cyano-1,5-dihydro-1-oxopyrido[1,2-*a*]benzimidazole-3-carboxylate, or methyl 1-cyano-4,9-dihydro-4-oxo-4a,9-diazafluorene-2-carboxylate

This is a good example of the sort of skeletal numbering problems that can arise. Consider first the name based on pyrido[1,2-a]benzimidazole. At first sight there are four possible numbering systems, a–d. All give the same locants to the N atoms (5 and 10). A choice among these is governed by IUPAC Rule B-3.4(c) (allowing carbon atoms common to two or more rings to receive lowest numbers). Thus system a is preferred (4a,5a,9a rather than 4a,5a,10a, 4a,9a,10a, or 5a,9a,10a). For the replacement name, only two numberings are possible, e and f, and system e is preferred since it gives lowest numbers to heteroatoms [Rule B-3.4(a); 4a,9 rather than 4b,9].

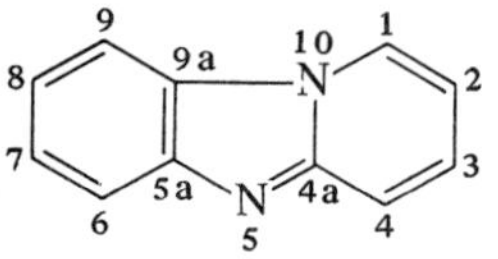

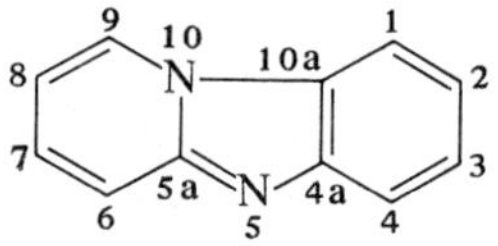

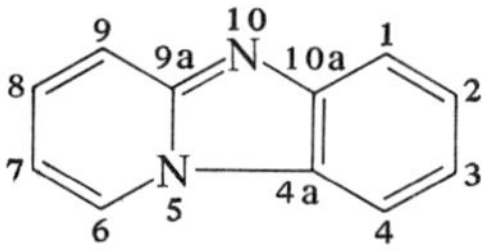

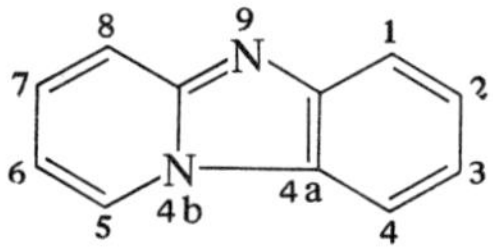

Example:

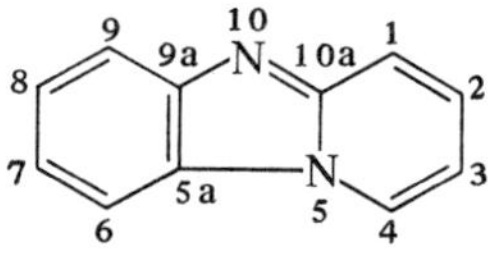

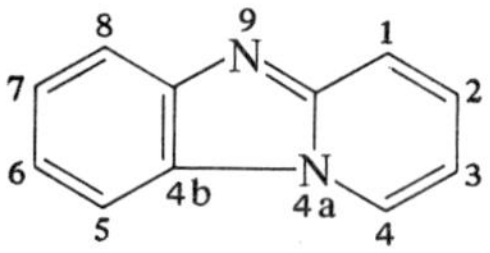

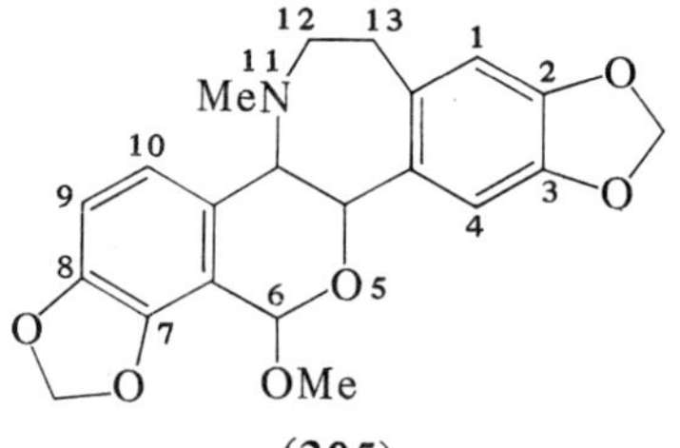

(205)

4b,6,10b,11,12,13-Hexahydro-6-methoxy-2,3:7,8-bismethylenedioxy-11-methyl-5-oxa-11-azabenzo[3,4]cyclohepta[1,2-a]naphthalene, or the correspondingly substituted [2]benzopyrano[3,4-a][3]benzazepine [rhoeadine (or rheadine)]

This is an example of a systematic name for an alkaloid. Chemical Abstracts would treat the dioxolen units as additional fused rings, but it seems preferable in cases of this type to regard them as substituents.

Example:

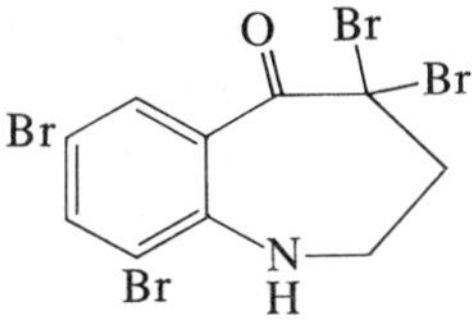

(206)

4,4,7,9-Tetrabromo-1,2,3,4-tetrahydro-1-benzazepin-5-one,
or 2,4,8,8-tetrabromo-5,6,7,8-tetrahydro-5-azabenzocyclohepten-9-one

This illustrates the correct way of naming a ketone derivative of a partially unsaturated fused heterocycle: as far as determining the mode of citation of the degree of hydrogenation is concerned, the parent is structure a, not b [which could yield incorrect names such as 4,4,7,9-tetrabromo-2,3,4,5-tetrahydro-1H-1-benzazepin-5-one or 4,4,7,9-tetrabromo-2,3-dihydro-1H-1-benzazepin-5(4H)-one] (IUPAC Rule C-315).

(a) 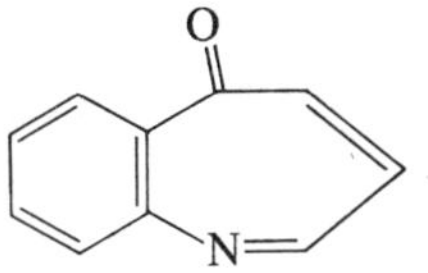   (b)

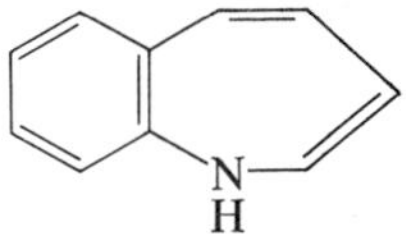

Example:

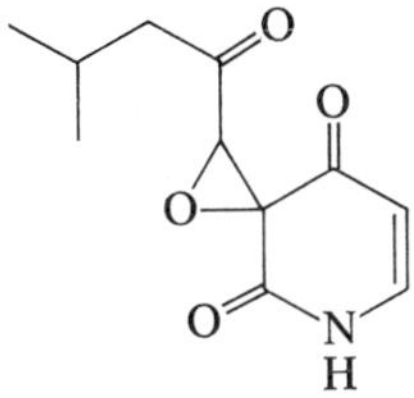

(207)

3-(3-Methylbutanoyl)spiro[oxiran-2,3'(2'H)-pyridine]-2',4'(1'H)-dione (flavipucine)

This again illustrates the problem of correct citation of hydrogenation. The unsubstituted spiro skeleton is structure a, named as spiro[oxiran-2,3'(2'H)-pyridine], it being necessary to introduce an "indicated hydrogen" (2'H) to allow for the spiro linkage. The ketone groups are

(a)

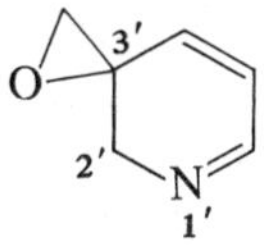

then formally introduced before final insertion of the maximum number of double bonds into the pyridine ring, and a further "indicated hydrogen" becomes necessary ($1'H$). The rules are not very clear with respect to the correct procedure when these two types of "indicated hydrogen" appear to require citation in the same name: one could regard citation of $2'H$ as superfluous, but it is probably safer to leave it in.

Example:

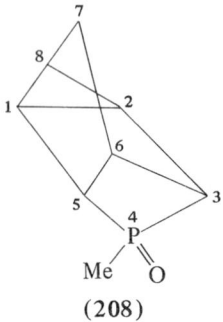

(208)

4-Methyl-4-phosphatetracyclo[3.3.0.0^{2,8}.0^{3,6}]octane 4-oxide

Complex bridged structures of this type are often incorrectly named in the literature. It may therefore be useful to describe the derivation of the above name in detail. The appropriate IUPAC Rule is A-32 (especially subrule A-32.31). The first operation is to identify the main ring, which must contain as many carbon atoms as possible. This is often awkward to visualize; in the present example, it is in fact possible to draw the skeleton as an eight-membered ring with three single-bond bridges (see representation a), and this ring is the basis for naming the compound, rather than one of the alternative representations: b (six-membered) and c (seven-membered). The main bridge is then identified: since all three

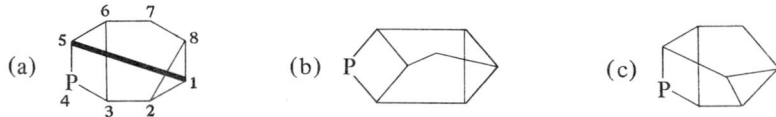

are of the same length, it is defined as the bridge that divides the main ring most equally (i.e., the one illustrated as a thickened line). Finally the skeleton is numbered, starting with one end of the main bridge. If the paths to the other end are of unequal length, the longer is followed first; if, as in the present case, they are equal in length, the numbering is chosen such as to give overall lowest locants to the other bridges. This latter criterion also decides here which end of the main bridge is numbered 1. (Thus the locants 2,3,6,8 are preferred to 3,4,6,8, 2,4,6,7, or 2,4,7,8.) In this example the position of the heteroatom does not affect the numbering.

ACKNOWLEDGMENTS

I thank the IUPAC authorities for permission to reproduce the extracts from the IUPAC Rules provided in the Appendix. I also thank my colleagues in the Chemical Society editorial office and at UKCIS, Nottingham, for advice and encouragement.

Appendix: IUPAC Rules A-11 to A-56, B-1 to B-15, C-14, and C-15

The following extracts from the 1969 IUPAC Definitive Rules for the Nomenclature of Organic Chemistry, Sections A–C,[6] are reproduced by permission of the IUPAC authorities.

MONOCYCLIC HYDROCARBONS

Rule A-11. Unsubstituted Compounds and Radicals

11.1—The names of saturated monocyclic hydrocarbons (with no side chains) are formed by attaching the prefix "cyclo" to the name of the acyclic saturated unbranched hydrocarbon with the same number of carbon atoms. The generic name of saturated monocyclic hydrocarbons (with or without side chains) is "cycloalkane".
Examples:

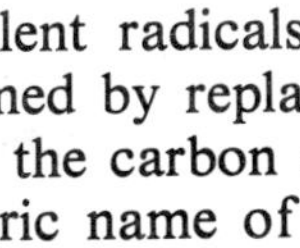

Cyclopropane 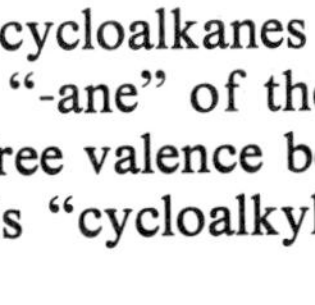   Cyclohexane

11.2—Univalent radicals derived from cycloalkanes (with no side chains) are named by replacing the ending "-ane" of the hydrocarbon name by "-yl", the carbon atom with the free valence being numbered as 1. The generic name of these radicals is "cycloalkyl".
Examples:

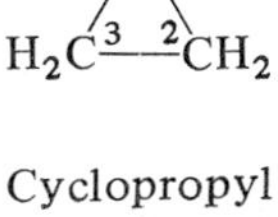

Cyclopropyl 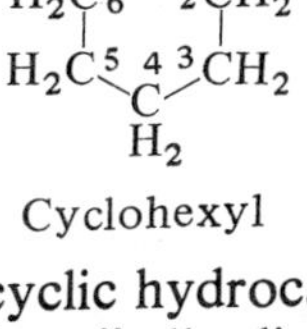   Cyclohexyl

11.3—The names of unsaturated monocyclic hydrocarbons (with no side chains) are formed by substituting "-ene", "-adiene", "-atriene", "-yne", "-adiyne", *etc.*, for "-ane" in the name of the corresponding cycloalkane. The double and triple bonds are given numbers as low as possible as in Rule **A-3.3**.

Examples:

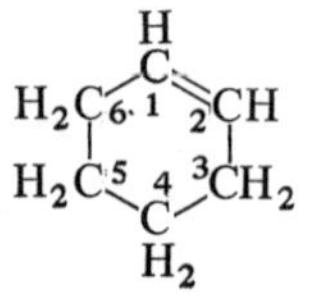

Cyclohexene

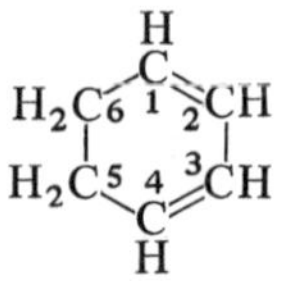

1,3-Cyclohexadiene

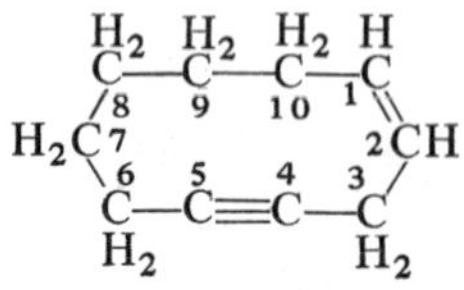

1-Cyclodecen-4-yne

The name "benzene" is retained.

11.4—The names of univalent radicals derived from unsaturated monocyclic hydrocarbons have the endings "-enyl", "-ynyl", "-dienyl", *etc.*, the positions of the double and triple bonds being indicated according to the principles of Rule **A-3.3**. The carbon atom with the free valence is numbered as 1, except as stated in the rules for terpenes (see Rules **A-72** to **A-75**).

Examples:

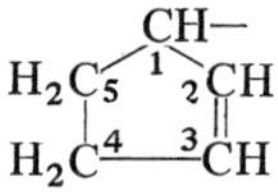

2-Cyclopenten-1-yl

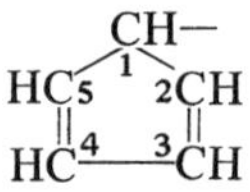

2,4-Cyclopentadien-1-yl

The radical name "phenyl" is retained.

11.5—Names of bivalent radicals derived from saturated or unsaturated monocyclic hydrocarbons by removal of two atoms of hydrogen from the same carbon atom of the ring are obtained by replacing the endings "-ane", "-ene", "-yne", by "-ylidene", "-enylidene" and "-ynylidene", respectively. The carbon atom with the free valences is numbered as 1, except as stated in the rules for terpenes.

Examples:

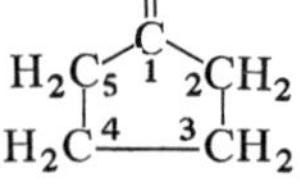

Cyclopentylidene

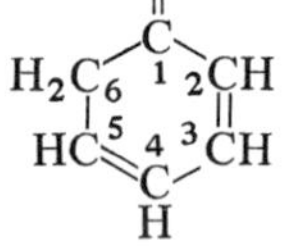

2,4-Cyclohexadien-1-ylidene

11.6—Bivalent radicals derived from saturated or unsaturated monocyclic hydrocarbons by removing a hydrogen atom from each of two different carbon atoms of the ring are named by replacing the endings

Appendix] MONOCYCLIC HYDROCARBONS 249

"-ane", "-ene", "-diene", "-yne", *etc.*, of the hydrocarbon name by "-ylene", "-enylene", "-dienylene", "-ynylene", *etc.*, the positions of the double and triple bonds and of the points of attachment being indicated. Preference in lowest numbers is given to the carbon atoms having the free valences.

Examples:

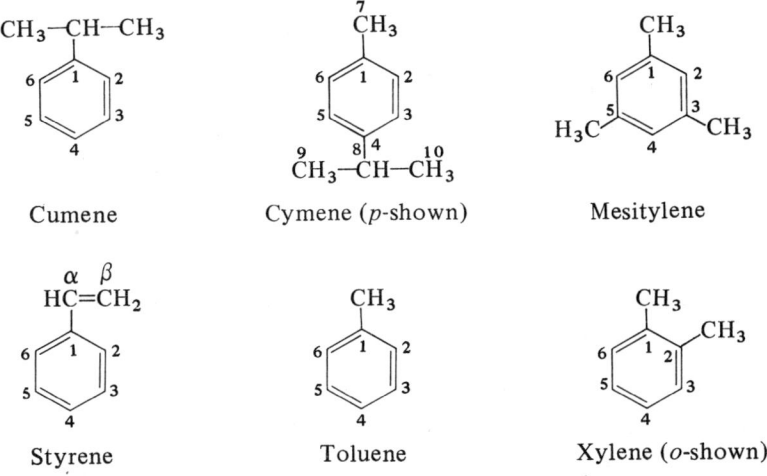

1,3-Cyclopentylene   3-Cyclohexen-1,2-ylene   2,5-Cyclohexadien-1,4-ylene

The following name is retained:

Phenylene (*p*-shown)

Rule A-12. Substituted Aromatic Compounds

12.1—The following names for monocyclic substituted aromatic hydrocarbons are retained:

Cumene   Cymene (*p*-shown)   Mesitylene

Styrene   Toluene   Xylene (*o*-shown)

12.2—Other monocyclic substituted aromatic hydrocarbons are named as derivatives of benzene or of one of the compounds listed in Part **.1** of this rule. However, if the substituent introduced into such a compound is identical with one already present in that compound, then the substituted compound is named as a derivative of benzene (see Rule **61.4**).

12.3—The position of substituents is indicated by numbers except that *o-* (*ortho*), *m-* (*meta*) and *p-* (*para*) may be used in place of 1,2-, 1,3-, and 1,4-, respectively, when only two substituents are present. The lowest numbers possible are given to substituents, choice between alternatives being governed by Rule **A-2** so far as applicable, except that when names are based on those of compounds listed in Part **.1** of this rule the first priority for lowest numbers is given to the substituent(s) already present in those compounds.

Examples:

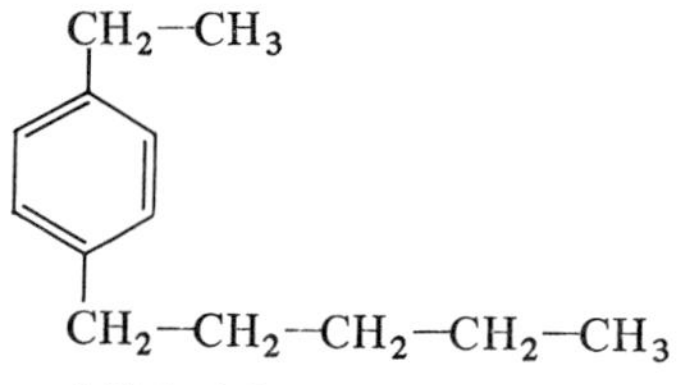

1-Ethyl-4-pentylbenzene
or *p*-Ethylpentylbenzene

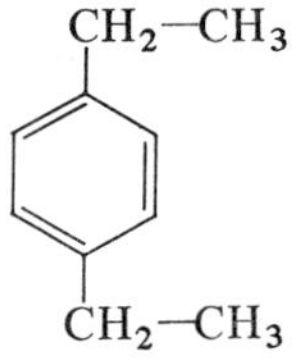

1,4-Diethylbenzene
or *p*-Diethylbenzene

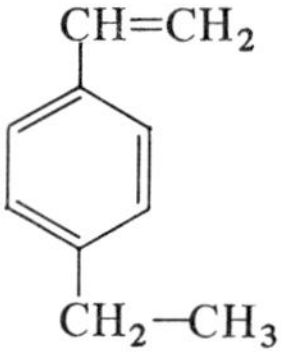

4-Ethylstyrene
or *p*-Ethylstyrene

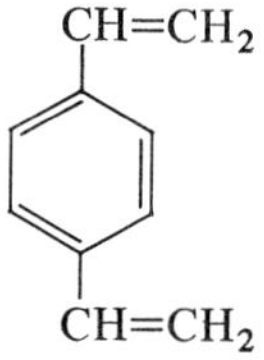

1,4-Divinylbenzene
or *p*-Divinylbenzene
not *p*-Vinylstyrene

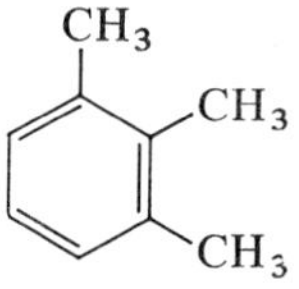

1,2,3-Trimethylbenzene
not Methylxylene
not Dimethyltoluene

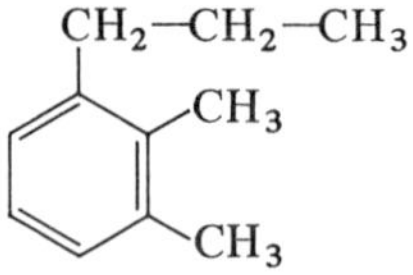

1,2-Dimethyl-
3-propylbenzene
or 3-Propyl-*o*-xylene

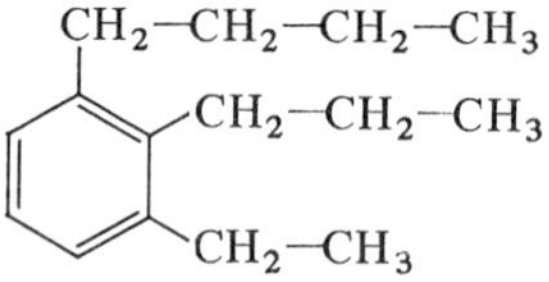

1-Butyl-3-ethyl-2-propylbenzene

Appendix] MONOCYCLIC HYDROCARBONS 251

12.4—The generic name of monocyclic and polycyclic aromatic hydrocarbons is "arene".

Rule A-13. Substituted Aromatic Radicals

13.1—Univalent radicals derived from monocyclic substituted aromatic hydrocarbons and having the free valence at a ring atom are given the names listed below. Such radicals not listed below are named as substituted phenyl radicals. The carbon atom having the free valence is numbered as 1.

Phenyl C_6H_5-

Cumenyl (*m*- shown)

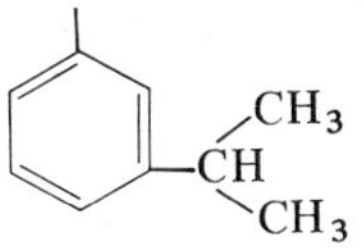

Mesityl

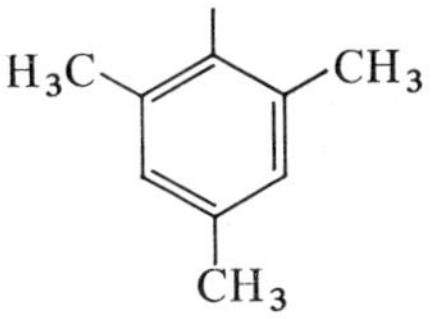

Tolyl (*o*- shown)

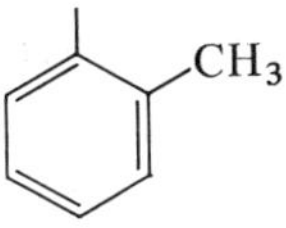

Xylyl (2,3- shown)

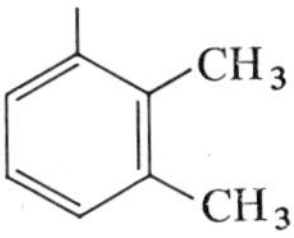

13.2—Since the name phenylene (*o*-, *m*- or *p*-) is retained for the radical $-C_6H_4-$ (exception to Rule **A-11.6**), bivalent radicals formed from substituted benzene derivatives and having the free valences at ring atoms are named as substituted phenylene radicals. The carbon atoms having the free valences are numbered 1,2-, 1,3 or 1,4- as appropriate.

13.3—The following trivial names for radicals having a single free valence in the side chain are retained:

Benzyl	$C_6H_5-\overset{\alpha}{C}H_2-$
Benzhydryl (alternative to Diphenylmethyl)	$(C_6H_5)_2\overset{\alpha}{C}H-$
Cinnamyl	$C_6H_5-\overset{\gamma}{C}H=\overset{\beta}{C}H-\overset{\alpha}{C}H_2-$
Phenethyl	$C_6H_5-\overset{\beta}{C}H_2-\overset{\alpha}{C}H_2-$
Styryl	$C_6H_5-\overset{\beta}{C}H=\overset{\alpha}{C}H-$
Trityl	$(C_6H_5)_3C-$

13.4—Multivalent radicals of aromatic hydrocarbons with the free valences in the side chain are named in accordance with Rule **A-4**. Examples:

Benzylidyne	$C_6H_5-C\equiv$
Cinnamylidene	$C_6H_5-\overset{\gamma}{C}H=\overset{\beta}{C}H-\overset{\alpha}{C}H=$

13.5—The generic names of univalent and bivalent aromatic hydrocarbon radicals are "aryl" and "arylene", respectively.

FUSED POLYCYCLIC HYDROCARBONS

Rule A-21. Trivial and Semi-trivial names

21.1—The names of polycyclic hydrocarbons with maximum number of non-cumulative* double bonds end in "-ene". The names listed on pp. 20 and 21 are retained.

---

* Cumulative double bonds are those present in a chain in which at least three contiguous carbon atoms are joined by double bonds; non-cumulative double bonds comprise every other arrangement of two or more double bonds in a single structure. The generic name "cumulene" is given to compounds containing three or more cumulative double bonds.
    Examples:

$CH_2=C=C=C=CH_2$
Cumulative

$CH_3-CH=CH-CH=CH-CH=CH_2$
or

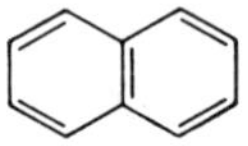

Non-cumulative

Appendix] FUSED POLYCYCLIC HYDROCARBONS 253

21.2—The names of hydrocarbons containing five or more fused benzene rings in a straight linear arrangement are formed from a numerical prefix as specified in Rule **A-1.1** followed by "-acene". [Examples on p. 255].

Examples (to Rule **A-21.2**):

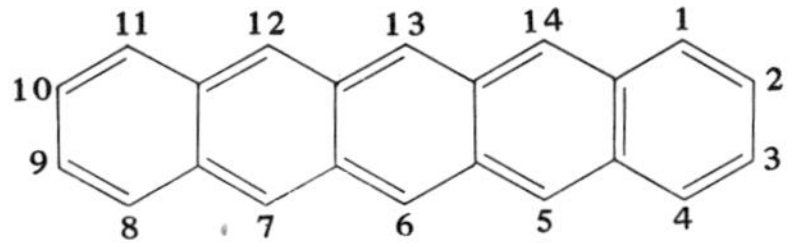

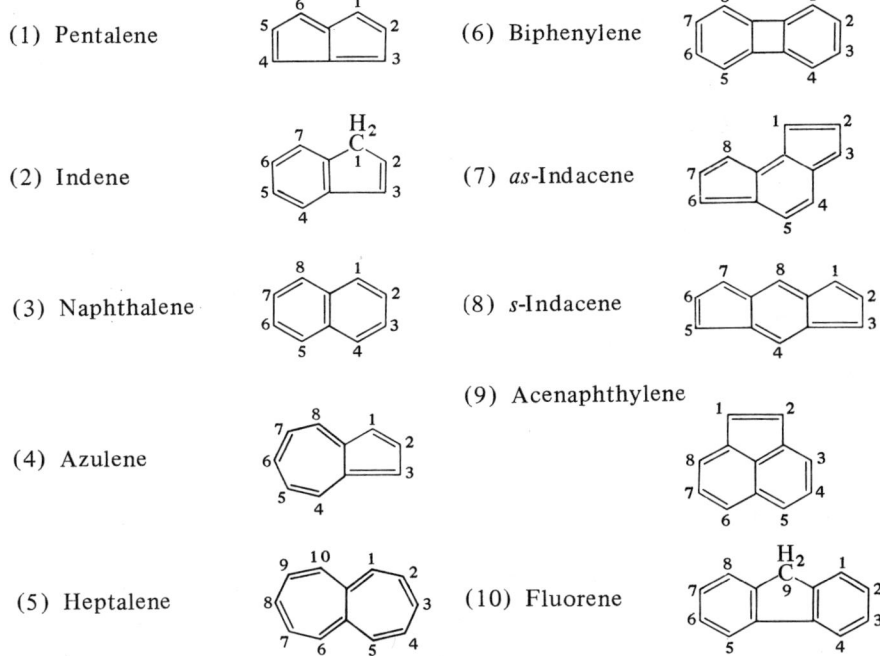

Pentacene

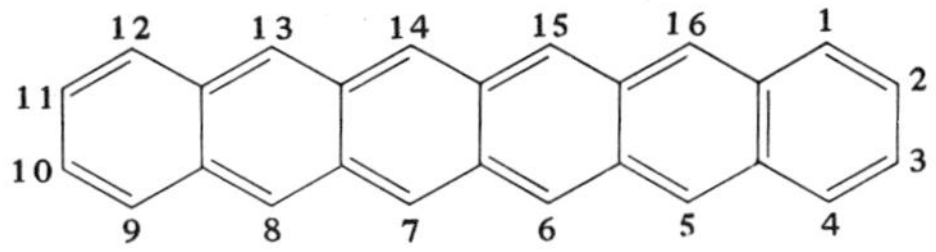

Hexacene

The following list contains the names of polycyclic hydrocarbons which are retained (see Rule **A-21.1**).

(1) Pentalene

(2) Indene

(3) Naphthalene

(4) Azulene

(5) Heptalene

(6) Biphenylene

(7) *as*-Indacene

(8) *s*-Indacene

(9) Acenaphthylene

(10) Fluorene

(11) Phenalene

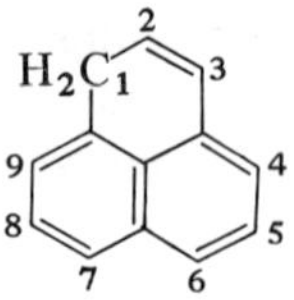

(12) Phenanthrene*

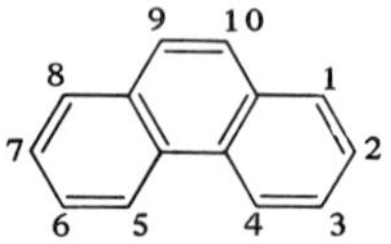

(13) Anthracene*

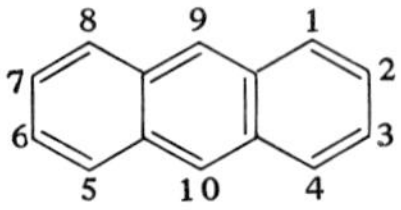

(14) Fluoranthene

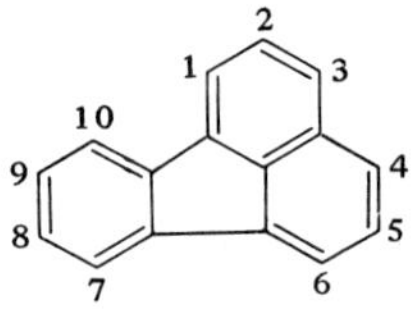

(15) Acephenanthrylene

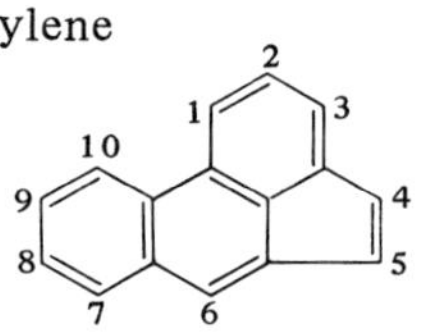

(16) Aceanthrylene

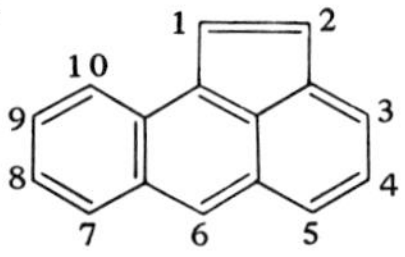

(17) Triphenylene

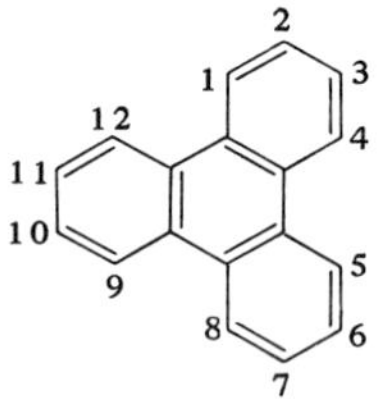

(18) Pyrene

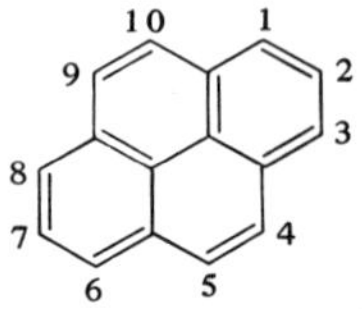

(19) Chrysene

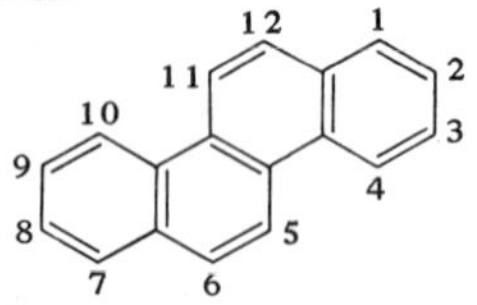

(20) Naphthacene

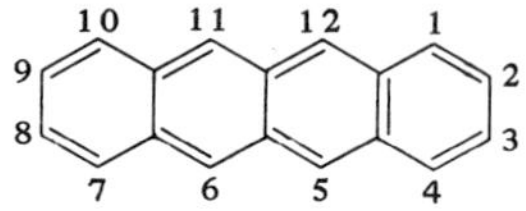

(21) Pleiadene

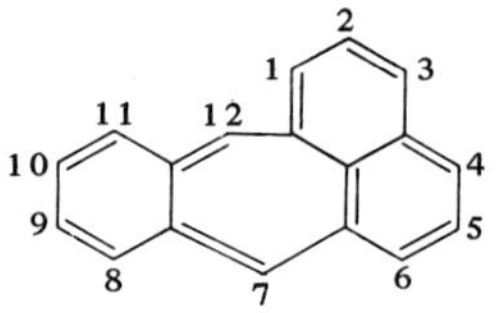

(22) Picene

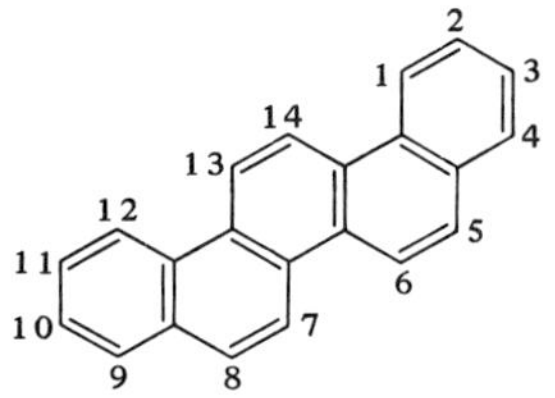

(23) Perylene

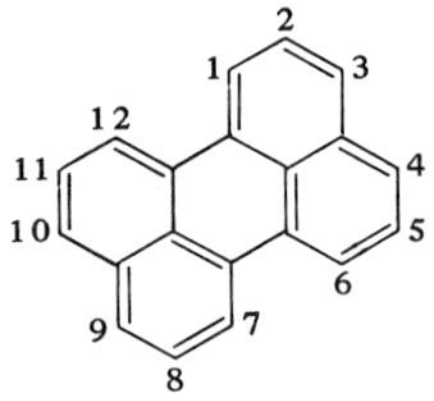

(24) Pentaphene

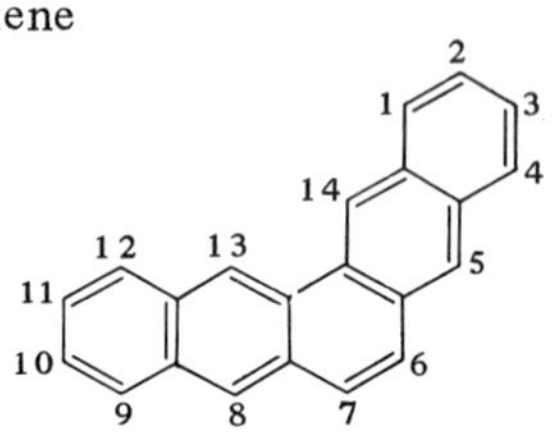

Appendix] FUSED POLYCYCLIC HYDROCARBONS 255

(25) Pentacene**

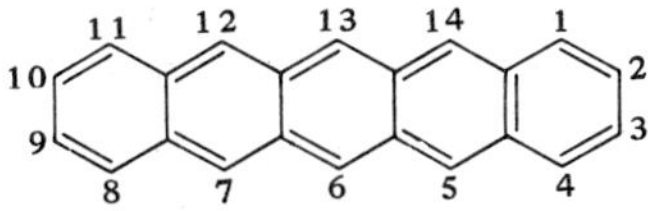

(26) Tetraphenylene***

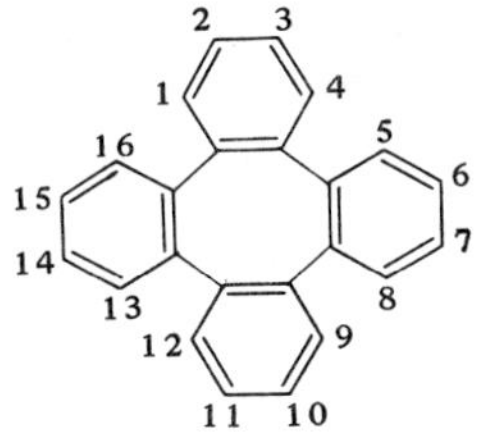

(27) Hexaphene

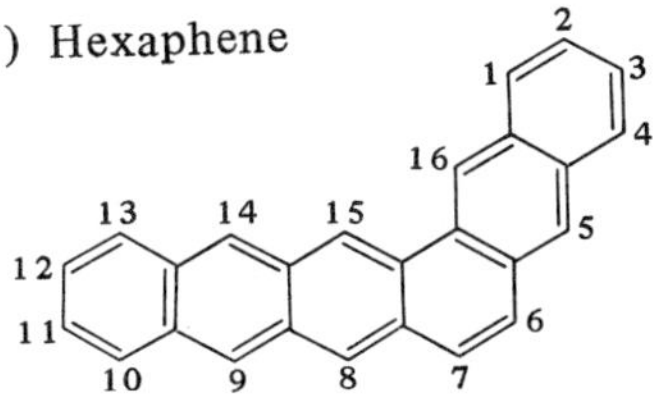

(28) Hexacene**

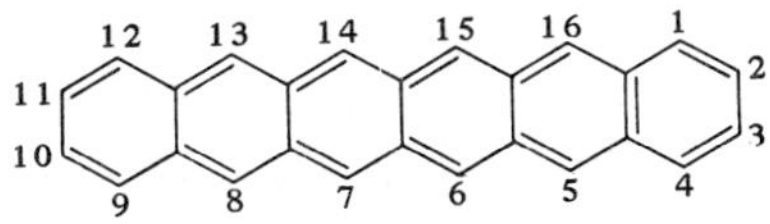

(29) Rubicene

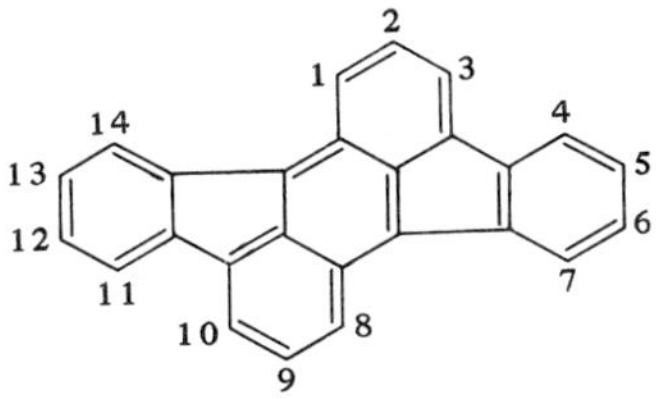

(30) Coronene

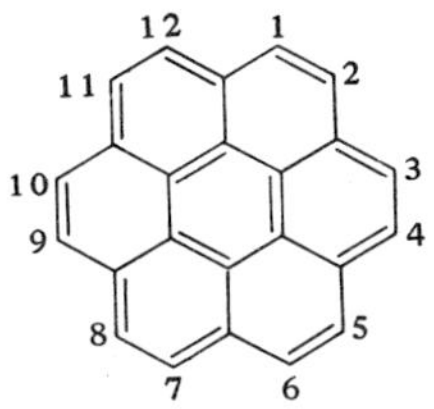

(31) Trinaphthylene***

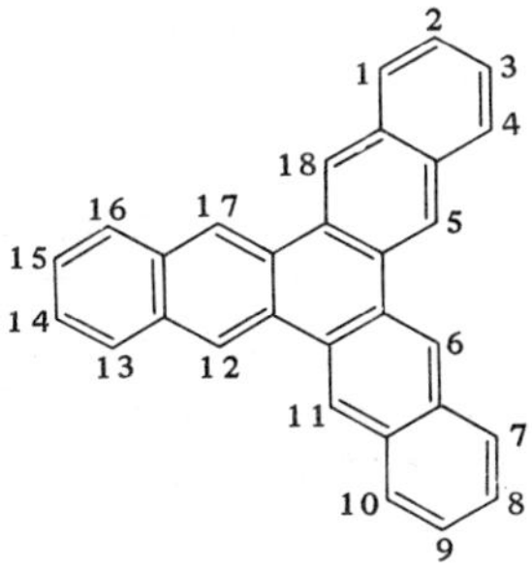

(32) Heptaphene

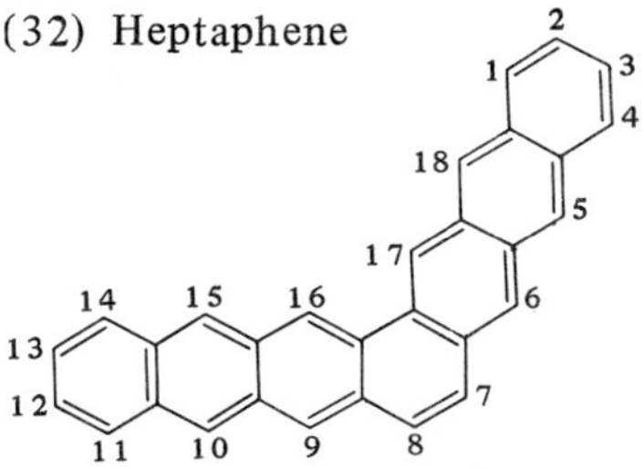

(33) Heptacene**

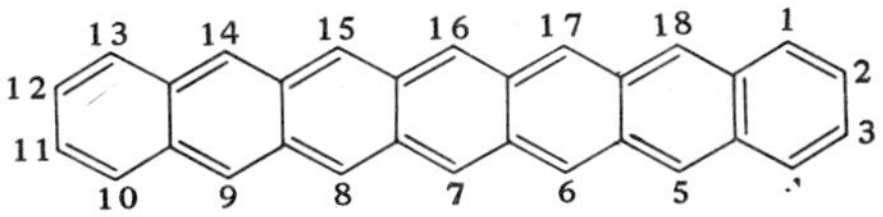

(34) Pyranthrene

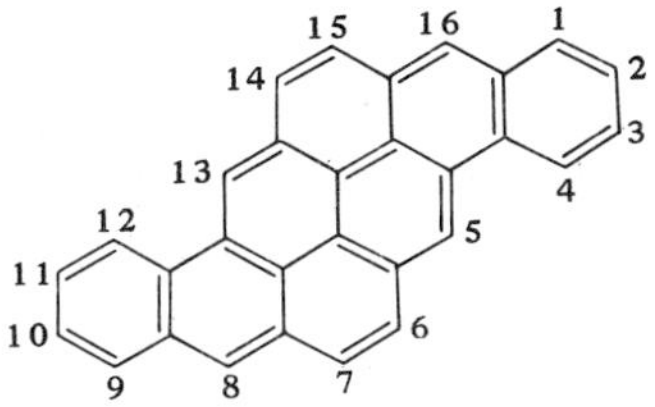

(35) Ovalene

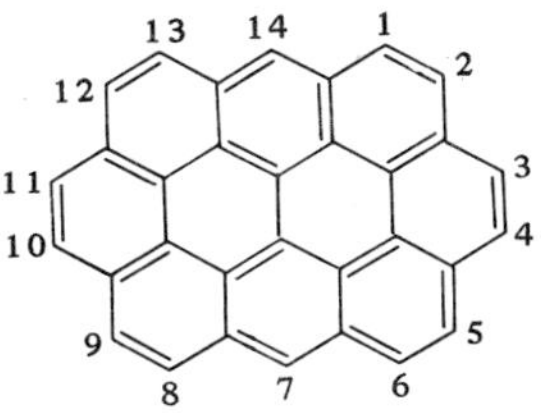

* Denotes exception to systematic numbering.
** See Rule A-21.2.
*** For isomer shown only.

256 IUPAC RULES [Appendix

21.3—"*Ortho*-fused"* or "*ortho*- and *peri*-fused"† polycyclic hydrocarbons with maximum number of non-cumulative double bonds which contain at least two rings of five or more members and which have no accepted trivial name such as those of Part .1 of this rule, are named by prefixing to the name of a component ring or ring system (the base component) designations of the other components. The base component should contain as many rings as possible (provided it has a trivial name), and should occur as far as possible from the beginning of the list of Rule A-21.1. The attached components should be as simple as possible.
Example:

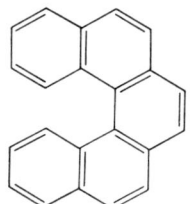

(not Naphthophenanthrene: benzo is "simpler" than naphtho, even though there are two benzo rings and only one naphtho)

Dibenzophenanthrene

21.4—The prefixes designating attached components are formed by changing the ending "-ene" of the name of the component hydrocarbon into "-eno"; *e.g.*, "pyreno" (from pyrene). When more than one prefix is present, they are arranged in alphabetical order. The following common

---

* Polycyclic compounds in which two rings have two, and only two, atoms in common are said to be "*ortho*-fused". Such compounds have n common faces and $2n$ common atoms (Example I).
† Polycyclic compounds in which one ring contains two, and only two, atoms in common with each of two or more rings of a contiguous series of rings are said to be "*ortho*- and *peri*-fused". Such compounds have n common faces and less than $2n$ common atoms (Examples II and III).
Examples:

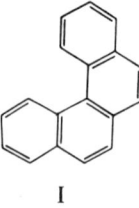

I	II	III
3 common faces	7 common faces	5 common faces
6 common atoms	8 common atoms	6 common atoms
"*Ortho*-fused" system	"*Ortho*- and *peri*-fused" systems	

Appendix] FUSED POLYCYCLIC HYDROCARBONS 257

abbreviated prefixes are recognized (see list in Part .1 of this rule):

Acenaphtho	from	Acenaphthylene	Naphtho	from	Naphthalene
Anthra	from	Anthracene	Perylo	from	Perylene
Benzo	from	Benzene	Phenanthro	from	Phenanthrene

For monocyclic prefixes other than "benzo", the following names are recognized, each to represent the form with the maximum number of non-cumulative double bonds: cyclopenta, cyclohepta, cycloocta, cyclonona, *etc.* When the base component is a monocyclic system, the ending "-ene" signifies the maximum number of non-cumulative double bonds, and thus does not denote one double bond only.

Examples:

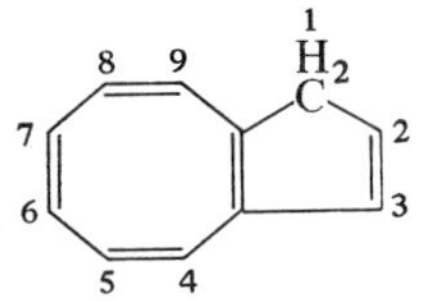

1H-Cyclopentacyclooctene 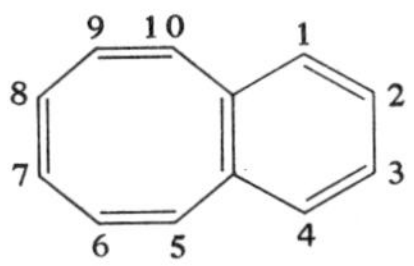   Benzocyclooctene

21.5—Isomers are distinguished by lettering the peripheral sides of the base component *a, b, c, etc.,* beginning with "*a*" for the side "1,2", "*b*" for "2,3" (or in certain cases "2,2*a*") and lettering every side around the periphery. To the letter as early in the alphabet as possible, denoting the side where fusion occurs, are prefixed, if necessary, the numbers of the positions of attachment of the other component. These numbers are chosen to be as low as is consistent with the numbering of the component, and their order conforms to the direction of lettering of the base component (see Examples II and IV). When two or more prefixes refer to equivalent positions so that there is a choice of letters, the prefixes are cited in alphabetical order according to Rule **A-21.4** and the location of the first cited prefix is indicated by a letter as early as possible in the alphabet (see Example V). The numbers and letters are enclosed in square brackets and placed immediately after the designation of the attached component. This expression merely defines the manner of fusion of the components.

Examples:

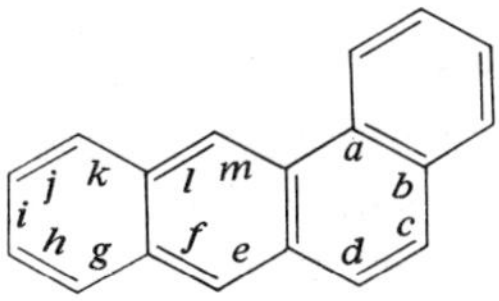

I
Benz[*a*]anthracene

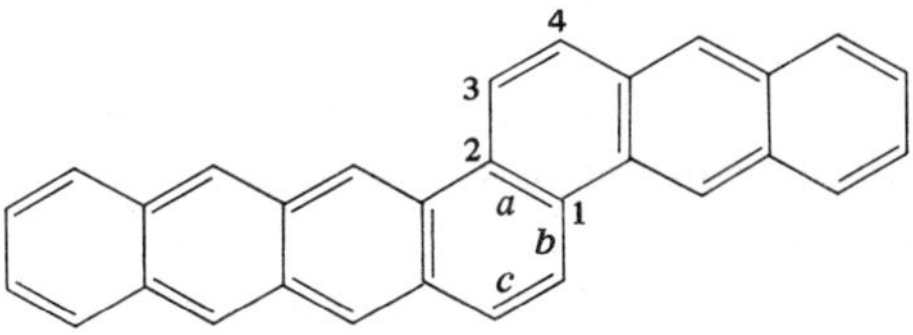

II
Anthra[2,1-*a*]naphthacene

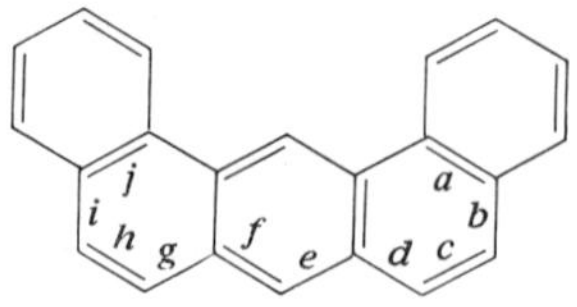

III
Dibenz[a,j]anthracene
(not Naphtho[2,1-b]phenanthrene)

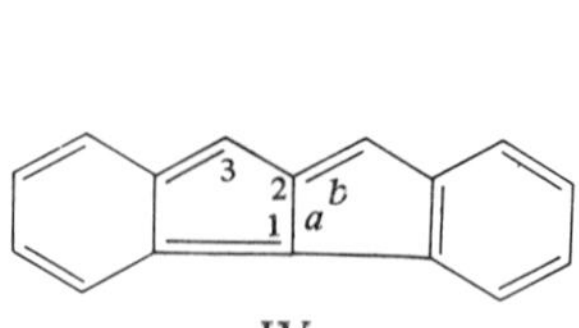

IV
Indeno[1,2-a]indene

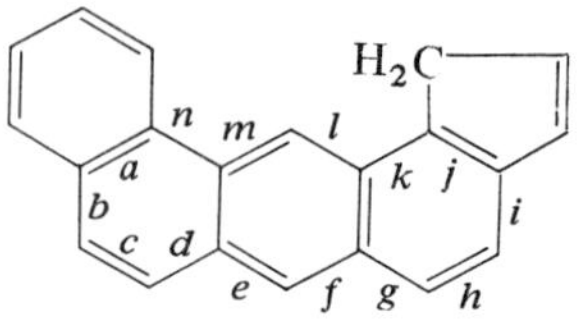

V
1H-Benzo[a]cyclopent[j]anthracene

The completed system consisting of the base component and the other components is then renumbered according to Rule **A-22**, the enumeration of the component parts being ignored.
Example:

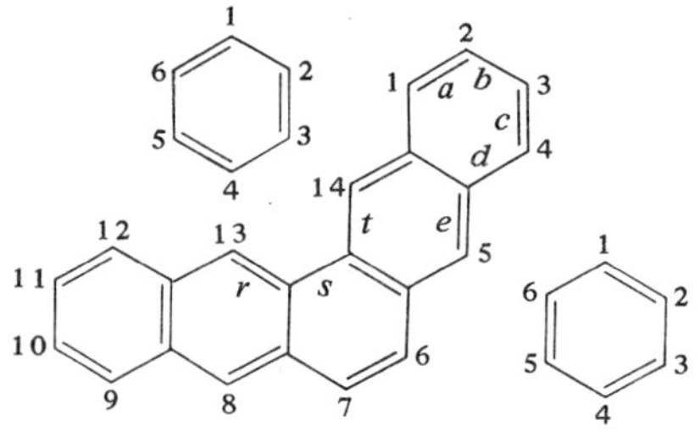

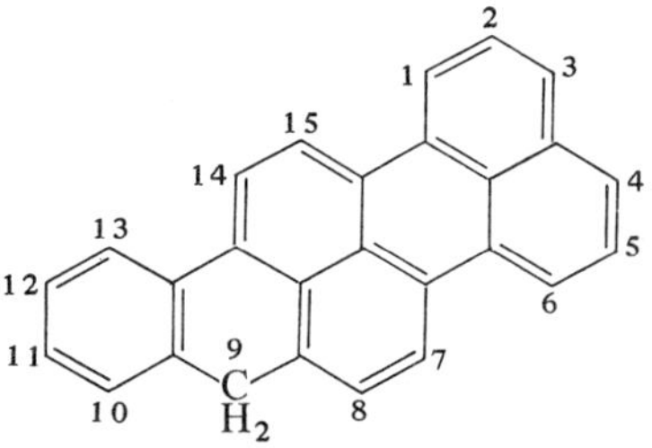

Benzene   Pentaphene   Benzene → 9H-Dibenzo[de,rst]pentaphene

21.6—When a name applies equally to two or more isomeric condensed parent ring systems with the maximum number of non-cumulative double bonds and when the name can be made specific by indicating the position of one or more hydrogen atoms in the structure, this is accomplished by modifying the name with a locant, followed by italic capital *H* for each of these hydrogen atoms. Such symbols ordinarily precede the name. The said atom or atoms are called "indicated hydrogen". The same principle is applied to radicals and compounds derived from these systems.

Appendix] FUSED POLYCYCLIC HYDROCARBONS 259

Examples:

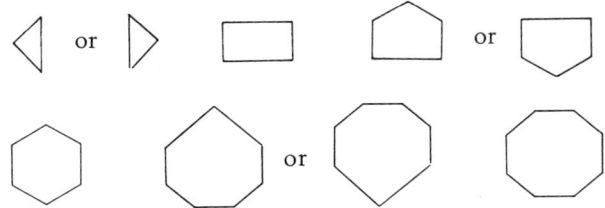

3H-Fluorene   2H-Indene

Rule A-22. Numbering

22.1—For the purposes of numbering, the individual rings of a polycyclic "*ortho*-fused" or "*ortho*- and *peri*-fused" hydrocarbon system are normally drawn as follows:

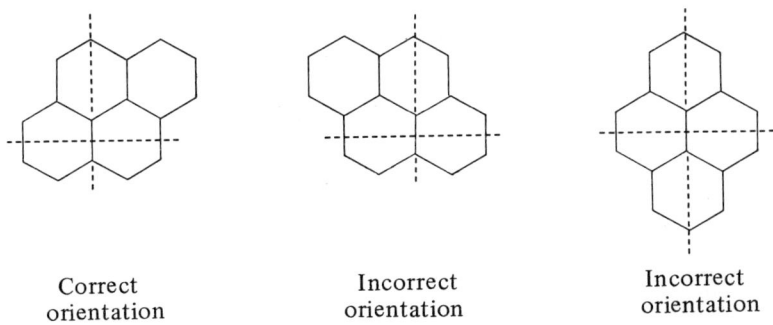

and the polycyclic system is oriented so that (*a*) the greatest number of rings are in a horizontal row and (*b*) a maximum number of rings are above and to the right of the horizontal row (upper right quadrant). If two or more orientations meet these requirements, the one is chosen which has as few rings as possible in the lower left quadrant.
Example:

Correct orientation   Incorrect orientation   Incorrect orientation

The system thus oriented is numbered in a clockwise direction commencing with the carbon atom not engaged in ring-fusion in the most counter-clockwise position of the uppermost ring, or if there is a choice,

of the uppermost ring farthest to the right, and omitting atoms common to two or more rings.
Example:

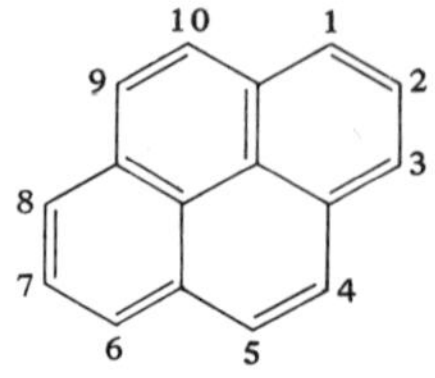

Correct 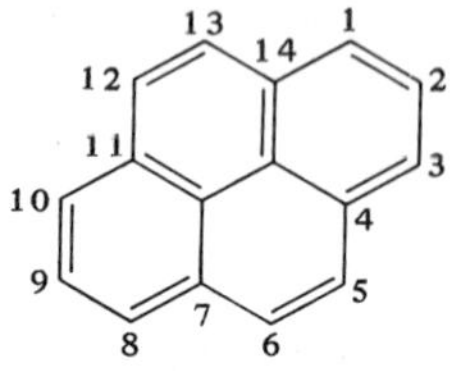 / Incorrect

22.2—Atoms common to two or more rings are designated by adding roman letters "a", "b", "c", *etc.*, to the number of the position immediately preceding. Interior atoms follow the highest number, taking a clockwise sequence wherever there is a choice.
Example:

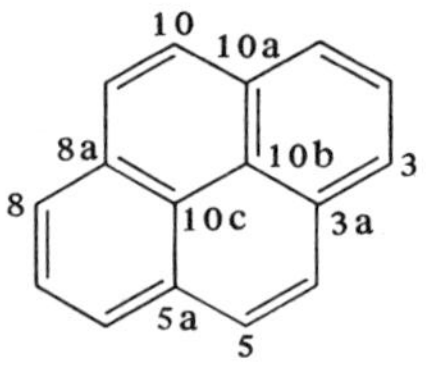

Correct 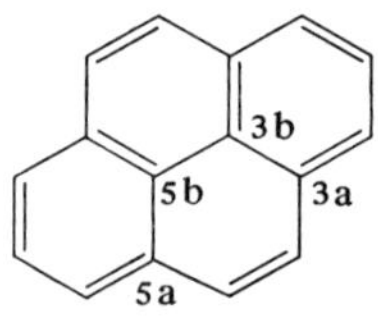 / Incorrect

22.3—When there is a choice, carbon atoms common to two or more rings follow the lowest possible numbers.
Examples [*cf.* Notes below]:

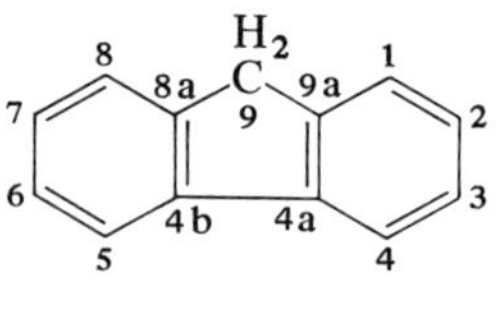

Correct 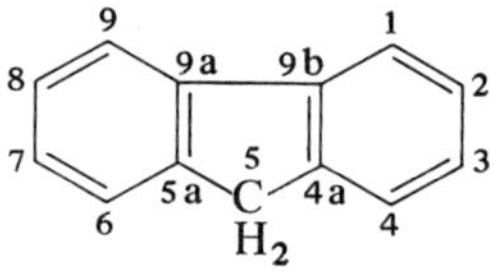 / Incorrect   I

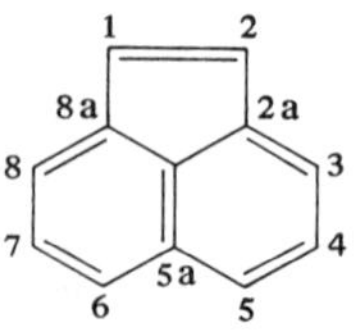

Correct 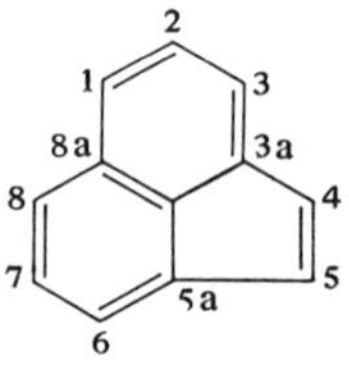 / Incorrect   II

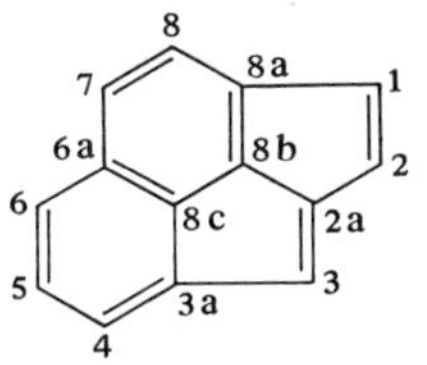

Correct

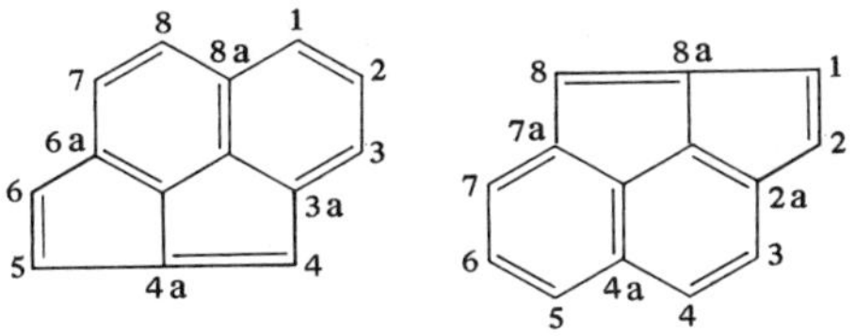

Incorrect

III

Notes: I. 4, 4, 8, 9 is lower than 4, 5, 9, 9.
II. 2, 5, 8 is lower than 3, 5, 8.
III. 2, 3, 6, 8 is lower than 3, 4, 6, 8 or 2, 4, 7, 8.

22.4—When there is a choice, the carbon atoms which carry an indicated hydrogen atom are numbered as low as possible.

Example:

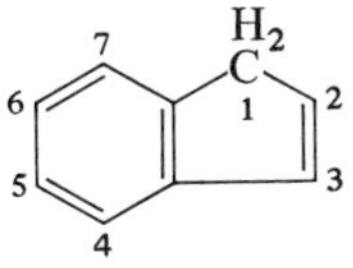

Correct

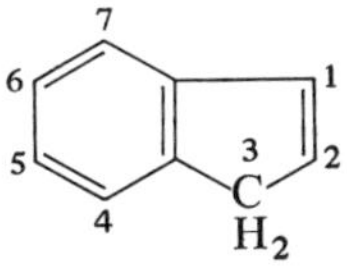

Incorrect

22.5—The following are recommended exceptions to the above rules on numbering:

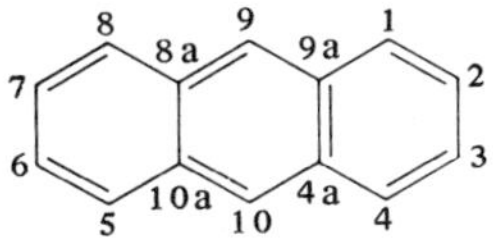

Anthracene

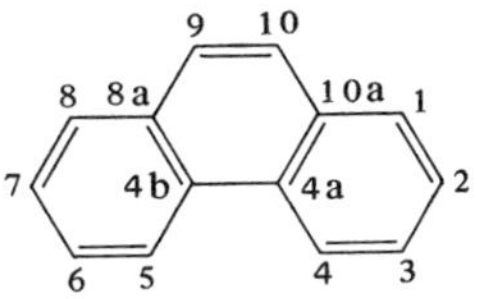

Phenanthrene

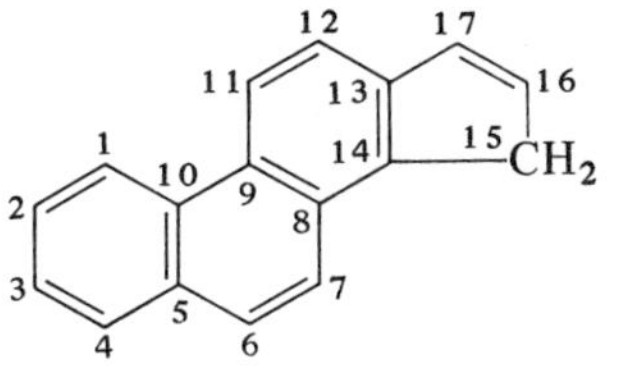

Cyclopenta[*a*]phenanthrene
(15*H*- shown)
See also rules on steroids

Rule A-23. Hydrogenated Compounds

23.1—The names of "*ortho*-fused" or "*ortho*- and *peri*-fused" polycyclic hydrocarbons with less than maximum number of non-cumulative double bonds are formed from a prefix "dihydro-", "tetrahydro-", *etc.*, followed by the name of the corresponding unreduced hydrocarbon. The prefix "perhydro-" signifies full hydrogenation. When there is a choice for *H* used for indicated hydrogen it is assigned the lowest available number.

Examples:

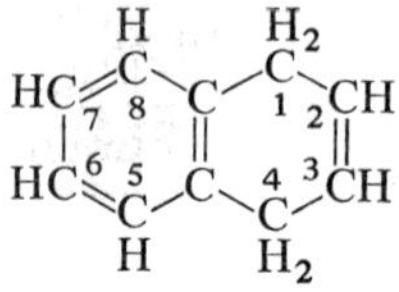

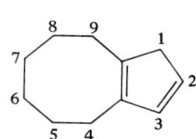

1,4-Dihydronaphthalene

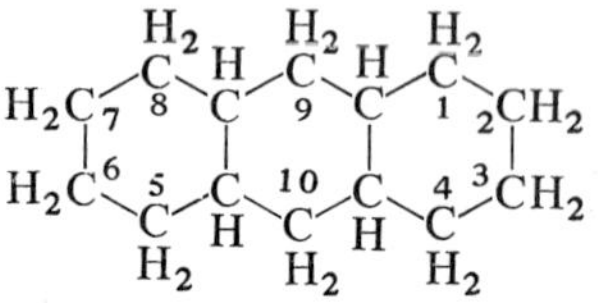

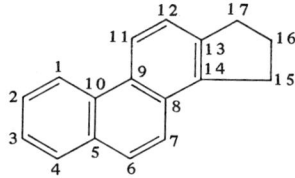

Tetradecahydroanthracene
or Perhydroanthracene

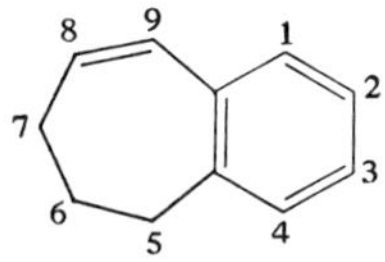

6,7-Dihydro-5*H*-benzo-, cycloheptene

4,5,6,7,8,9-Hexahydro-1*H*-cyclopentacyclooctene

16,17-Dihydro-15*H*-cyclopenta[*a*]phenanthrene

Exceptions:

The following names are retained:

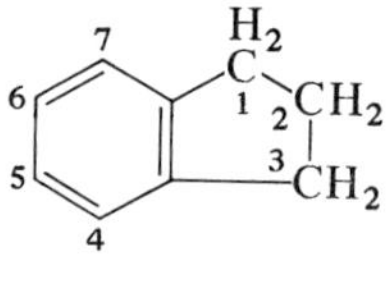

Indan

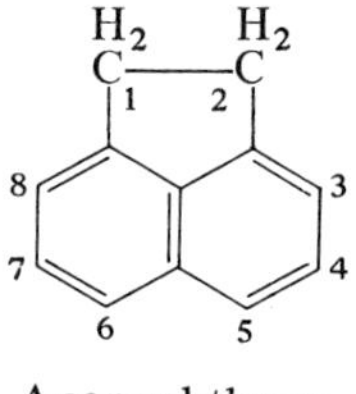

Acenaphthene

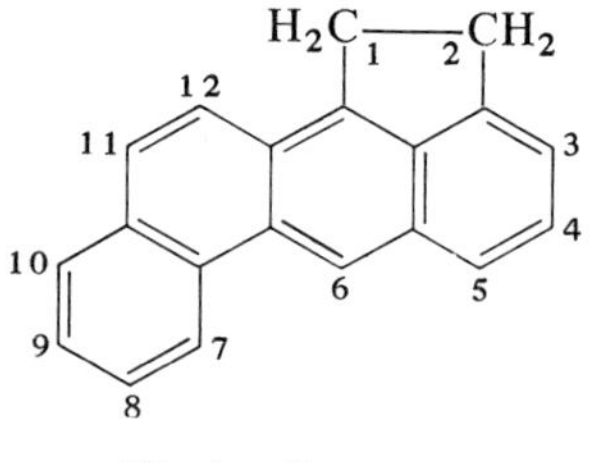

Cholanthrene

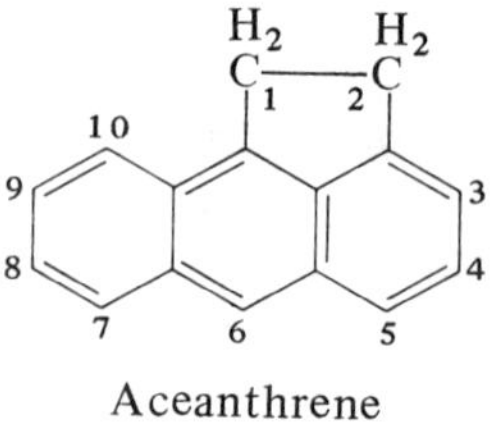

Aceanthrene

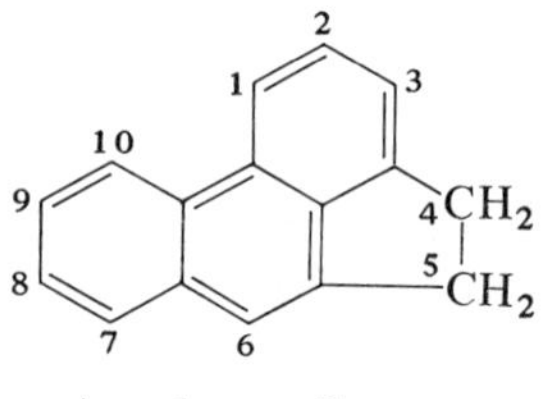

Acephenanthrene

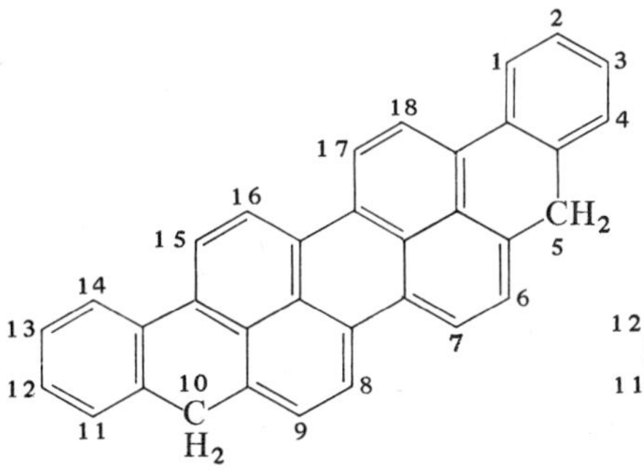

Violanthrene

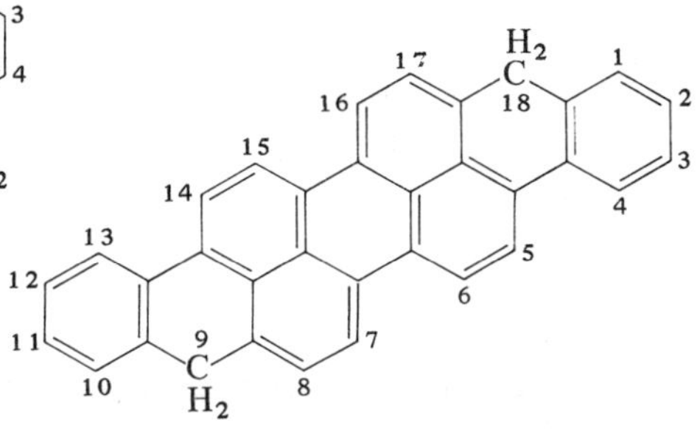

Isoviolanthrene

23.2—When there is a choice, the carbon atoms to which hydrogen atoms are added are numbered as low as possible.
Example:

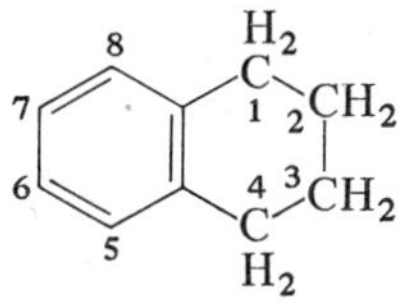

Correct

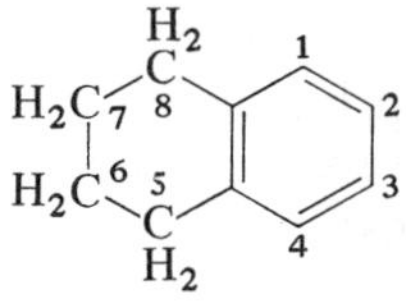

Incorrect

23.3—Substituted polycyclic hydrocarbons are named according to the same principles as substituted monocyclic hydrocarbons (see Rules **A-12** and **A-61**).

23.5 (Alternate to part of Rule **A-23.1**)—The names of "*ortho*-fused" polycyclic hydrocarbons which have (*a*) less than the maximum number of non-cumulative double bonds, (*b*) at least one terminal unit which is most conveniently named as an unsaturated cycloalkane derivative, and (*c*) a double bond at the positions where rings are fused together, may be derived by joining the name of the terminal unit to that of the other component by means of a letter "o" with elision of a terminal "e". The abbreviations for fused aromatic systems laid down in Rule **A-21.4** are used, and the exceptions of Rule **A-23.1** apply.
Examples:

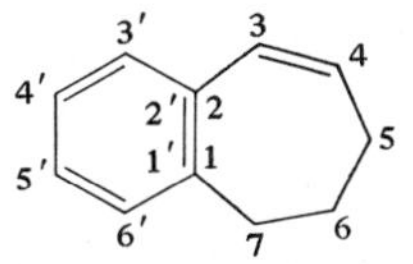

1,2-Benzo-1,3-cycloheptadiene

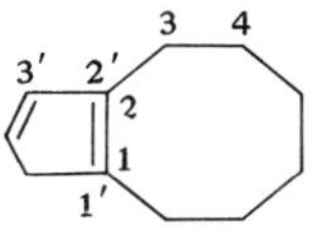

1,2-Cyclopenta-1′,3′-dienocyclooctene

1,2-Cyclopentenophenanthrene

Rule A-24. Radical Names from Trivial and Semi-trivial Names

24.1—For radicals derived from polycyclic hydrocarbons, the numbering of the hydrocarbon is retained. The point or points of attachment are given numbers as low as is consistent with the fixed numbering of the hydrocarbon.

24.2—Univalent radicals derived from "*ortho*-fused" or "*ortho*- and *peri*-fused" polycyclic hydrocarbons with names ending in "-ene" by removal of a hydrogen atom from an aromatic or alicyclic ring are named in principle by changing the ending "-ene" of the names of the hydrocarbons to "-enyl".

Examples:

2-Indenyl   1-Pyrenyl   1-Acenaphthenyl

Exceptions:

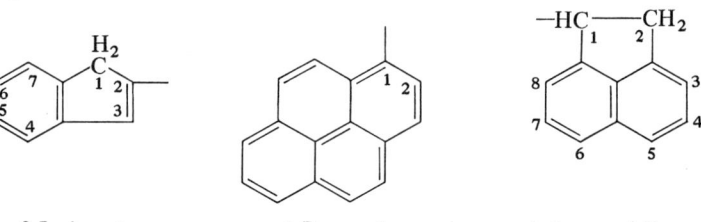

Naphthyl (2- shown)   Anthryl (2- shown)

Phenanthryl (2-shown)   5,6,7,8-Tetrahydro-2-naphthyl

24.3—Bivalent radicals derived from univalent polycyclic hydrocarbon radicals whose names end in "-yl" by removal of one hydrogen atom from the carbon atom with the free valence are named by adding "-idene" to the name of the corresponding univalent radical.

Examples:

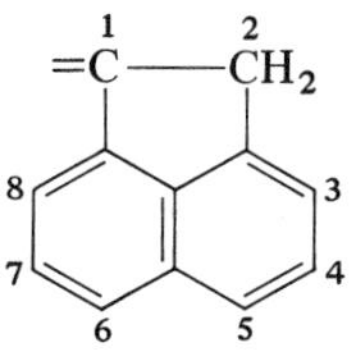

1-Acenaphthenylidene

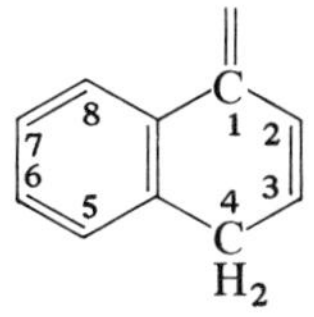

1(4*H*)-Naphthylidene
(for 4*H* see Rule A-21.6)
or 1,4-Dihydro-1-naphthylidene

24.4—Bivalent radicals derived from "*ortho*-fused" or "*ortho*- and *peri*-fused" polycyclic hydrocarbons by removal of a hydrogen atom from each of two different carbon atoms of the ring are named by changing the ending "-yl" of the univalent radical name to "-ylene" or "-diyl". Multivalent radicals, similarly derived, are named by adding "-triyl", "-tetrayl", *etc.*, to the name of the ring system.

Examples:

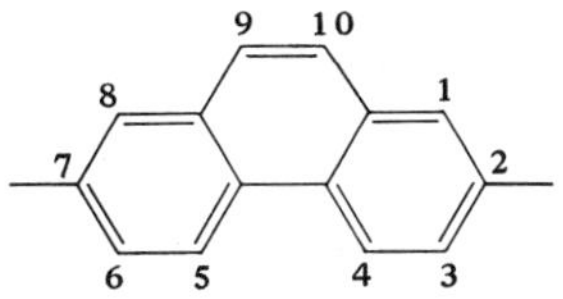

2,7-Phenanthrylene
or 2,7-Phenanthrenediyl

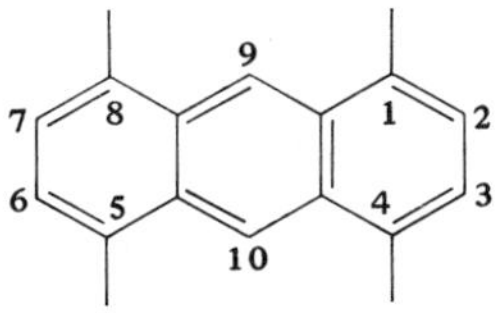

1,4,5,8-Anthracenetetrayl

Rule A-28. Radical Names for Fused Cyclic Systems with Side Chains

28.1—Radicals formed from hydrocarbons consisting of polycyclic systems and side chains are named according to the principles of the preceding rules.

BRIDGED HYDROCARBONS

EXTENSION OF THE VON BAEYER SYSTEM

Rule A-31. Bicyclic Systems

31.1—Saturated alicyclic hydrocarbon systems consisting of two rings only, having two or more atoms in common, take the name of an open chain hydrocarbon containing the same total number of carbon atoms preceded by the prefix "bicyclo-". The number of carbon atoms in each of the three bridges* connecting the two tertiary carbon atoms is indicated in brackets in descending order.

Examples:

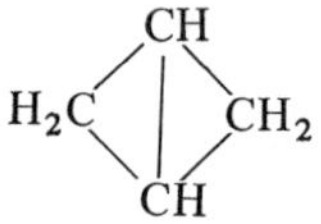

Bicylo[1.1.0]butane 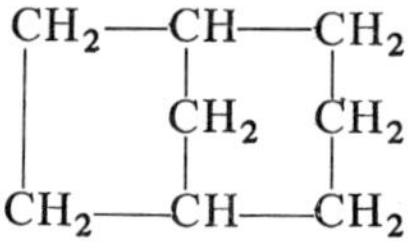   Bicyclo[3.2.1]octane 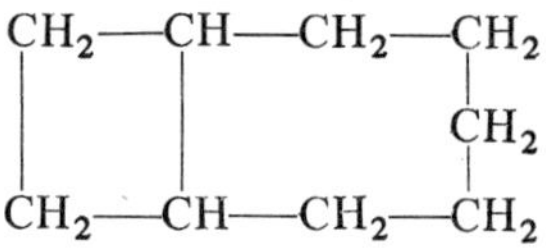   Bicyclo[5.2.0]nonane

31.2—The system is numbered commencing with one of the bridgeheads, numbering proceeding by the longest possible path to the second bridgehead; numbering is then continued from this atom by the longer unnumbered path back to the first bridgehead and is completed by the shortest path.

Examples:

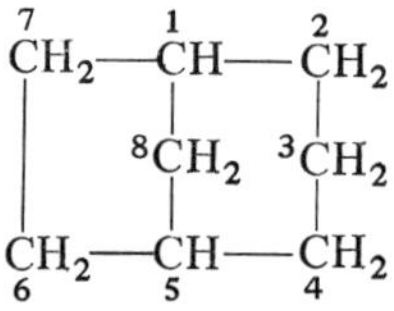

Bicyclo[3.2.1]octane 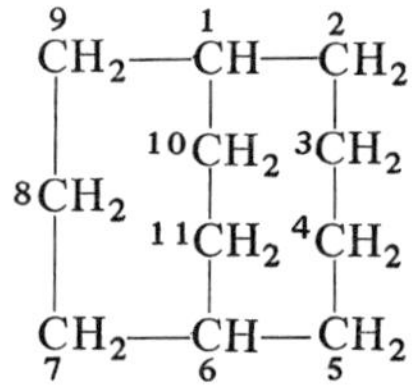   Bicyclo[4.3.2]undecane

Note: Longest path 1, 2, 3, 4, 5
　　　Next longest path 5, 6, 7, 1
　　　Shortest path 1, 8, 5

31.3—Unsaturated hydrocarbons are named in accordance with the principles set forth in Rule **A-11.3.** When there is a choice in numbering, unsaturation is given the lowest numbers.

---

* A bridge is a valence bond or an atom or an unbranched chain of atoms connecting two different parts of a molecule. The two tertiary carbon atoms connected through the bridge are termed "bridgeheads".

Examples:

$^6CH_2-^1CH-^2CH$
$\quad\quad|\quad\quad\|$
$\quad ^7CH_2\quad\|$
$\quad\quad|\quad\quad\|$
$\quad ^8CH_2\quad\|$
$\quad\quad|\quad\quad\|$
$^5CH_2-^4CH-^3CH$

Bicyclo[2.2.2]oct-2-ene

$^{16}CH=^1C-(CH_2)_5-^7CH_2$
$\quad\quad\quad|\quad\quad\quad\quad\quad|$
$\quad\quad ^{17}CH\quad\quad\quad\quad|$
$\quad\quad\quad|\quad\quad\quad\quad\quad|$
$\quad\quad ^{18}CH\quad\quad\quad\quad|$
$\quad\quad\quad|\quad\quad\quad\quad\quad|$
$^{15}CH=^{14}C-(CH_2)_5-^8CH_2$

Bicyclo[12.2.2]octadeca-1(16),14,17-triene
or Bicyclo[12.2.2]octadeca-14,16(1),17-triene
(See Rule **A-3.1** for double locants)

31.4—Radicals derived from bridged hydrocarbons are named in accordance with the principles set forth in Rule **A-11**. The numbering of the hydrocarbon is retained and the point or points of attachment are given numbers as low as is consistent with the fixed numbering of the saturated hydrocarbon.

Examples:

$^7CH_2-^1CH-^2CH-$
$\quad\quad|\quad\quad\quad|$
$\quad\quad ^8CH_2\ ^3CH_2$
$\quad\quad|\quad\quad\quad|$
$^6CH_2-^5CH-^4CH_2$

Bicyclo[3.2.1]oct-2-yl

$^6CH-^1CH-^2CH-$
$\quad|\quad\quad\quad\|$
$\quad ^7CH_2\quad\|$
$\quad|\quad\quad\quad\|$
$\quad ^8CH_2\quad\|$
$\quad|\quad\quad\quad\|$
$^5CH-^4CH-^3CH_2$

Bicyclo[2.2.2]oct-5-en-2-yl

$^{11}CH_2-^{12}CH=^1C\!-\!\!-^2CH_2-^3CH-$
$\quad|\quad\quad\quad\quad|\quad\quad\quad\quad|$
$^{10}CH_2\quad\quad ^{13}CH_2\quad\quad ^4CH_2$
$\quad|\quad\quad\quad\quad\quad\quad\quad\quad\quad|$
$^9CH_2\!-\!\!-^8CH_2-^7CH-^6CH_2-^5CH_2$

Bicyclo[5.5.1]tridec-1(12)-en-3-yl
or Bicyclo[5.5.1]tridec-12(1)-en-3-yl
(See Rule **A-3.1** for double locants)

Rule A-32. Polycyclic Systems

32.11—Cyclic hydrocarbon systems consisting of three or more rings may be named in accordance with the principles stated in Rule **A-31**. The appropriate prefix "tricyclo-", "tetracyclo-", *etc.*, is substituted for "bicyclo-" before the name of the open-chain hydrocarbon containing the same total number of carbon atoms. Radicals derived from these hydrocarbons are named according to the principles set forth in Rule **A-31.4**.

32.12—A polycyclic system is regarded as containing a number of rings equal to the number of scissions required to convert the system into an open-chain compound.

32.13—The word "cyclo" is followed by brackets containing, in decreasing order, numbers indicating the number of carbon atoms in:

the two branches of the main ring,
the main bridge,
the secondary bridges.

Examples:

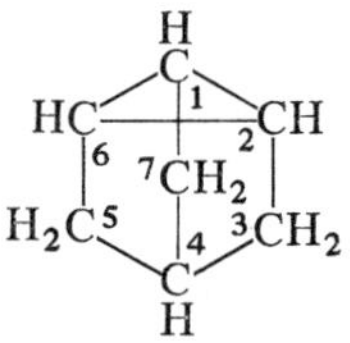

Tricyclo[2.2.1.0*]heptane 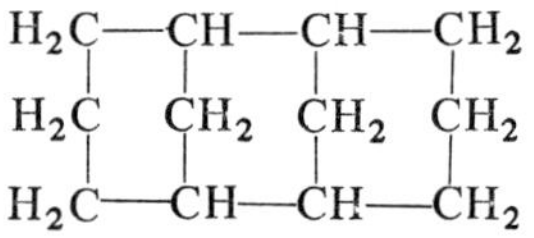   Tricyclo[5.3.1.1*]dodecane

* For location and numbering of the secondary bridge see Rules **A-32.22, A-32.23, A-32.31**.

32.21—The main ring and the main bridge form a bicyclic system whose numbering is made in compliance with Rule **A-31**.

32.22—The location of the other or so-called secondary bridges is shown by superscripts following the number indicating the number of carbon atoms in the said bridges.

32.23—For the purpose of numbering, the secondary bridges are considered in decreasing order. The numbering of any bridge follows from the part already numbered, proceeding from the highest-numbered bridgehead. If equal bridges are present, the numbering begins at the highest-numbered bridgehead.

32.31—When there is a choice, the following criteria are considered in turn until a decision is made:

(*a*) The main ring shall contain as many carbon atoms as possible, two of which must serve as bridgeheads for the main bridge.

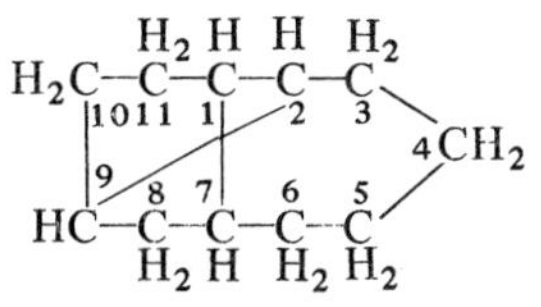

Tricyclo[5.4.0.0^{2,9}]undecane
Correct numbering

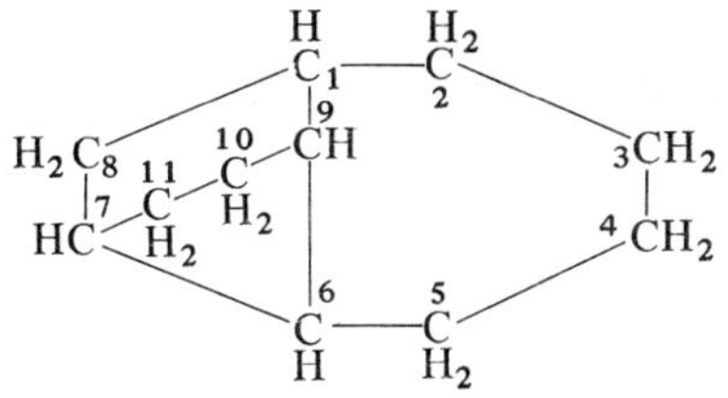

Tricyclo[4.2.1.2^{7,9}]undecane
Incorrect numbering

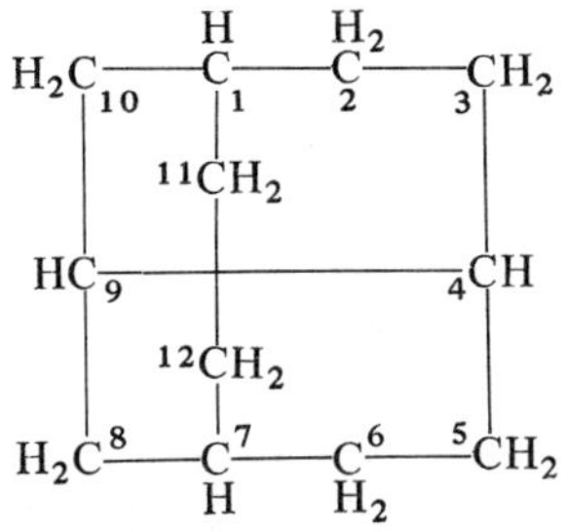

Tricyclo[5.3.2.0^{4,9}]dodecane
Correct numbering

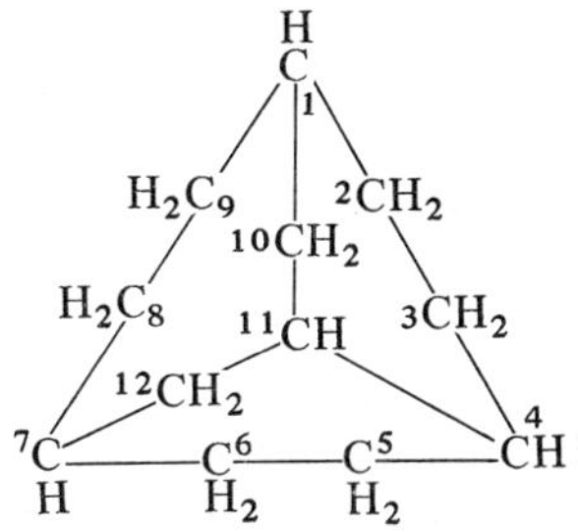

Tricyclo[5.2.3.0^{4,11}]dodecane
Incorrect numbering

(*b*) The main bridge shall be as large as possible.

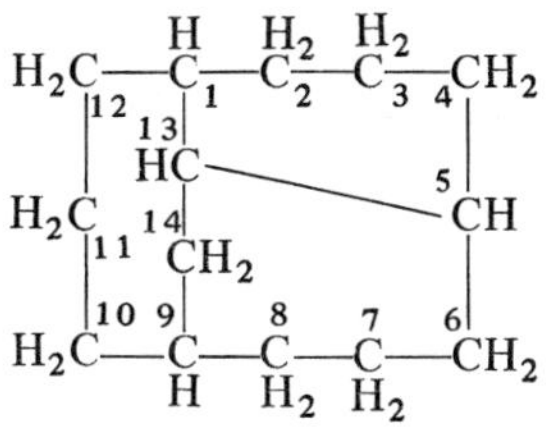

Tricyclo[7.3.2.0^{5,13}]tetradecane
Correct numbering

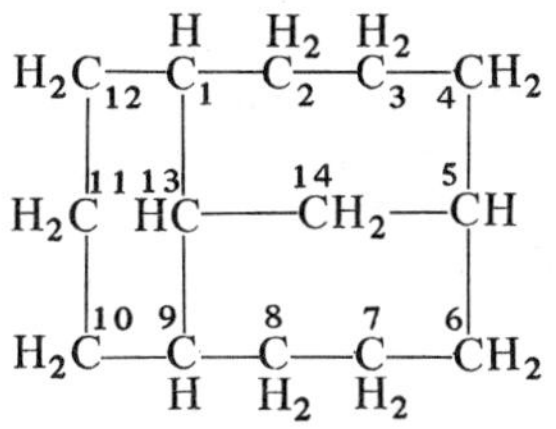

Tricyclo[7.3.1.1^{5,13}]tetradecane
Incorrect numbering

(*c*) The main ring shall be divided as symmetrically as possible by the main bridge.

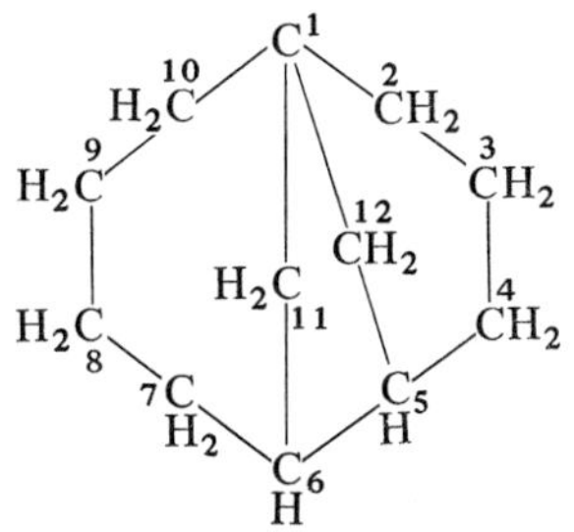

Tricyclo[4.4.1.1^{1,5}]dodecane
Correct numbering

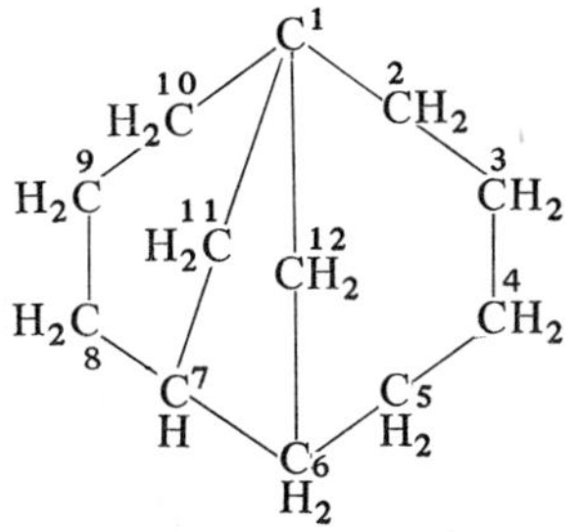

Tricyclo[5.3.1.1^{1,6}]dodecane
Incorrect numbering

(d) The superscripts locating the other bridges shall be as small as possible (in the sense indicated in Rule A-2.2).

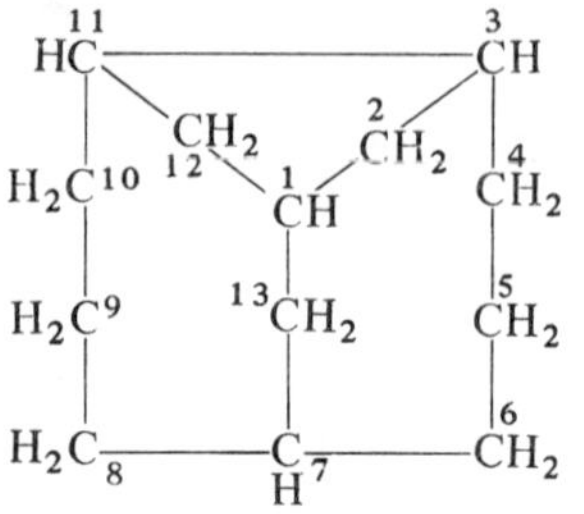

Tricyclo[5.5.1.0^{3,11}]tridecane
Correct numbering

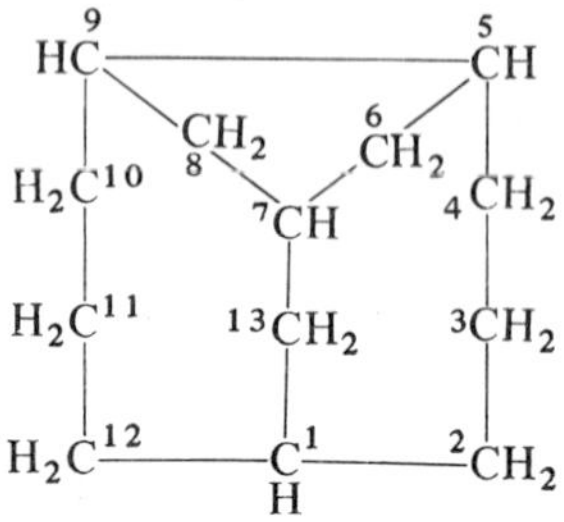

Tricyclo[5.5.1.0^{5,9}]tridecane
Incorrect numbering

Rule A-34. Hydrocarbon Bridges

34.1—Polycyclic hydrocarbon systems which can be regarded as "*ortho*-fused" or "*ortho*- and *peri*-fused" systems according to Rule **A-21** and which, at the same time, have other bridges*, are first named as "*ortho*-fused" or "*ortho*- and *peri*-fused" systems. The other bridges are then indicated by prefixes derived from the name of the corresponding hydrocarbon by replacing the final "-ane", "-ene", *etc.*, by "-ano", "-eno", *etc.*, and their positions are indicated by the points of attachment in the parent compound. If bridges of different types are present, they are cited in alphabetical order.

Examples of bridge names:

Butano	—CH$_2$—CH$_2$—CH$_2$—CH$_2$—
Benzeno (*o*-, *m*-, *p*-)	—C$_6$H$_4$—
Ethano	—CH$_2$—CH$_2$—
Etheno	—CH=CH—
Methano	—CH$_2$—
Propano	—CH$_2$—CH$_2$—CH$_2$—

Examples:

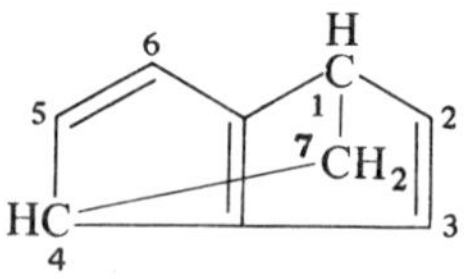

1,4-Dihydro-1,4-methanopentalene

* The term "bridge", when used in connection with an "*ortho*-fused" or "*ortho*- and *peri*-fused" polycyclic system as defined in the note to Rule **A-31.1** also includes "bivalent cyclic systems".

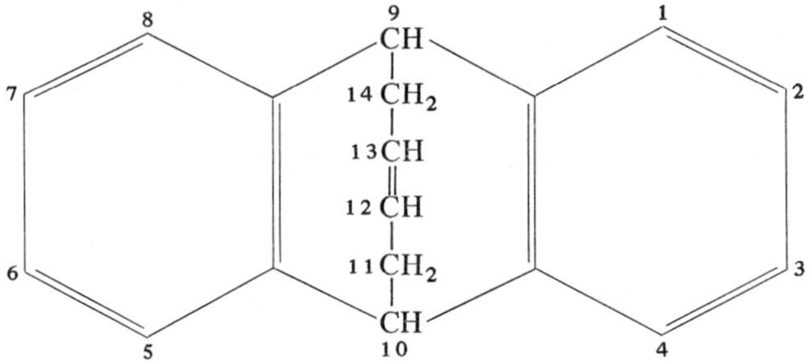

BRIDGED HYDROCARBONS

9,10-Dihydro-9,10-(2-buteno)anthracene

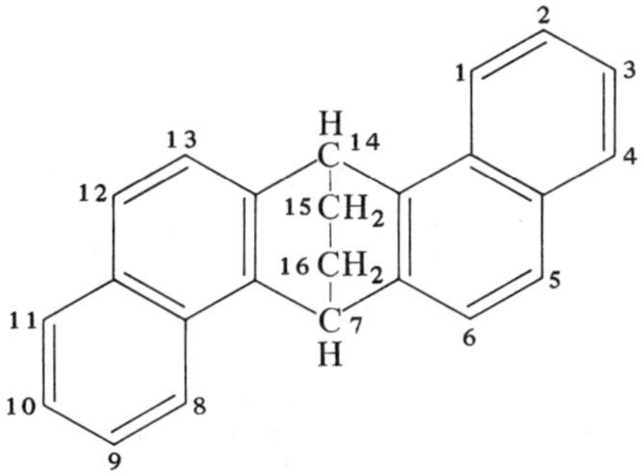

7,14-Dihydro-7,14-ethanodibenz[*a*,*h*]anthracene

34.2—The parent "*ortho*-fused" or "*ortho*- and *peri*-fused" system is numbered as prescribed in Rule **A-22**. Where there is a choice, the position numbers of the bridgeheads should be as low as possible. The remaining bridges are then numbered in turn starting each time with the bridge atom next to the bridgehead possessing the highest number.

Example:

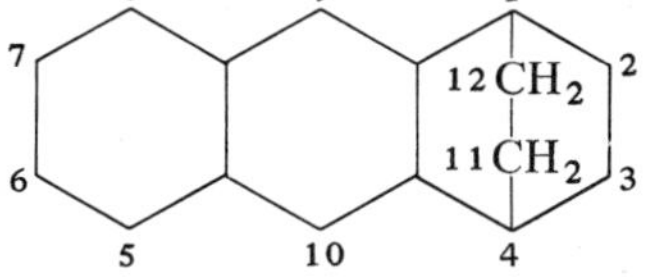

Perhydro-1,4-ethanoanthracene

not 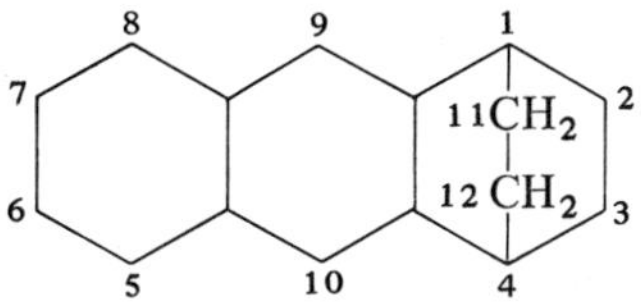   or

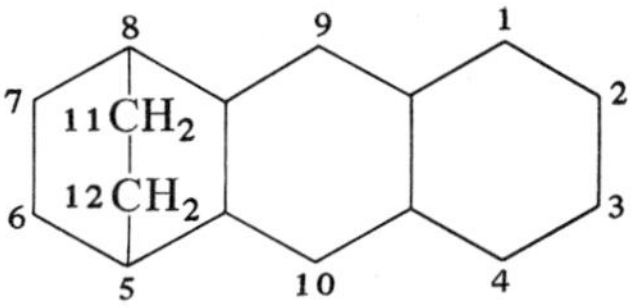

34.3—When there is a choice of position numbers for the points of attachment for several individual bridges, the lowest numbers are assigned to the bridgeheads in the order of citation of the bridges and the bridge atoms are numbered according to the preceding rule.

Example:

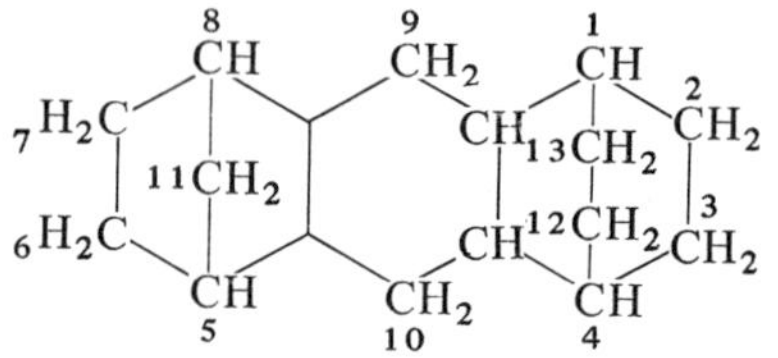

Perhydro-1,4-ethano-5,8-methanoanthracene

34.4—When the bridge is formed from a bivalent cyclic hydrocarbon radical, low numbers are given to the carbon atoms constituting the shorter bridge and numbering proceeds around the ring.

Example:

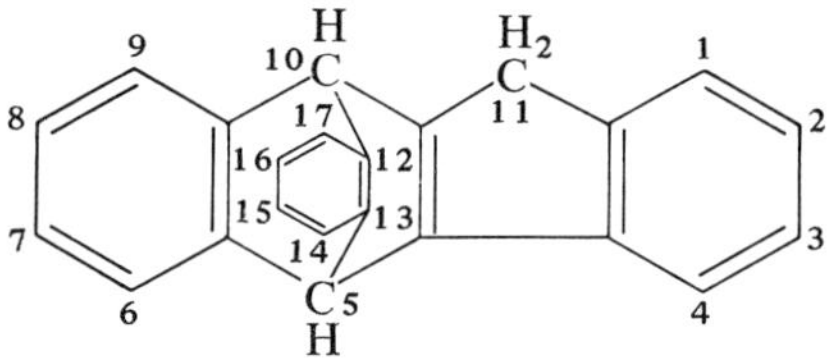

10,11-Dihydro-5,10-*o*-benzeno-5*H*-benzo[*b*]fluorene

34.5—Names for radicals derived from the bridged hydrocarbons considered in Rule **A-34.1** are constructed in accordance with the principles set forth in Rule **A-24**. The abbreviated radical names naphthyl, anthryl, phenanthryl, naphthylene, etc., permitted as exceptions to Rules **A-24.2** and **A-24.4**, are replaced in such cases by the regularly formed names naphthalenyl, anthracenyl, phenanthrenyl, naphthalenediyl, *etc.*

Examples:

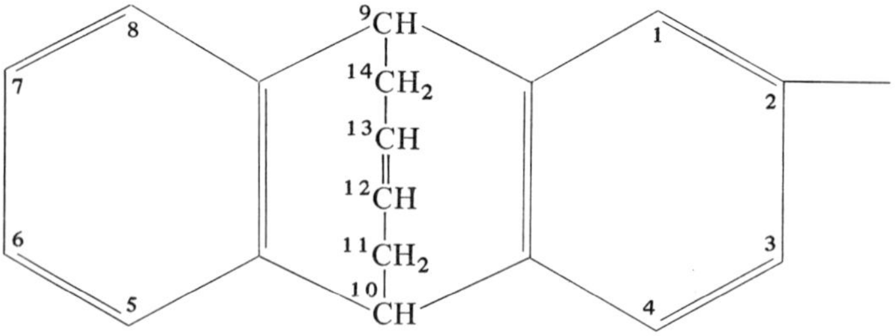

9,10-Dihydro-9,10-(2-buteno)anthracen-2-yl

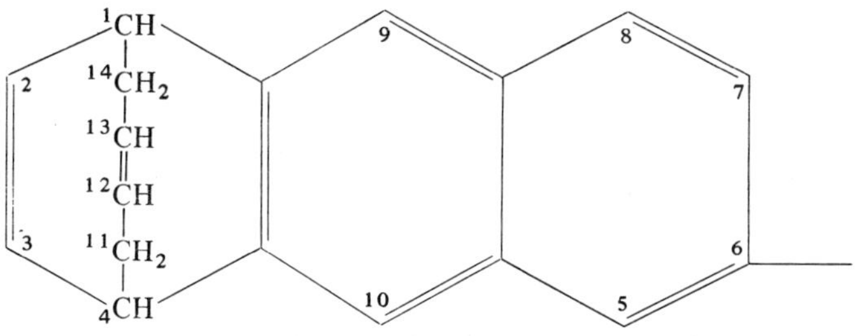

Appendix] SPIRO HYDROCARBONS 273

1,4-Dihydro-1,4-(2-buteno)anthracen-6-yl

SPIRO HYDROCARBONS

A "spiro union" is one formed by a single atom which is the only common member of two rings. A "free spiro union" is one constituting the only union direct or indirect between two rings*. The common atom is designated as the "spiro atom". According to the number of spiro atoms present, the compounds are distinguished as monospiro-, dispiro-, trispiro-compounds, *etc.* The following rules apply to the naming of compounds containing free spiro unions.

Rule A-41. Compounds: Method 1

41.1—Monospiro compounds consisting of only two alicyclic rings as components are named by placing "spiro" before the name of the normal acyclic hydrocarbon of the same total number of carbon atoms. The number of carbon atoms linked to the spiro atom in each ring is indicated in ascending order in brackets placed between the spiro prefix and the hydrocarbon name.

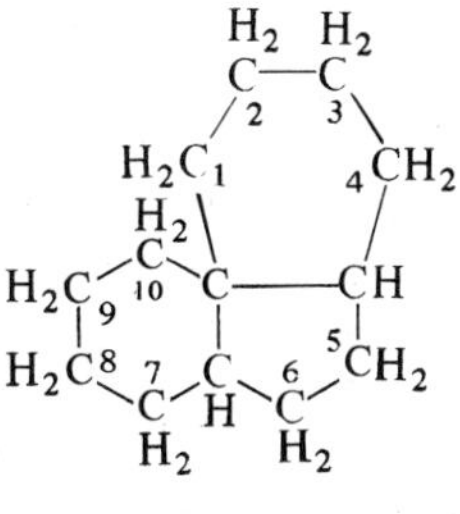

This compound is named by previous rules as dodecahydrobenz[*c*]indene.

* An example of a compound where the spiro union is *not* free is:

Examples:

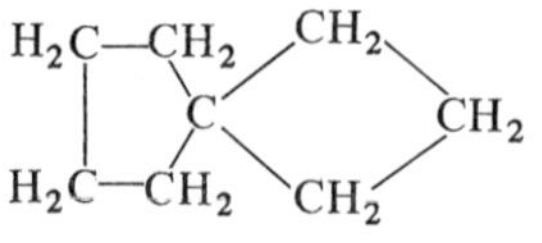

$$\text{Spiro[3,4]octane}$$

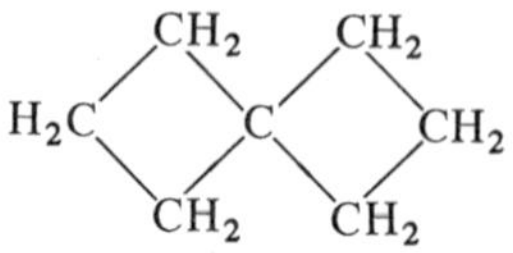

$$\text{Spiro[3.3]heptane}$$

41.2—The carbon atoms in monospiro hydrocarbons are numbered consecutively starting with a ring atom next to the spiro atom, first through the smaller ring (if such be present) and then through the spiro atom and around the second ring.

Example:

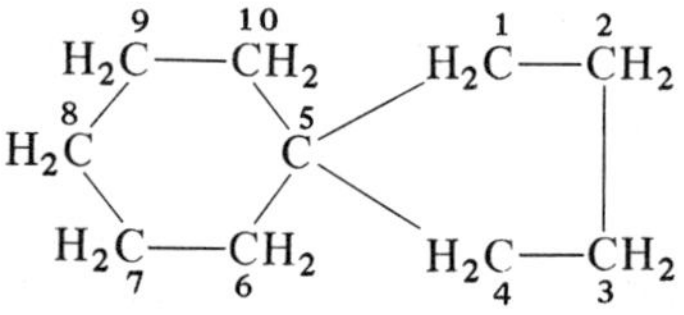

$$\text{Spiro[4.5]decane}$$

41.3—When unsaturation is present, the same enumeration pattern is maintained, but in such a direction around the rings that the double and triple bonds receive numbers as low as possible in accordance with Rule A-11.

Example:

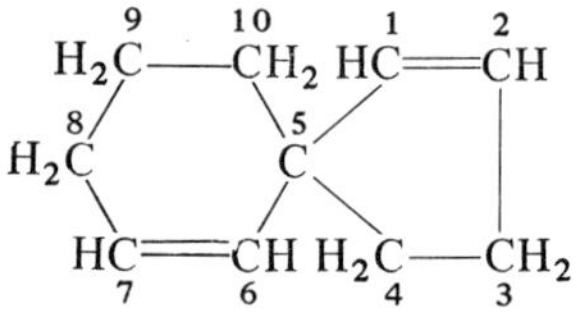

$$\text{Spiro[4.5]deca-1,6-diene}$$

41.4—If one or both components of the monospiro compound are fused polycyclic systems, "spiro" is placed before the names of the components arranged in alphabetical order and enclosed in brackets. Established numbering of the individual components is retained. The lowest possible number is given to the spiro atom, and the numbers of the second component are marked with primes. The position of the spiro atom is indicated by placing the appropriate numbers between the names of the two components.

Example:

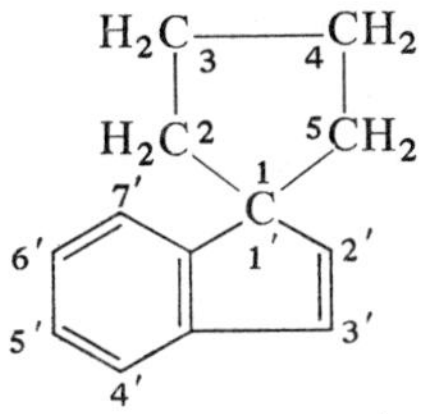

Spiro[cyclopentane-1,1'-indene]

41.5—Monospiro compounds containing two similar polycyclic components are named by placing the prefix "spirobi" before the name of the component ring system. Established enumeration of the polycyclic system is maintained and the numbers of one component are distinguished by primes. The position of the spiro atom is indicated in the name of the spiro compound by placing the appropriate locants before the name.

Example:

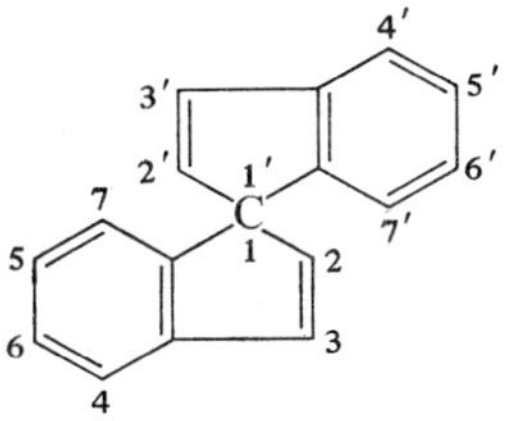

1,1'-Spirobiindene

41.6—Polyspiro compounds consisting of a linear assembly of three or more alicyclic systems are named by placing "dispiro-", "trispiro-", "tetraspiro-", etc., before the name of the unbranched-chain acyclic hydrocarbon of the same total number of carbon atoms. The numbers of carbon atoms linked to the spiro atoms in each ring are indicated in brackets in the same order as the numbering proceeds about the ring. Numbering starts with a ring atom next to a terminal spiro atom and proceeds in such a way as to give the spiro atoms as low numbers as possible.

Example:

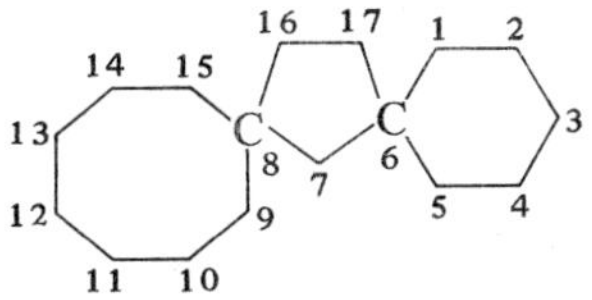

Dispiro[5.1.7.2]heptadecane

41.7—Polycyclic compounds containing more than one spiro atom and at least one fused polycyclic component are named in accordance with Part .4 of this rule by replacing "spiro" with "dispiro", "trispiro", *etc.*, and choosing the end components by alphabetical order.

Example:

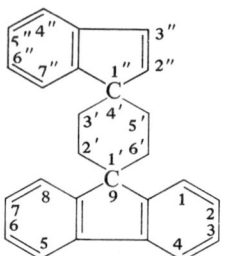

Dispiro[fluorene-9,1'-cyclohexane-4',1"-indene]

Rule A-42. Compounds: Method 2

42.1 (Alternate to Rules **A-41.1** and **A-41.2**)—When two dissimilar cyclic components are united by a spiro union, the name of the larger component is followed by the affix "spiro" which, in turn, is followed by the name of the smaller component. Between the affix "spiro" and the name of each component system is inserted the number denoting the spiro position in the appropriate ring system, these numbers being as low as permitted by any fixed enumeration of the component. The components retain their respective enumerations but numerals for the component mentioned second are primed. Numerals 1 may be omitted when a free choice is available for a component.

Examples:

Cyclopentanespirocyclobutane   Cyclohexanespirocyclopentane

2*H*-Indene-2-spiro-1'-cyclopentane

42.2 (Alternate to **A-41.3**)—Rule **A-41.3** applies also with appropriate different enumeration, where nomenclature is according to Rule

A-42.1, but the spiro junction has priority for lowest numbers over unsaturation.

Example:

2-Cyclohexenespiro-(2′-cyclopentene)

42.3 (Alternate to **A-41.5**)—The nomenclature of Rule **A-41.5** is applied also to monocyclic components with identical saturation, the spiro union being numbered 1.

Example:

Spirobicyclohexane   but   2-Cyclohexenespiro-(3′-cyclohexene)

42.4 (Alternate to **A-41.6** and **A-41.7**)—Polycyclic compounds containing more than one spiro atom are named in accordance with Rule **A-42.1** starting from the senior* end-component irrespective of whether the components are simple or fused rings.

Examples:

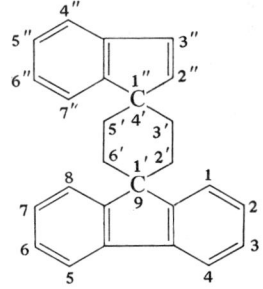

Cyclooctanespirocyclopentane-3′-spirocyclohexane

Fluorene-9-spiro-1′-cyclohexane-4′-spiro-1″-indene

* "Seniority" in respect to spiro compounds is based on the principles:

(i) an aggregate is senior to a monocycle;
(ii) of aggregates, the senior is that containing the largest number of individual rings;
(iii) of aggregates containing the same number of individual rings, the senior is that containing the largest ring;
(iv) if aggregates consist of equal numbers of equal rings the senior is the first occurring in the alphabetical list of names.

Rule A-43. Radicals

43.1—Radicals derived from spiro hydrocarbons are named according to the principles set forth in Rules **A-11** and **A-24**.

Examples:

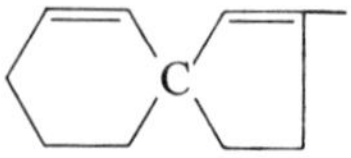

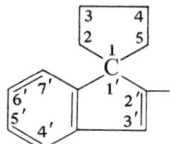

Spiro[4.5]deca-1,6-dien-2-yl
(cf. Rules **A-41.3** and **A-11**)
or 2-Cyclohexenespiro-2'-cyclopenten-3'-yl
(cf. Rule **A-42.2**)

Spiro[cyclopentane-1,1'-inden]-2'-yl
(cf. Rules **A-41.4** and **A-24**)

HYDROCARBON RING ASSEMBLIES

Rule A-51. Definition

51.1—Two or more cyclic systems (single rings or fused systems) which are directly joined to each other by double or single bonds are named "ring assemblies" when the number of such direct ring junctions is one less than the number of cyclic systems involved.

Examples:

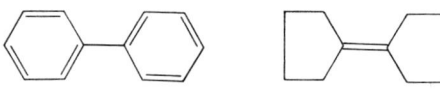

Ring assemblies

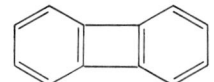

Fused polycyclic system

Rule A-52. Two Identical Ring Systems

52.1—Assemblies of two identical cyclic hydrocarbon systems are named in either of two ways: (*a*) by placing the prefix "bi-" before the name of the corresponding radical, or (*b*) for systems joined by a single bond by placing the prefix "bi-" before the name of the corresponding hydrocarbon. In each case, the numbering of the assembly is that of the corresponding radical or hydrocarbon, one system being assigned unprimed numbers and the other primed numbers. The points of attachment are indicated by placing the appropriate locants before the name.

Examples:

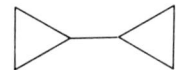   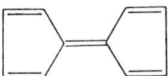

1,1'-Bicyclopropyl
or 1,1'-Bicyclopropane

1,1'-Bicyclopentadienylidene
or $\Delta^{1,1'}$-Bicyclopentadienylidene
(cf. Footnote to Rule B-1.2)

52.2—If there is a choice in numbering, unprimed numbers are assigned to the system which has the lower-numbered point of attachment.

Example:

1,2'-Binaphthyl
or 1,2'-Binaphthalene

52.3—If two identical hydrocarbon systems have the same point of attachment and contain substituents at different positions, the locants of these substituents are assigned according to Rule **A-2.2**; for this purpose an unprimed number is considered lower than the same number when primed. Assemblies of primed and unprimed numbers are arranged in ascending numerical order.

Examples:

2,3,3',4',5'-Pentamethylbiphenyl
(not 2',3,3',4,5-Pentamethylbiphenyl)

2-Ethyl-2'-propylbiphenyl

52.4—The name "biphenyl" is used for the assembly consisting of two benzene rings.

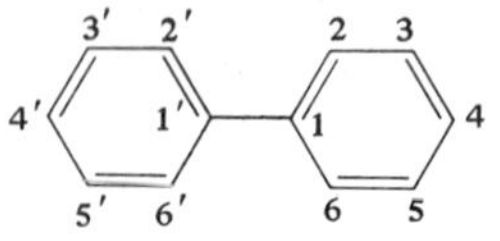

Biphenyl

Rule A-53. Non-identical Ring Systems

53.1—Other hydrocarbon ring assemblies are named by selecting one ring system as the base component and considering the other systems as substituents of the base component. Such substituents are arranged in alphabetical order. The base component is assigned unprimed numbers and the substituents are assigned numbers with primes.

53.2—The base component is chosen by considering the following characteristics in turn until a decision is reached:

(*a*) The system containing the larger number of rings.

Examples:

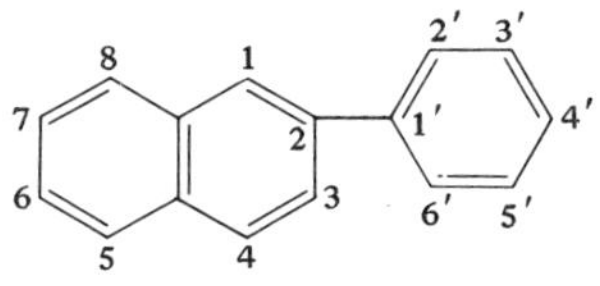

2-Phenylnaphthalene

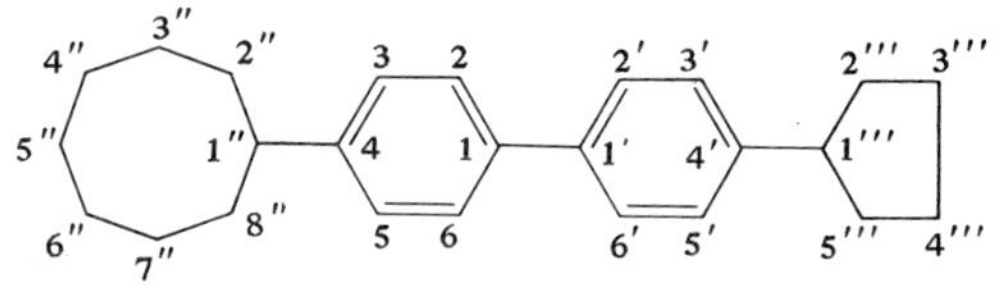

4-Cyclooctyl-4'-cyclopentylbiphenyl

(*b*) The system containing the larger ring.

Examples:

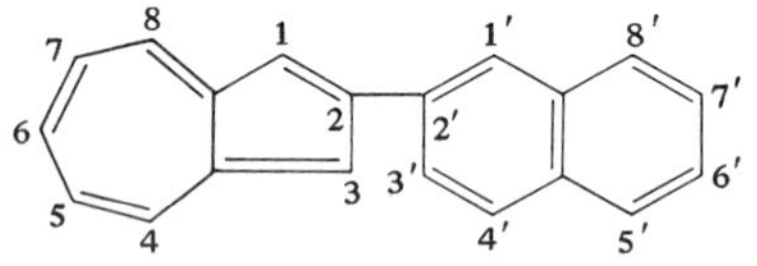

2-(2'-Naphthyl)azulene

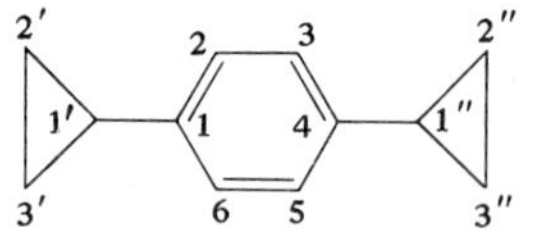

1,4-Dicyclopropylbenzene
or *p*-Dicyclopropylbenzene

(*c*) The system in the lowest state of hydrogenation (see also Part .3 of this rule).

Example:

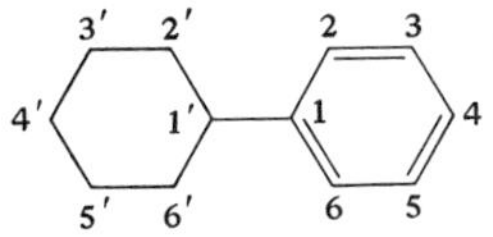

Cyclohexylbenzene

(d) The order of ring systems as set forth in the list of Rule **A-21.1**.
53.3—Compounds covered by Part .2(c) of this rule may also be named as hydrogenation products according to Rule **A-23**.

Example:

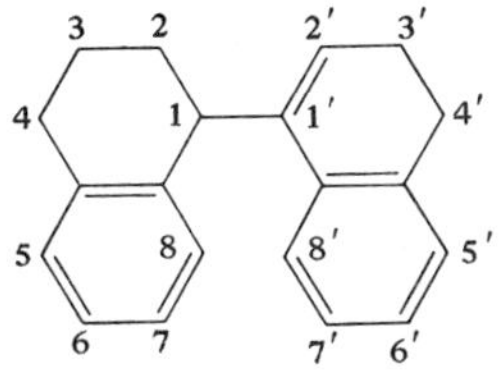

1,2,3,3′,4,4′-Hexahydro-1,1′-binaphthyl
or 1,2,3,3′,4,4′-Hexahydro-1,1′-binaphthalene

Rule A-54. Three or More Identical Ring Systems

54.1—Unbranched assemblies consisting of three or more identical hydrocarbon ring systems are named by placing an appropriate numerical prefix before the name of the hydrocarbon corresponding to the repetitive unit. The following numerical prefixes are used:

>   3. ter-   7. septi-
>   4. quater-   8. octi-
>   5. quinque-   9. novi-
>   6. sexi-   10. deci-

Example:

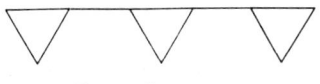

Tercyclopropane

54.2—Unprimed numbers are assigned to one of the terminal systems, the other systems being primed serially. Points of attachment are assigned the lowest numbers possible.

Examples:

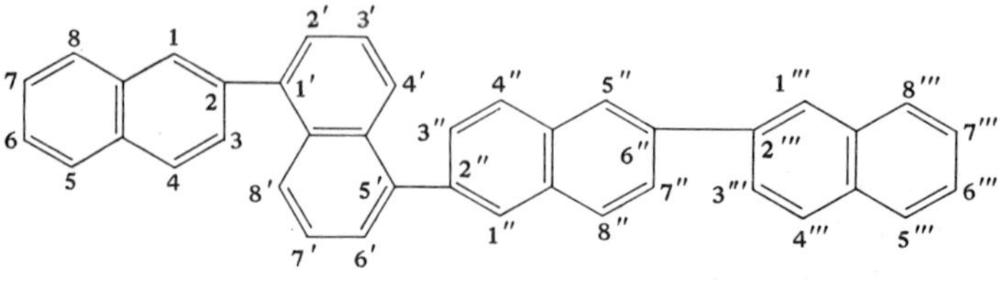

2,1′:5′,2″:6″,2‴-Quaternaphthalene

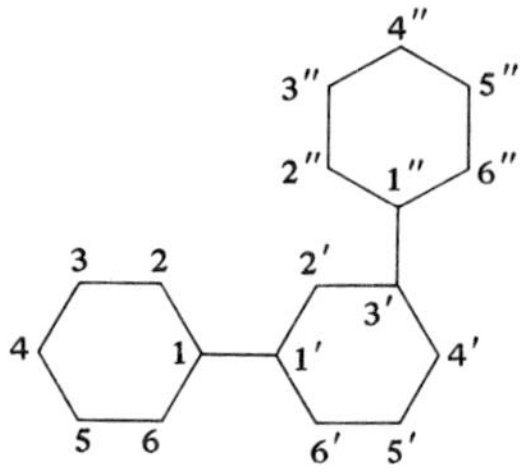

1,1′:3′,1″-Tercyclohexane

54.3—As exceptions, unbranched assemblies consisting of benzene rings are named by using the appropriate prefix with the radical name "phenyl".

Examples:

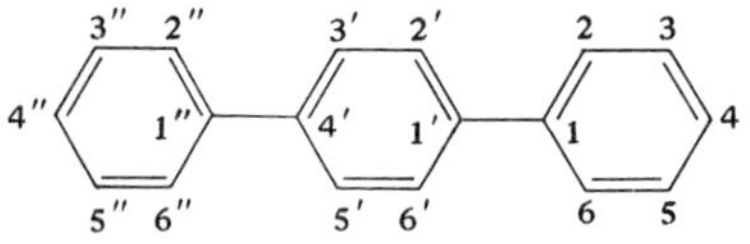

p-Terphenyl
or 1,1′:4′,1″-Terphenyl

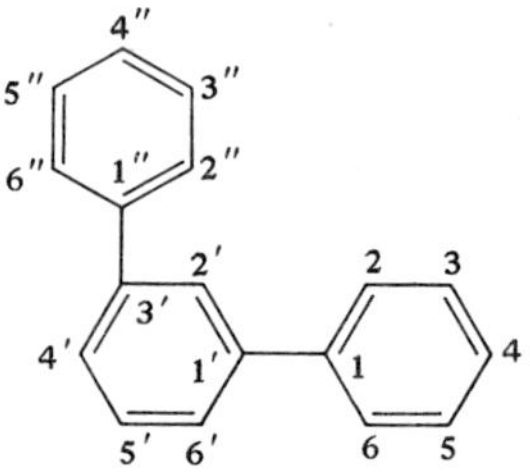

m-Terphenyl
or 1,1′:3′,1″-Terphenyl

Rule A-55. Radicals for Identical Ring Systems (Alternative in part to Rule A-56.1)

55.1—Univalent and multivalent radicals derived from assemblies of identical hydrocarbon ring systems are named by adding "-yl", "-ylene" or "-diyl", "-triyl", *etc.*, to the name of the ring assembly.

Examples:

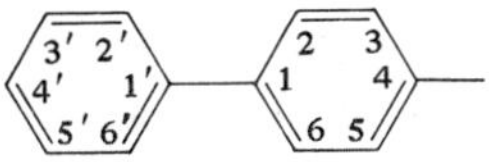

4-Biphenylyl

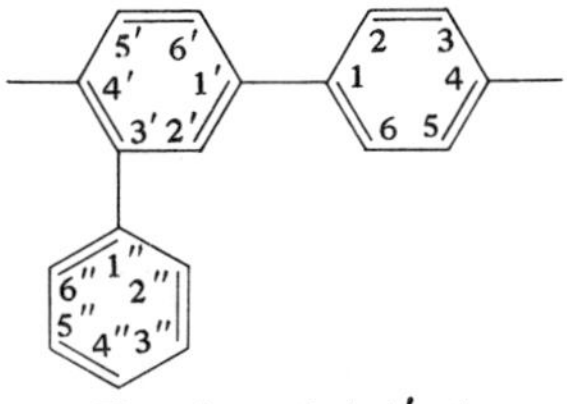

m-Terphenyl-4,4′-ylene
or *m*-Terphenyl-4,4′-diyl

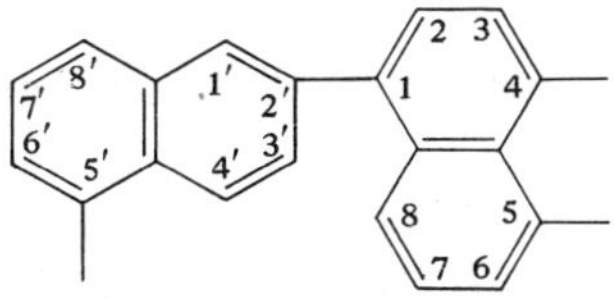

[1,2′-Binaphthalene]-4,5,5′-triyl

Rule A-56. Radicals for Non-benzenoid Ring Systems (Alternative in part to Rule A-55.1)

56.1—Radicals derived from hydrocarbon ring assemblies other than benzene ring assemblies by removal of one or more hydrogen atoms from only one ring are named with that ring as the parent radical, the remaining rings being named as substituents.

Examples:

6-(2-Anthryl)-2,3-naphthalenediyl
or 6-(2-Anthryl)-2,3-naphthylene

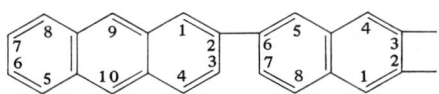

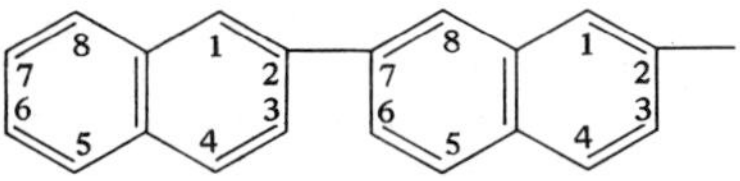

7-(2-Naphthyl)-2-naphthyl

Note: This method is used for assemblies of non-identical systems; also it is sometimes preferable to that of Rule **A-55** for assemblies of identical systems when a group to be specified as a suffix or as a separate word is present in a chain attached to the ring assembly.

FUNDAMENTAL HETEROCYCLIC SYSTEMS

SPECIALIST HETEROCYCLIC NOMENCLATURE

Rule B-1. Extension of Hantzsch–Widman System

1.1—Monocyclic compounds containing one or more hetero atoms in a three- to ten-membered ring are named by combining the appropriate prefix or prefixes from Table I (eliding "a" where necessary) with a stem from Table II. The state of hydrogenation is indicated either in the stem, as shown in Table II, or by the prefixes "dihydro-", "tetrahydro-", *etc.*, according to Rule **B-1.2**.

TABLE I

Element	Valence	Prefix	Element	Valence	Prefix
Oxygen	II	Oxa	Antimony	III	Stiba*
Sulfur	II	Thia	Bismuth	III	Bismutha
Selenium	II	Selena	Silicon	IV	Sila
Tellurium	II	Tellura	Germanium	IV	Germa
Nitrogen	III	Aza	Tin	IV	Stanna
Phosphorus	III	Phospha*	Lead	IV	Plumba
Arsenic	III	Arsa*	Boron	III	Bora
			Mercury	II	Mercura

* When immediately followed by "-in" or "-ine", "phospha-" should be replaced by "phosphor-", "arsa-" should be replaced by "arsen-" and "stiba-" should be replaced by "antimon-". In addition, the saturated six-membered rings corresponding to phosphorin and arsenin are named phosphorinane and arsenane.

Appendix] SPECIALIST HETEROCYCLIC NOMENCLATURE   285

TABLE II

No. of members in the ring	Rings containing nitrogen		Rings containing no nitrogen	
	Unsaturation (a)	Saturation	Unsaturation (a)	Saturation
3	-irine	-iridine	-irene	-irane (e)
4	-ete	-etidine	-ete	-etane
5	-ole	-olidine	-ole	-olane
6	-ine (b)	(c)	-in (b)	-ane (d)
7	-epine	(c)	-epine	-epane
8	-ocine	(c)	-ocin	-ocane
9	-onine	(c)	-onin	-onane
10	-ecine	(c)	-ecin	-ecane

(a) Corresponding to the maximum number of non-cumulative double bonds, the hetero elements having the normal valences shown in Table I.
(b) For phosphorus, arsenic, antimony, see the special provisions of Table I.
(c) Expressed by prefixing "perhydro" to the name of the corresponding unsaturated compound.
(d) Not applicable to silicon, germanium, tin and lead. In this case, "perhydro-" is prefixed to the name of the corresponding unsaturated compound.
(e) The syllables denoting the size of rings containing 3, 4 or 7-10 members are derived as follows: "ir" from t*ri*, "et" from t*etr*a, "ep" from h*ep*ta, "oc" from *oc*ta, "on" from n*on*a, and "ec" from d*ec*a.

Examples:

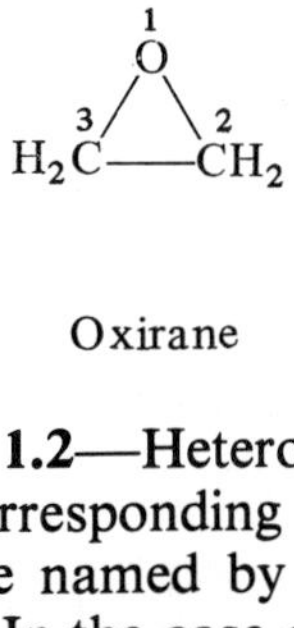

Oxirane 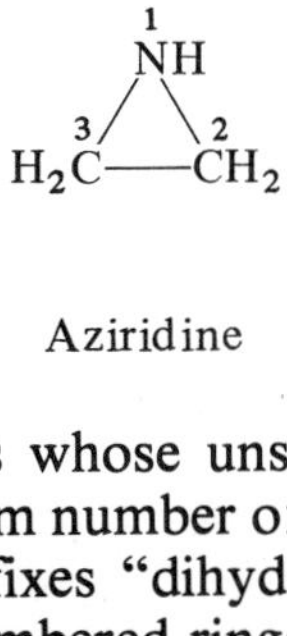   Aziridine 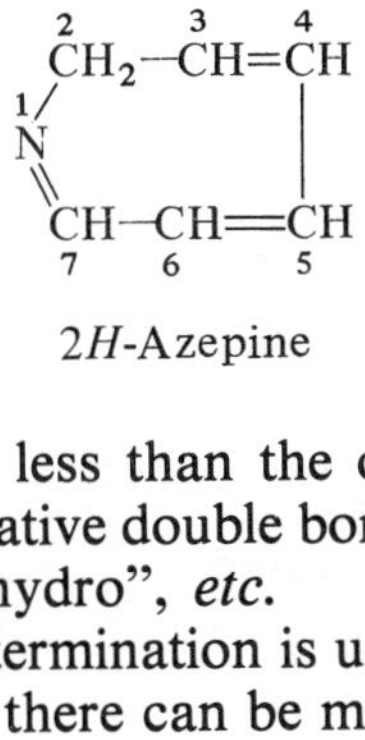   2*H*-Azepine

1.2—Heterocyclic systems whose unsaturation is less than the one corresponding to the maximum number of non-cumulative double bonds are named by using the prefixes "dihydro-", "tetrahydro", *etc.*

In the case of 4- and 5-membered rings, a special termination is used for the structures containing one double bond, when there can be more than one non-cumulative double bond.

No. of members of the partly saturated rings	Rings containing nitrogen	Rings containing no nitrogen
4	-etine	-etene
5	-oline	-olene

Examples:

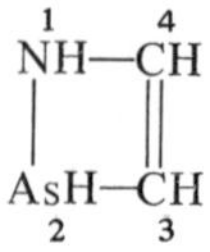

Δ^3-1,2-Azarsetine*

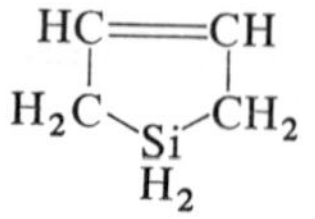

Silolene

1.3—Multiplicity of the same hetero atom is indicated by a prefix "di-", "tri-", etc., placed before the appropriate "a" term (Table I).

Example:

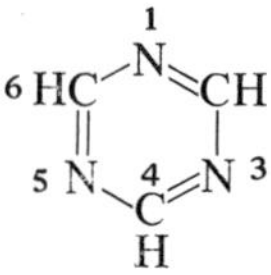

1,3,5-Triazine

1.4—If two or more kinds of "a" terms occur in the same name, their order of citation is by descending group number of the Periodic Table and increasing atomic number in the group as illustrated by the sequence in Table I.

Examples:

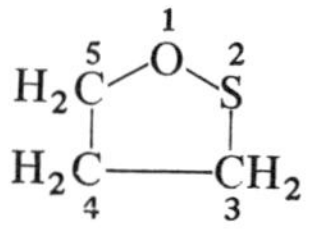

1,2-Oxathiolane

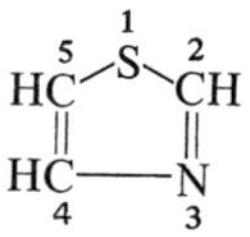

1,3-Thiazole

1.51—The position of a single hetero atom determines the numbering in a monocyclic compound.

Example:

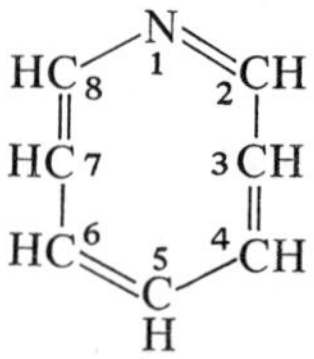

Azocine

---

* As exceptions, Greek capital delta (Δ), followed by superscript locant(s), is used to denote a double bond in a compound named according to Rule **B-1.2** if its name is preceded by locants for hetero atoms; and also to denote a double bond uniting components in an assembly of rings (cf. Examples to Rules **A-52.1** and **C-71.1**) or in conjunctive names (cf. Rule **C-55.1**).

1.52—When the same hetero atom occurs more than once in a ring, the numbering is chosen to give the lowest locants to the hetero atoms.

Example:

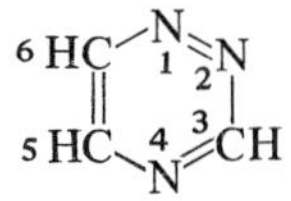

1,2,4-Triazine

1.53—When hetero atoms of different kinds are present, the locant 1 is given to a hetero atom which is as high as possible in Table I. The numbering is then chosen to give the lowest locants to the hetero atoms.

Examples:

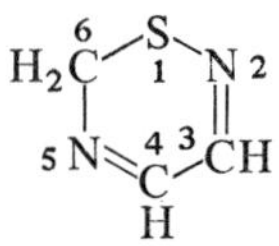

6H-1,2,5-Thiadiazine
(not: 2,1,4-Thiadiazine)
(not: 1,3,6-Thiadiazine)

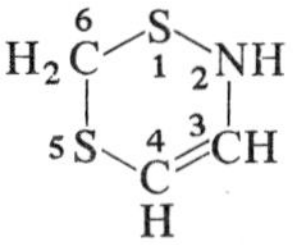

2H,6H-1,5,2-Dithiazine
(not: 1,3,4-Dithiazine)
(not: 1,3,6-Dithiazine)
(not: 1,5,4-Dithiazine)

The numbering must begin with the sulfur atom. This condition eliminates 2,1,4-thiadiazine. Then the nitrogen atoms receive the lowest possible locant, which eliminates 1,3,6-thiadiazine.

The numbering has to begin with a sulfur atom. The choice of this atom is determined by the set of locants which can be attributed to the remaining hetero atoms of any kind.
As the set 1,2,5 is lower than 1,3,4 or 1,3,6 or 1,5,4 in the usual sense, the name is 1,5,2-dithiazine.

Rule B-2. Trivial and Semi-trivial Names

2.11—The following trivial and semi-trivial names constitute a partial list of such names which are retained for the compound and as a basis of fusion names. The names of the radicals shown are formed according to Rule **B-5**.

	Parent Compound		Radical Name
(1)	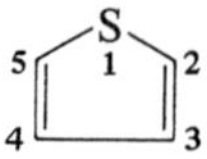 Thiophene		Thienyl (2- shown) 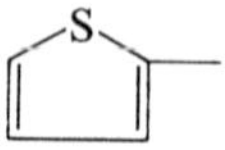
(2)	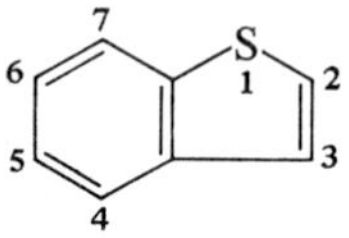 Benzo[b]thiophene (replacing thianaphthene)		Benzo[b]thienyl (2- shown) 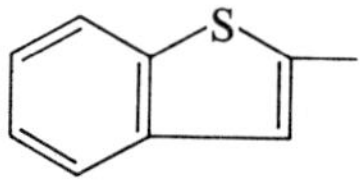
(3)	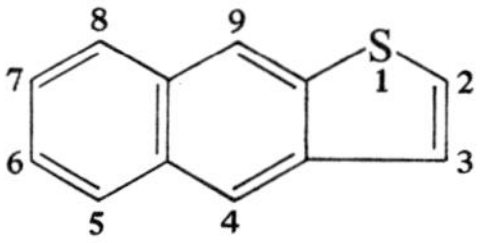 Naphtho[2,3-b]thiophene (replacing thiophanthrene)		Naphtho[2,3-b]thienyl (2- shown) 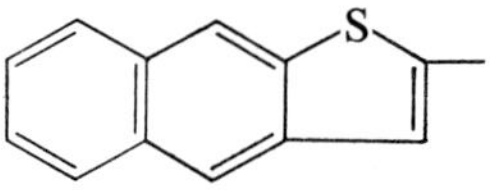
(4)	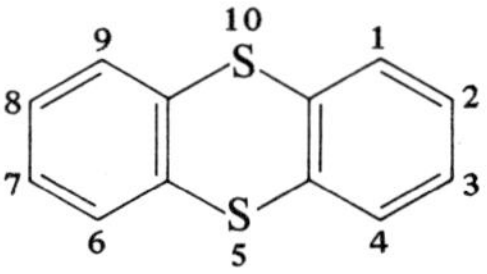 Thianthrene		Thianthrenyl (2- shown) 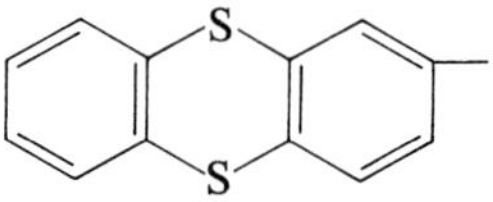
(5)	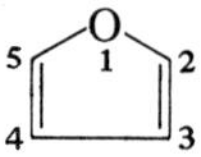 Furan		Furyl (3- shown) 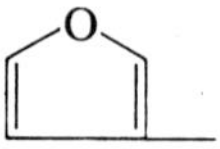
(6)	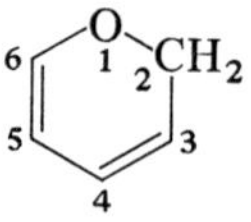 Pyran (2H- shown)		Pyranyl (2H-Pyran-3-yl shown) 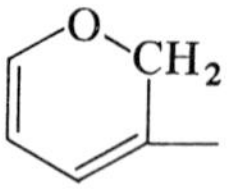
(7)	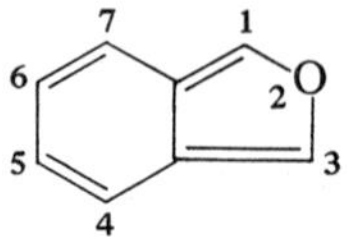 Isobenzofuran		Isobenzofuranyl (1- shown) 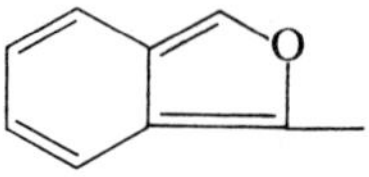

Appendix] SPECIALIST HETEROCYCLIC NOMENCLATURE 289

	Parent Compound		Radical Name
(8)	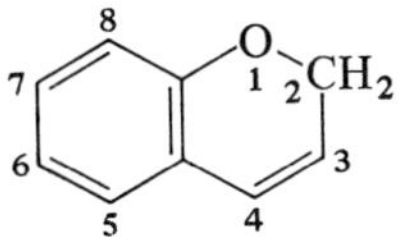	Chromene (2H- shown)	Chromenyl (2H-Chromen-3-yl shown) 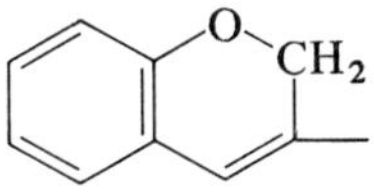
(9)	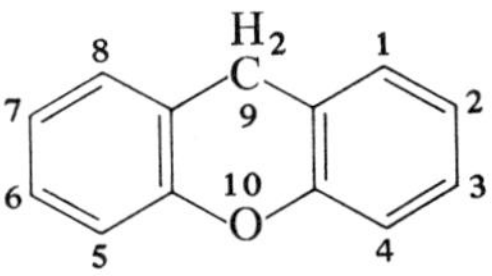	Xanthene*	Xanthenyl* (2- shown) 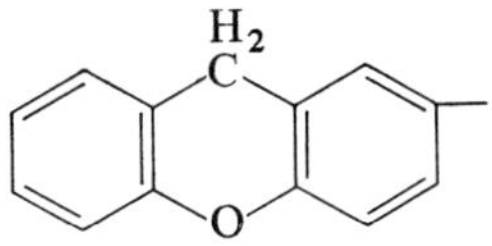
(10)	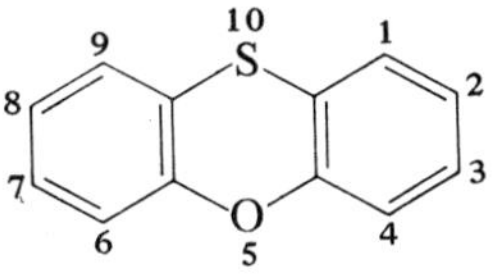	Phenoxathiin	Phenoxathiinyl (2- shown) 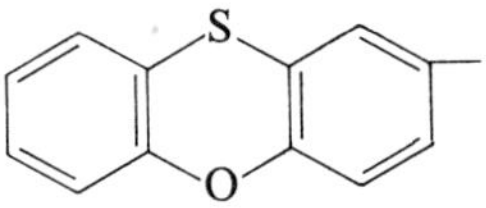
(11)	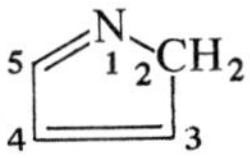	2H-Pyrrole	2H-Pyrrolyl (2H-Pyrrol-3-yl shown) 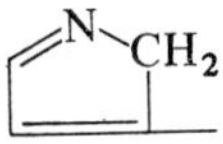
(12)	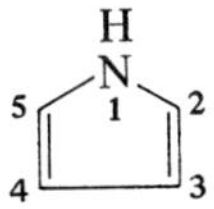	Pyrrole	Pyrrolyl (3- shown) 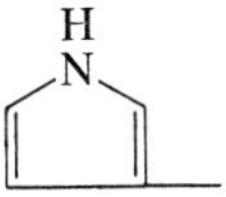
(13)	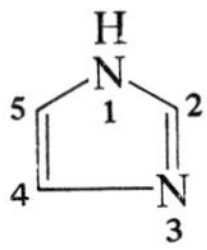	Imidazole	Imidazolyl (2- shown) 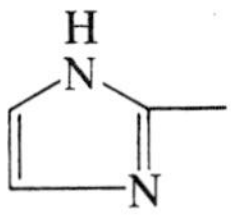

* Denotes exceptions to systematic numbering.

	Parent Compound		Radical Name
(14)	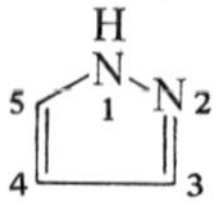	Pyrazole	Pyrazolyl (1-shown) 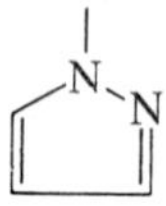
(15)	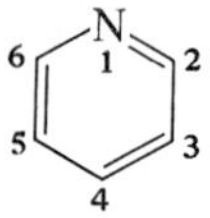	Pyridine	Pyridyl (3-shown) 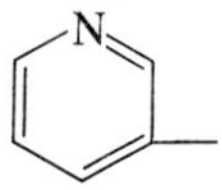
(16)	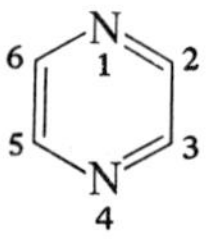	Pyrazine	Pyrazinyl 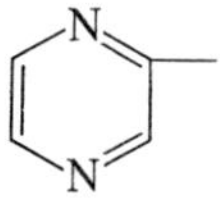
(17)	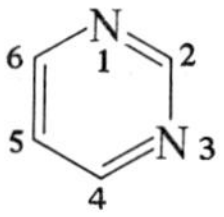	Pyrimidine	Pyrimidinyl (2-shown) 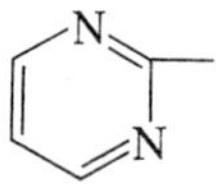
(18)	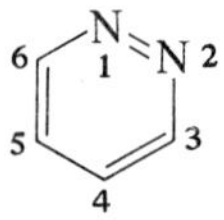	Pyridazine	Pyridazinyl (3-shown) 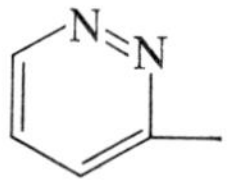
(19)	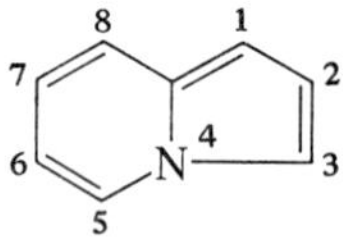	Indolizine	Indolizinyl (2-shown) 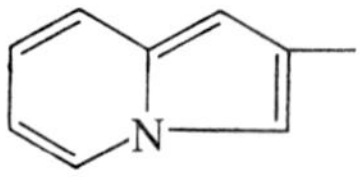
(20)	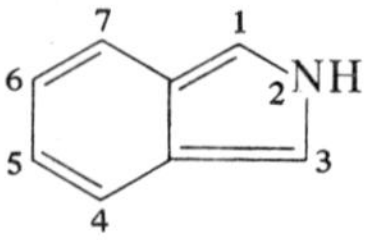	Isoindole	Isoindolyl (2-shown) 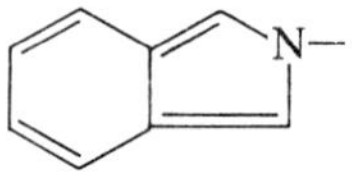

Appendix] SPECIALIST HETEROCYCLIC NOMENCLATURE

	Parent Compound		Radical Name
(21)	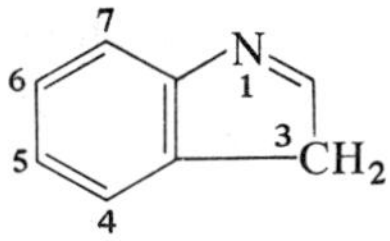	3H-Indole	3H-Indolyl (3H-Indol-2-yl shown) 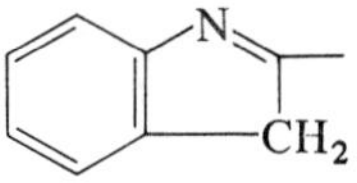
(22)	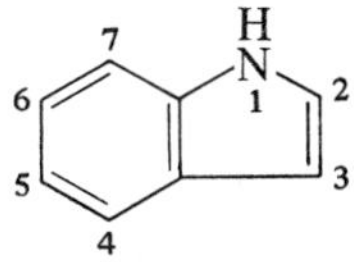	Indole	Indolyl (1-shown) 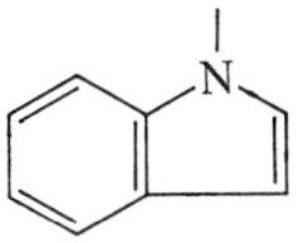
(23)	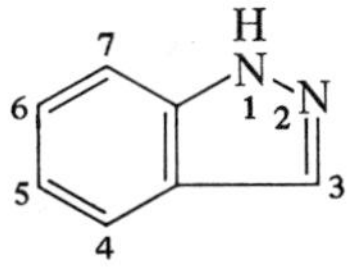	1H-Indazole	Indazolyl (1H-Indazol-3-yl shown) 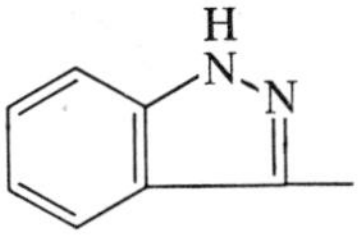
(24)	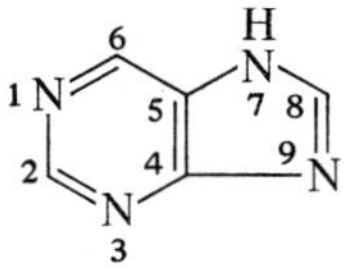	Purine*	Purinyl* (8-shown) 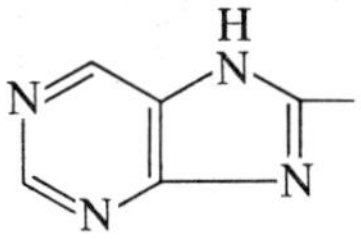
(25)	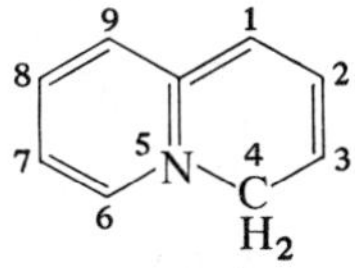	4H-Quinolizine	4H-Quinolizinyl (4H-Quinolizin-2-yl shown) 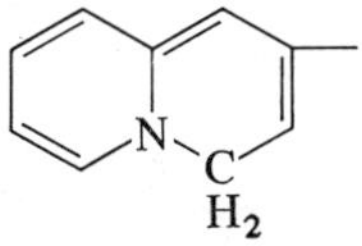
(26)	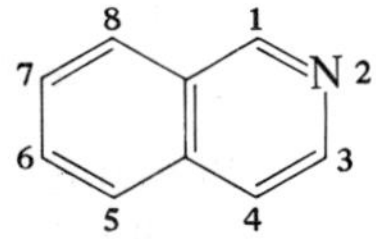	Isoquinoline	Isoquinolyl (3-shown) 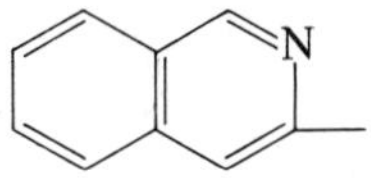

* Denotes exceptions to systematic numbering.

Parent Compound		Radical Name
(27)	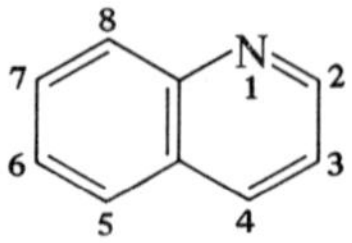 Quinoline	Quinolyl (2- shown) 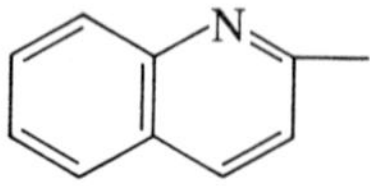
(28)	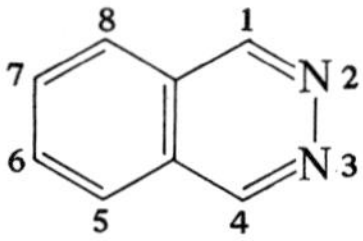 Phthalazine	Phthalazinyl (1- shown) 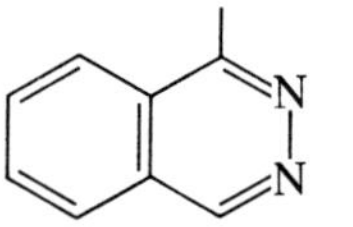
(29)	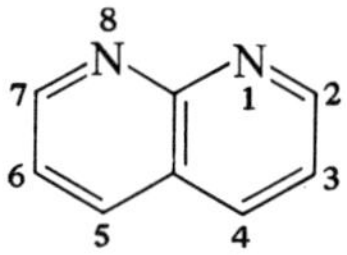 Naphthyridine (1,8- shown)	Naphthyridinyl (1,8-Naphthyridin-2-yl shown) 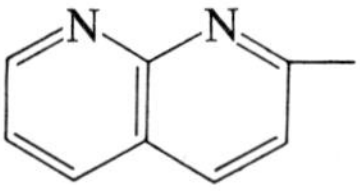
(30)	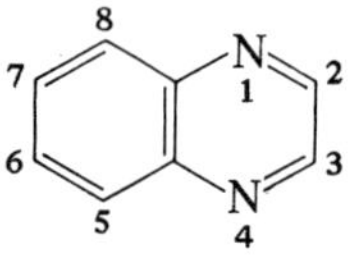 Quinoxaline	Quinoxalinyl (2- shown) 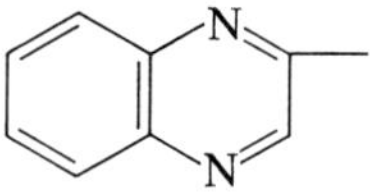
(31)	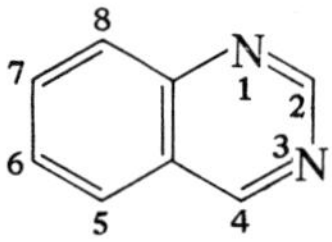 Quinazoline	Quinazolinyl (2- shown) 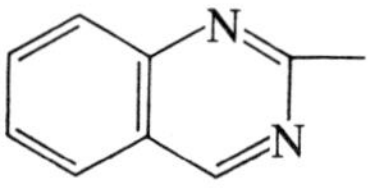
(32)	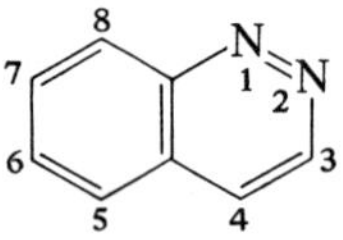 Cinnoline	Cinnolinyl (3- shown) 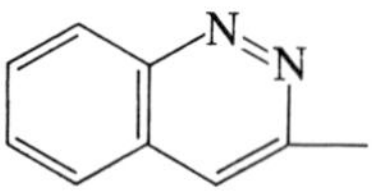
(33)	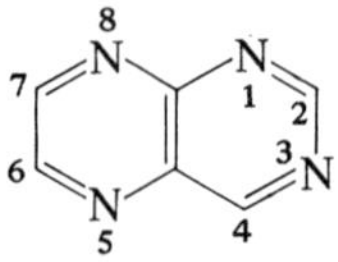 Pteridine	Pteridinyl (2- shown) 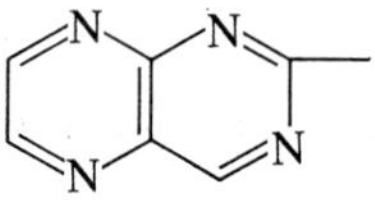

Appendix] SPECIALIST HETEROCYCLIC NOMENCLATURE 293

	Parent Compound		Radical Name
(34)	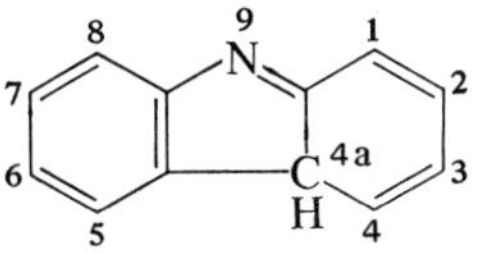	4aH-Carbazole*	4aH-Carbazolyl* (4aH-Carbazol-2-yl shown) 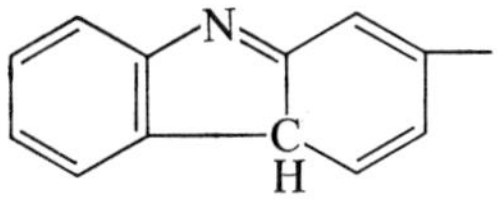
(35)	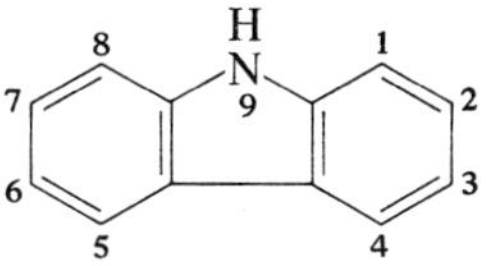	Carbazole*	Carbazolyl* (2- shown) 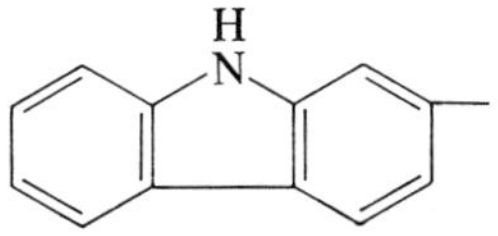
(36)	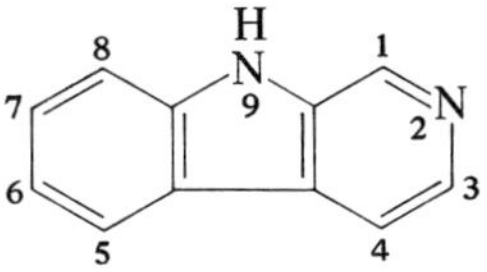	β-Carboline	β-Carbolinyl (β-Carbolin-3-yl shown) 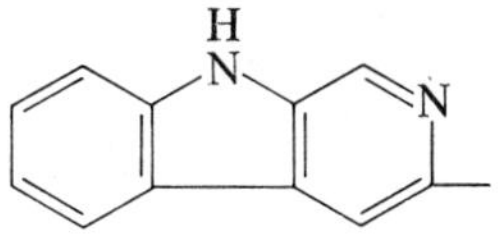
(37)	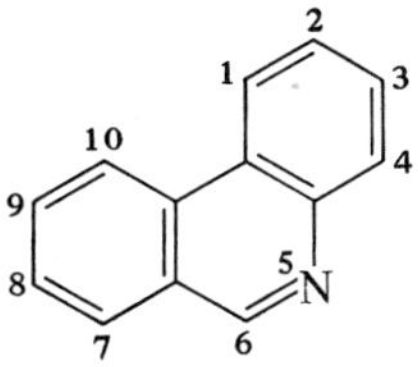	Phenanthridine	Phenanthridinyl (3- shown) 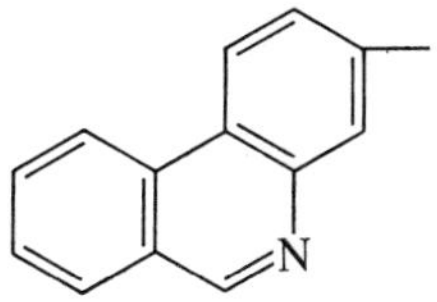
(38)	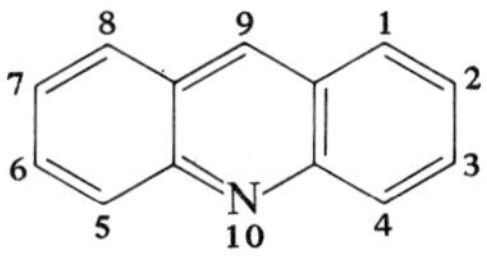	Acridine*	Acridinyl* (2- shown) 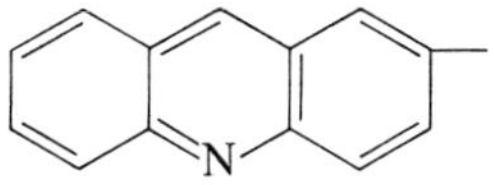
(39)	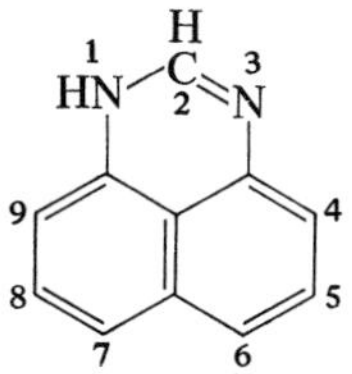	Perimidine	Perimidinyl (2- shown) 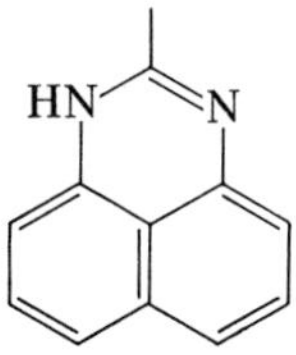

* Denotes exceptions to systematic numbering.

Parent Compound		Radical Name
(40)	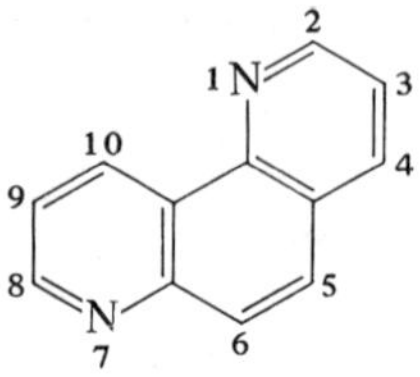 Phenanthroline (1,7- shown)	Phenanthrolinyl (1,7-Phenanthrolin-3-yl shown) 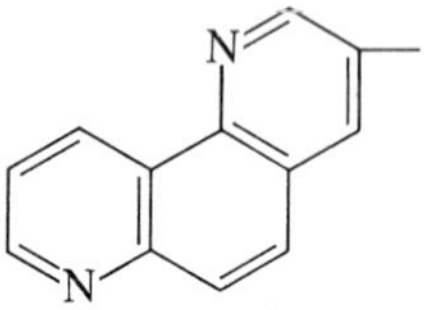
(41)	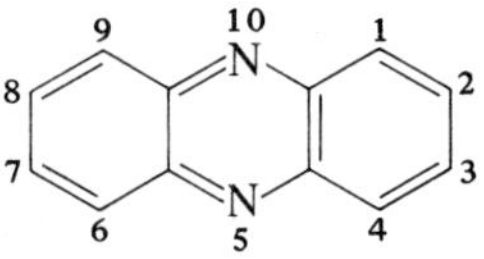 Phenazine	Phenazinyl (1- shown) 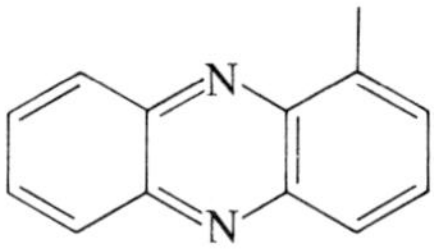
(42)	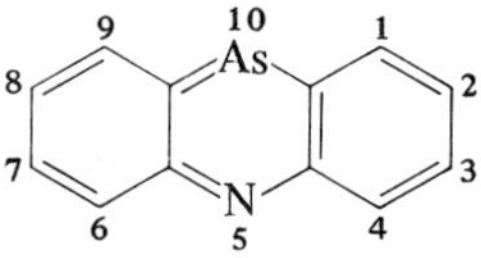 Phenarsazine	Phenarsazinyl (2- shown) 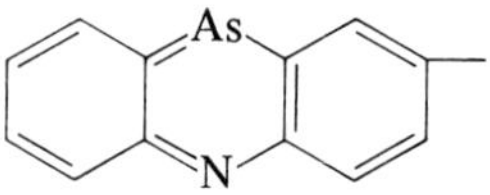
(43)	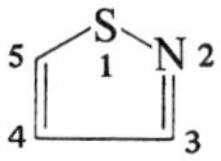 Isothiazole	Isothiazolyl (3- shown) 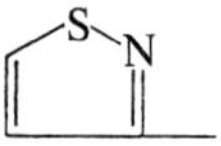
(44)	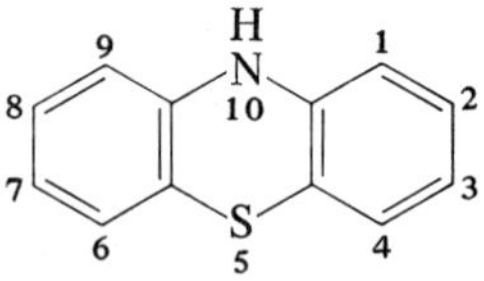 Phenothiazine	Phenothiazinyl (2- shown) 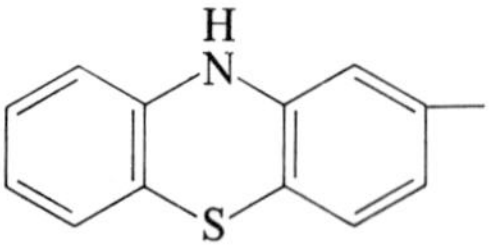
(45)	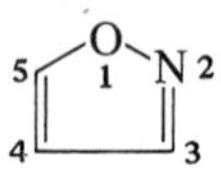 Isoxazole	Isoxazolyl (3- shown) 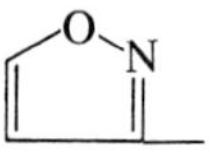

Appendix]   SPECIALIST HETEROCYCLIC NOMENCLATURE   295

Parent Compound		Radical Name
(46)	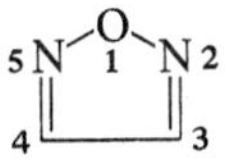 Furazan	Furazanyl (3- shown) 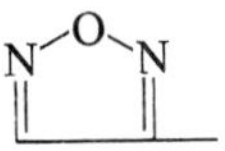
(47)	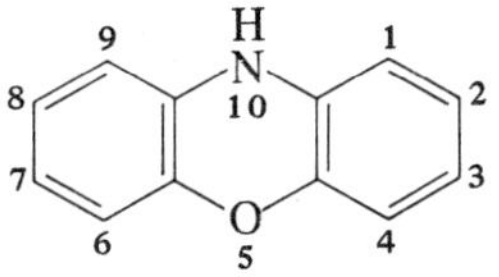 Phenoxazine	Phenoxazinyl (2- shown) 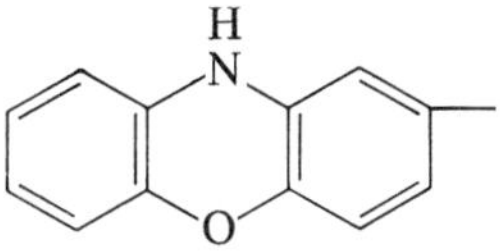

B-2.12—The following trivial and semi-trivial names are retained but are not recommended for use in fusion names. The names of the radicals shown are formed according to Rule **B-5**.

Parent Compound		Radical Name
(1)	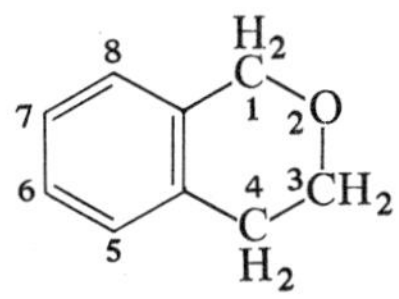 Isochroman	Isochromanyl (3- shown) 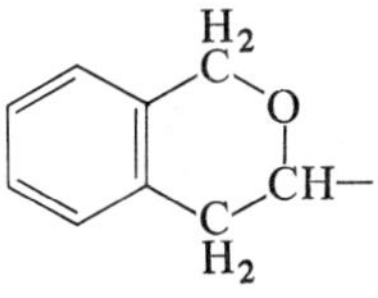
(2)	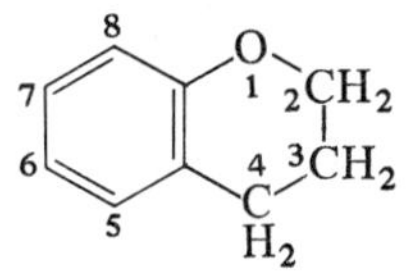 Chroman	Chromanyl (7- shown) 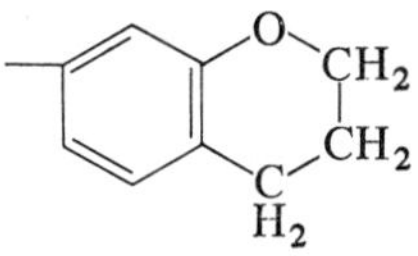
(3)	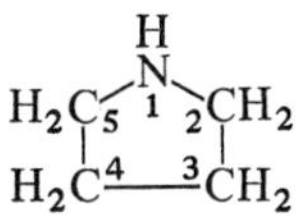 Pyrrolidine	Pyrrolidinyl (2- shown) 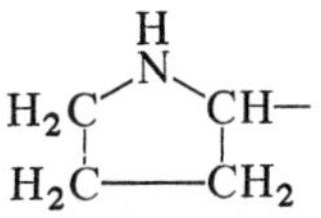

Parent Compound		Radical Name
(4)	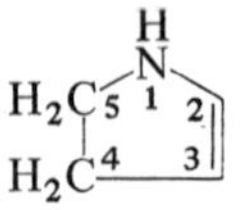 Pyrroline (2- shown*)	Pyrrolinyl (2-Pyrrolin-3-yl* shown) 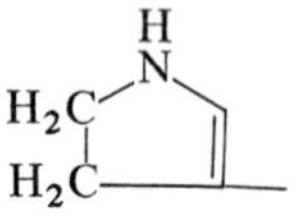
(5)	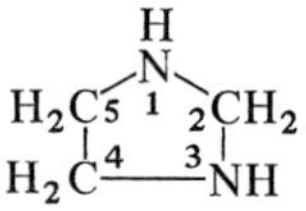 Imidazolidine	Imidazolidinyl (2-shown) 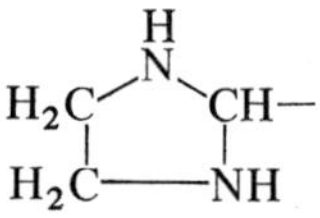
(6)	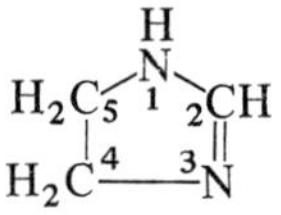 Imidazoline (2- shown*)	Imidazolinyl (2-Imidazolin-4-yl* shown) 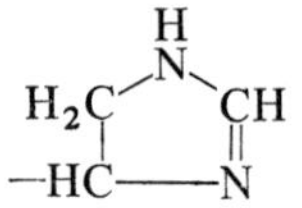
(7)	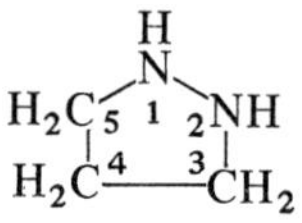 Pyrazolidine	Pyrazolidinyl (2- shown) 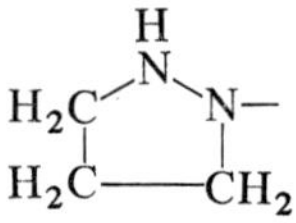
(8)	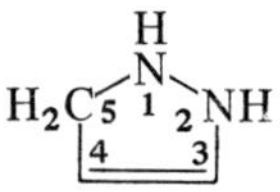 Pyrazoline (3- shown*)	Pyrazolinyl (3-Pyrazolin-2-yl* shown) 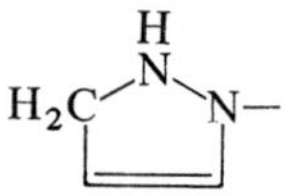
(9)	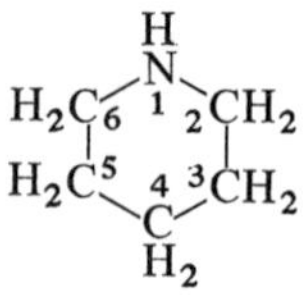 Piperidine	Piperidyl† (2- shown) 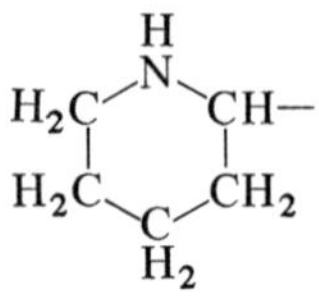

* The "2-" denotes the position of the double bond.

Appendix] **SPECIALIST HETEROCYCLIC NOMENCLATURE** 297

	Parent Compound		Radical Name
(10)	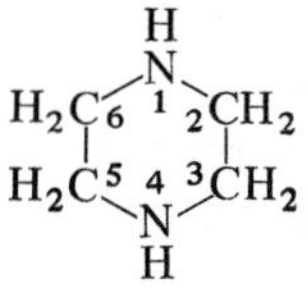 Piperazine		Piperazinyl (1- shown) 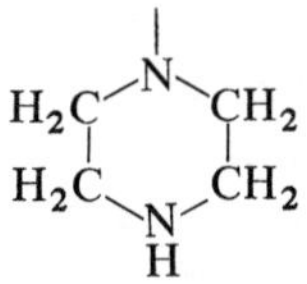
(11)	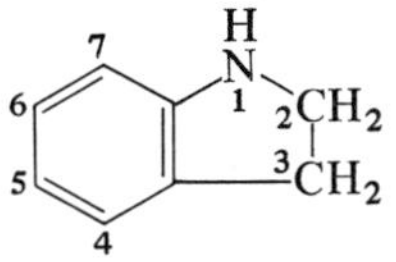 Indoline		Indolinyl (1- shown) 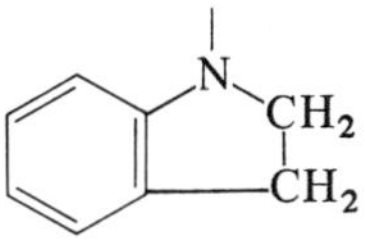
(12)	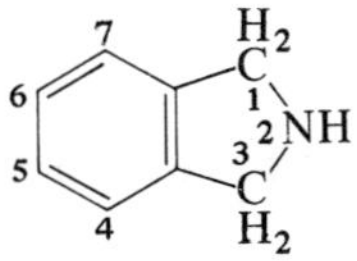 Isoindoline		Isoindolinyl (1- shown) 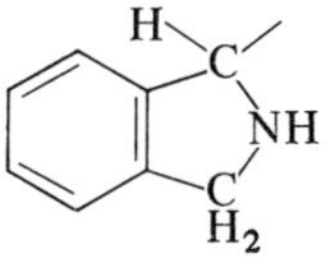
(13)	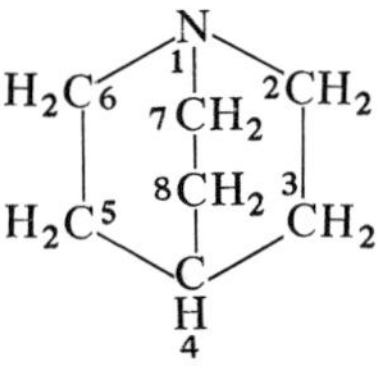 Quinuclidine		Quinuclidinyl (2- shown) 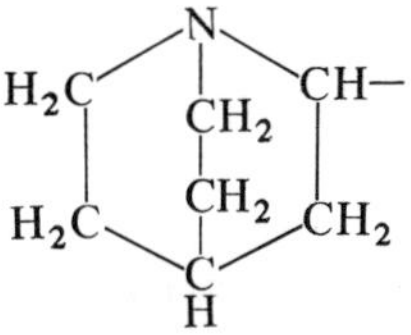
(14)	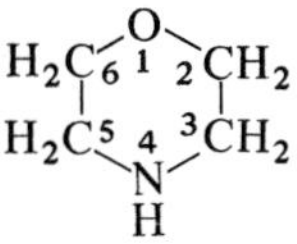 Morpholine		Morpholinyl‡ (3- shown) 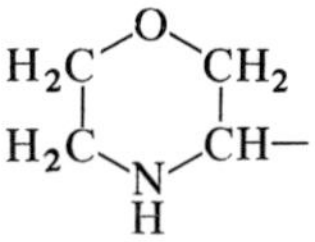

* The "3-" denotes the position of the double bond.
† For 1-Piperidyl use piperidino.
‡ For 4-Morpholinyl use morpholino.

Rule B-3. Fused Heterocyclic Systems

3.1—"*Ortho*-fused" and "*ortho*- and *peri*-fused" ring compounds containing hetero atoms are named according to the fusion principle described in Rule A-21 for hydrocarbons. The components are named according to Rules A-21, B-1 and B-2. When the name of a component in a fusion name contains locants (numerals or letters) that do not apply also to the numbering of the fused system, these locants are placed in square brackets (as are also the locants for fusion positions required by Rule A-21.5). The base component should be a heterocyclic system. If there is a choice, the base component should be, by order of preference:

(*a*) A nitrogen-containing component.

Example:

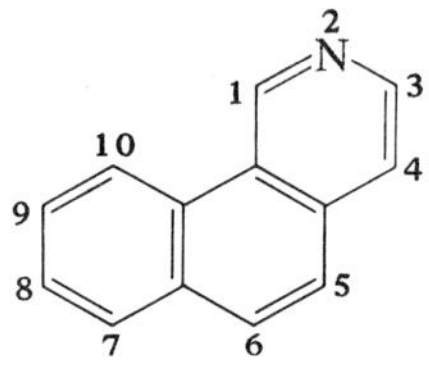

Benzo[*h*]isoquinoline
not Pyrido[3,4-*a*]naphthalene

(*b*) A component containing a hetero atom (other than nitrogen) as high as possible in Table I.*

Example:

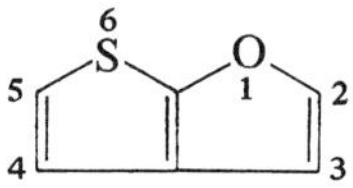

Thieno[2,3-*b*]furan
not Furo[2,3-*b*]thiophene

(*c*) A component containing the greatest number of rings.

Example:

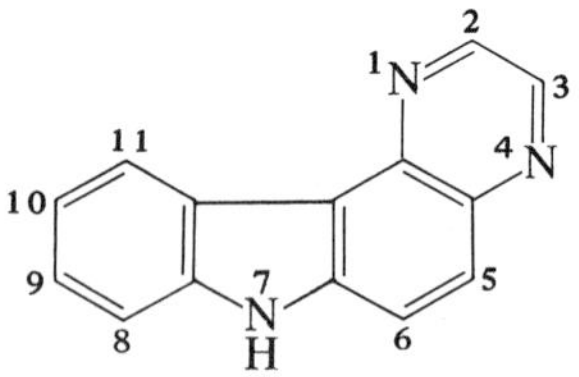

7*H*-Pyrazino[2,3-*c*]carbazole
not 7*H*-Indolo[3,2-*f*]quinoxaline

---
* Author's note. The wording has been known to cause confusion. Rule **B3.1**(*b*) is applicable only in the *absence* of a nitrogen-containing component.

(d) A component containing the largest possible individual ring.

Example:

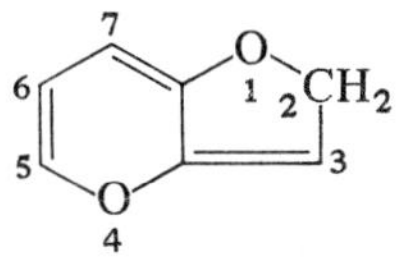

2H-Furo[3,2-b]pyran
not 2H-Pyrano[3,2-b]furan

(e) A component containing the greatest number of hetero atoms of any kind.

Example:

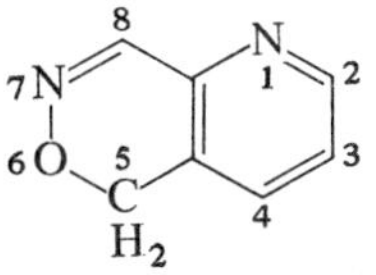

5H-Pyrido[2,3-d]-o-oxazine
not o-Oxazino[4,5-b]pyridine

(f) A component containing the greatest variety of hetero atoms.

Examples:

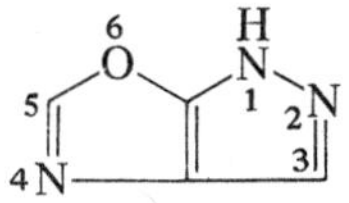

1H-Pyrazolo[4,3-d]oxazole
not 1H-Oxazolo[5,4-c]pyrazole

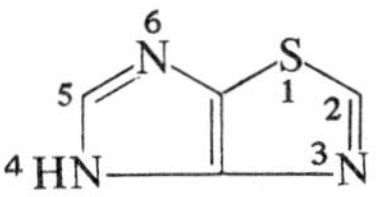

4H-Imidazo[4,5-d]thiazole
not 4H-Thiazolo[4,5-d]imidazole

(g) A component containing the greatest number of hetero atoms first listed in Table I.

Example:

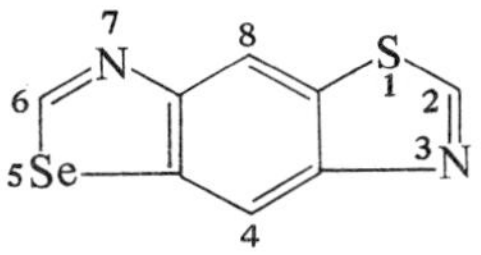

Selenazolo[5,4-f]benzothiazole*
not Thiazolo[5,4-f]benzoselenazole

* In this example the hetero atom first listed in Table I is sulfur and the greatest number of sulfur atoms in a ring is one.

(*h*) If there is a choice between components of the same size containing the same number and kind of hetero atoms choose as the base component that one with the lower numbers for the hetero atoms before fusion.

Example:

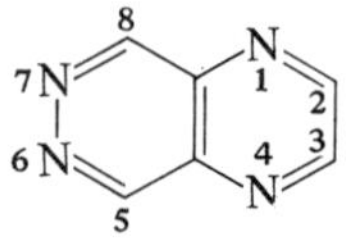

Pyrazino[2,3-*d*]pyridazine

3.2—If a position of fusion is occupied by a hetero atom, the names of the component rings to be fused are so chosen as both to contain the hetero atom.

Example:

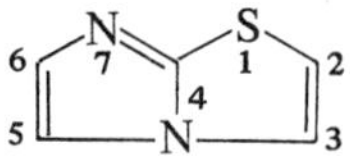

Imidazo[2,1-*b*]thiazole

3.3—The following contracted fusion prefixes may be used: furo, imidazo, isoquino, pyrido, quino and thieno.

Examples:

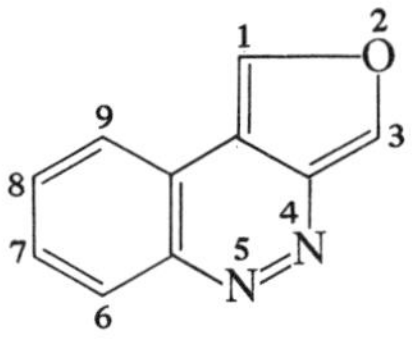

Furo[3,4-*c*]cinnoline 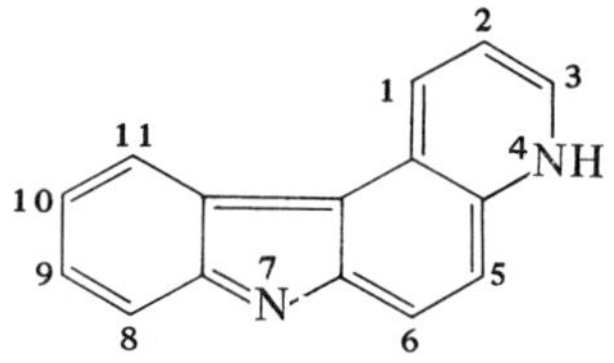   4*H*-Pyrido[2,3-*c*]carbazole

3.4—In peripheral numbering of the complete fused systems, the ring system is oriented and numbered according to the principles of Rule A-22. When there is a choice of orientations, it is made in the following sequence in order to:

(*a*) Give low numbers to hetero atoms, thus:

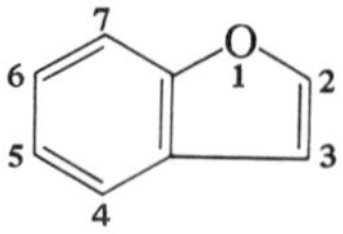

Benzo[*b*]furan 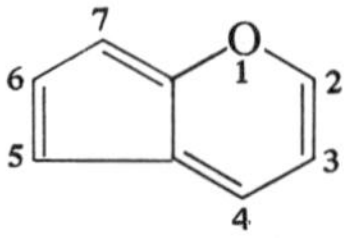   Cyclopenta[*b*]pyran 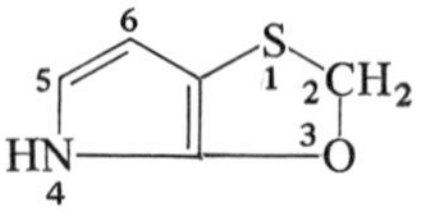   4*H*[1,3]-Oxathiolo[5,4-*b*]pyrrole
(Note: 1,3,4 lower than 1,3,6)

(b) Give low numbers to hetero atoms in order of Table I, thus:

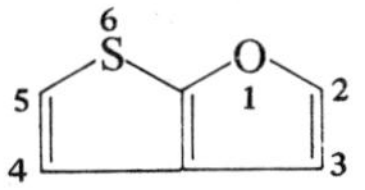

Thieno[2,3-b]furan

(c) Allow carbon atoms common to two or more rings to follow the lowest possible numbers (see Rules **A-22.2** and **A-22.3**). [A hetero atom common to two rings is numbered according to Rule **B-3.4(e)**], thus:

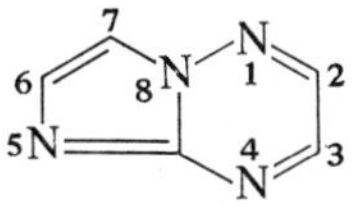

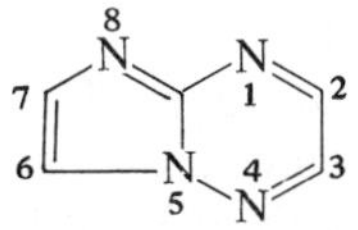

Imidazo[1,2-b][1,2,4]triazine   not   or

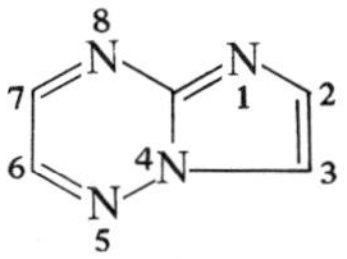

In a compound name for a fusion prefix (*i.e.*, when more than one pair of square brackets is required), the points of fusion in the compound prefix are indicated by the use of unprimed and primed numbers, the unprimed numbers being assigned to the ring attached directly to the base component, thus:

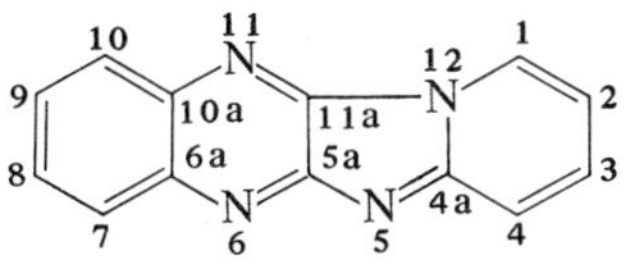

Pyrido[1',2':1,2]imidazo-[4,5-b]quinoxaline   not

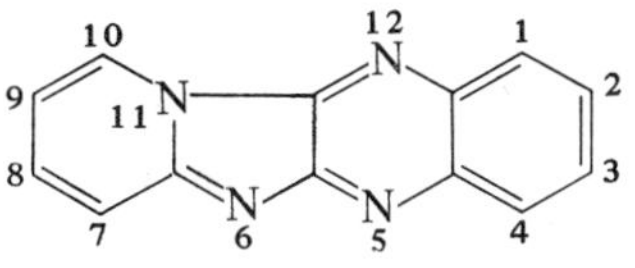

or

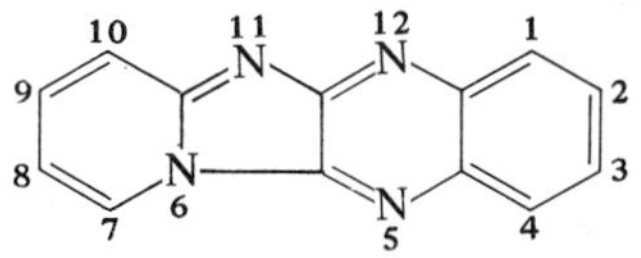

or

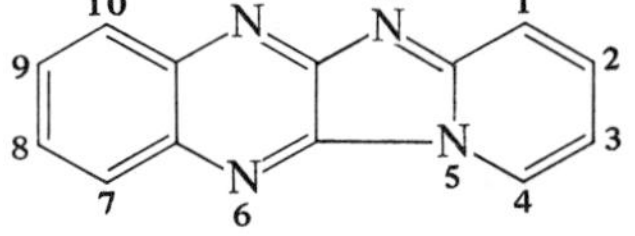

(d) Give hydrogen atoms lowest numbers possible, thus:

4H-1,3-Dioxolo[4,5-d]imidazole

(e) The ring is numbered as for hydrocarbons but numbers are given to all hetero atoms even when common to two or more rings. Interior hetero atoms are numbered last following the shortest path from the highest previous number.

3.5—As exceptions, two-ring systems in which a benzene ring is fused to a hetero ring may be named by prefixing numbers indicating the positions of the hetero atoms to benzo followed by the name of the heterocyclic component. Numbering is assigned by the principles set forth in Rule **B-3.4**(a), (b), and (d). The names provided by this rule may also be used for components of more complex fused systems.

Examples:

3-Benzoxepin
(not Benz[d]oxepin)

4H-3,1-Benzoxazine
(not 4H-Benz[d][1,3]oxazine]

1H-Pyrrolo[1,2-b][2]benzazepine
(not 1H-Benzo[e]pyrrolo[1,2-a]azepine)

Rule B-4. Replacement Nomenclature (also known as "a" Nomenclature)*

4.1—Names of monocyclic hetero compounds may be formed by prefixing "a" terms (see Table I of Rule **B-1.1**), preceded by their locants, to the name of the corresponding hydrocarbon†. Numbering is

---

* The Stelzner Method is abandoned.

† The "corresponding hydrocarbon" is obtained from the heterocyclic compound by formally replacing each hetero atom with >CH₂, ≥CH, or —C— in accord with the valence 2, 3 or 4 of the hetero atom replaced.

Appendix] SPECIALIST HETEROCYCLIC NOMENCLATURE 303

assigned so as to give lowest numbers in the following order: first to hetero atoms in the order of Table I, next to hetero atoms as a complete set, next to multiple bonds, next to substituents as a complete set, and then to substituents in alphabetical order.

Examples:

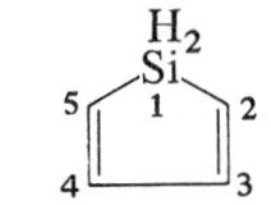

Sila-2,4-cyclopentadiene

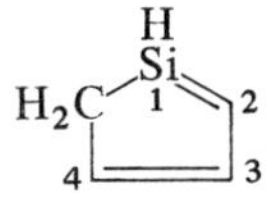

Sila-1,3-cyclopentadiene

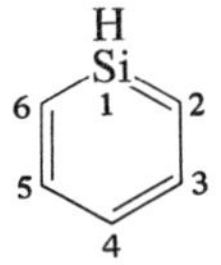

Silabenzene

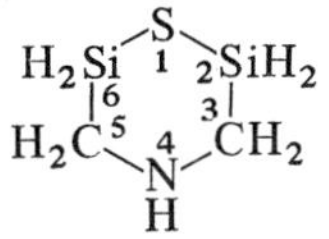

1-Thia-4-aza-2,6-disilacyclohexane

4.2—Fused heterocyclic systems may be named by prefixing "a" terms preceded by their locants, to the name of the corresponding hydrocarbon. The numbering of the corresponding hydrocarbon is retained, irrespective of the position of the hetero atoms; where there is a choice, low numbers are assigned in the following order: first to hetero atoms as a complete set, next to hetero atoms in order of Table I, and then to multiple bonds in the heterocyclic compound according to the principles of Rule A-11.3. These principles are applied in one of two ways, as follows:

(*a*) When the corresponding hydrocarbon does not contain the maximum number of non-cumulative double bonds and can be named without the use of hydro prefixes, as for indan, then the hydrocarbon is named in that state of hydrogenation.

Example:

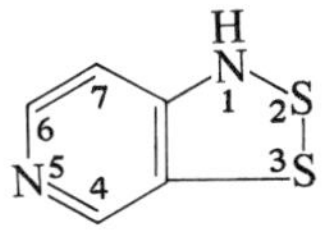

2,3-Dithia-1,5-diazaindan

(b) When the two conditions of paragraph (a) are not fulfilled, positions in the skeleton of the corresponding hydrocarbon that are occupied by hetero atoms are denoted by "a" prefixes, and the parent heterocyclic compound is considered to be that which contains the maximum number of conjugated or isolated* double bonds, but the corresponding hydrocarbon is named in the form in which it contains the maximum number of non-cumulative double bonds. Hydrogen additional to that present in the parent heterocyclic compound is named by hydro prefixes and/or as H in front of the "a" terms.

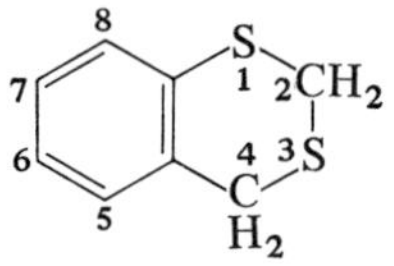

4H-1,3-Dithianaphthalene

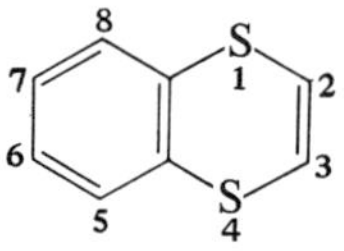

1,4-Dithianaphthalene

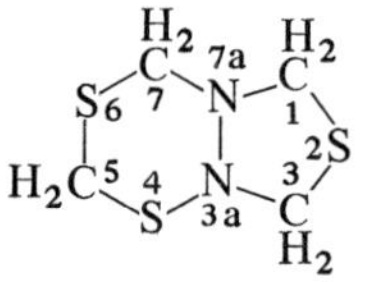

2,4,6-Trithia-3a,7a-diazaindene

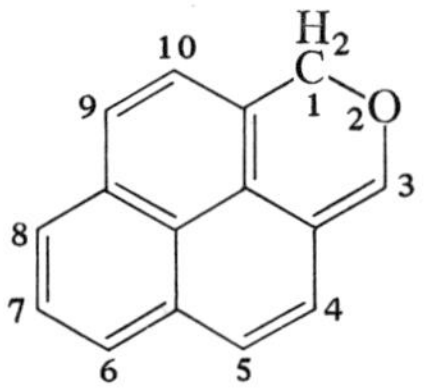

1H-2-Oxapyrene

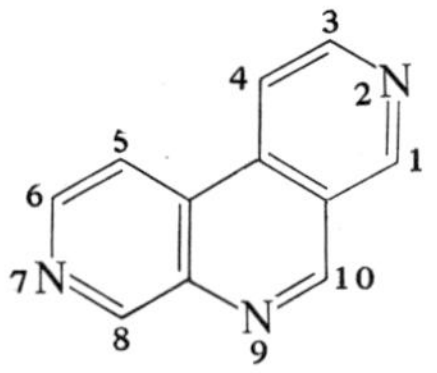

2,7,9-Triazaphenanthrene

* Isolated double bonds are those which are neither conjugated nor cumulative as in

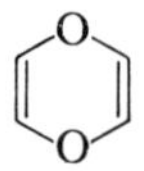

or the B ring of

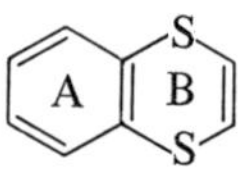

Appendix] SPECIALIST HETEROCYCLIC NOMENCLATURE 305

4.3—In fusion names, the "a" terms precede the complete name of the parent hydrocarbon. Prefixes denoting ordinary substitution precede the "a" terms.

Example:

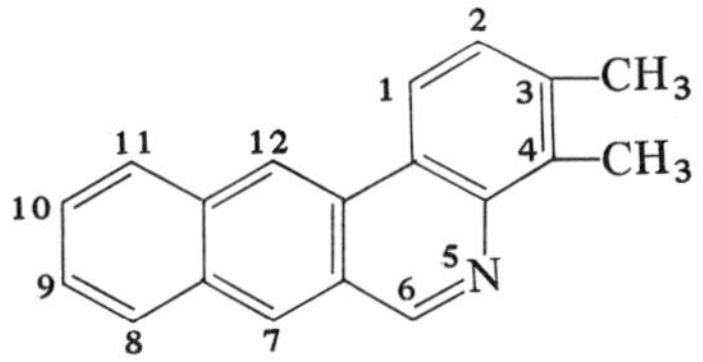

3,4-Dimethyl-5-azabenz[a]anthracene

Rule B-5. Radicals

5.11—Univalent radicals derived from heterocyclic compounds by removal of hydrogen from a ring are in principle named by adding "yl" to the names of the parent compounds (with elision of final "e", if present).

Examples:

>Indolyl   from indole
>Pyrrolinyl from pyrroline
>Triazolyl from triazole
>Triazinyl from triazine

(For further examples see Rule **B-2.11**.)

The following exceptions are retained: furyl, pyridyl, piperidyl, quinolyl, isoquinolyl and thienyl (from thiophene) (see also Rule **B-2.12**). Also retained are furfuryl (for 2-furylmethyl), furfurylidene (for 2-furylmethylene), furfurylidyne (for 2-furylmethylidyne), thenyl (for thienylmethyl), thenylidene (for thienylmethylene) and thenylidyne (for thienylmethylidyne).

As exceptions, the names "piperidino" and "morpholino" are preferred to "1-piperidyl" and "4-morpholinyl".

5.12—Bivalent radicals derived from univalent heterocyclic radicals whose names end in "-yl" by removal of one hydrogen atom from the atom with the free valence are named by adding "-idene" to the name of the corresponding univalent radical.

Examples:

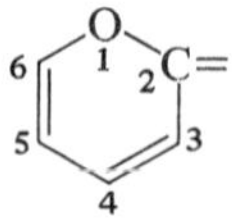

2H-Pyran-2-ylidene

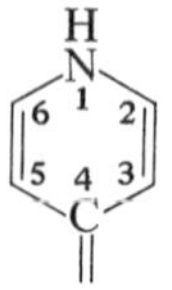

4(1H)-Pyridylidene
or 1,4-Dihydro-4-pyridylidene

5.13—Multivalent radicals derived from heterocyclic compounds by removal of two or more hydrogen atoms from different atoms in the ring are named by adding "-diyl", "-triyl", *etc.*, to the name of the ring system.

Example:

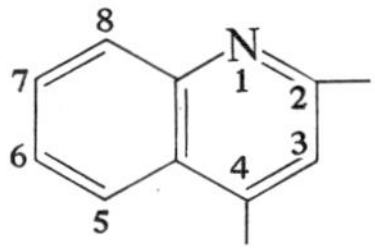

2,4-Quinolinediyl

5.21—The use of "a" terms (Rule **B-4**) does not affect the formation of radical names. Such names are strictly analogous to those of the hydrocarbon analogs except that the "a" terms establish enumeration in whole or in part.

Examples:

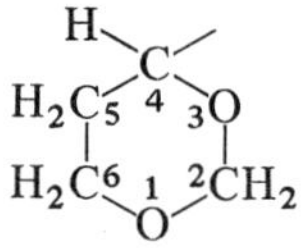

1,3-Dioxa-4-cyclohexyl

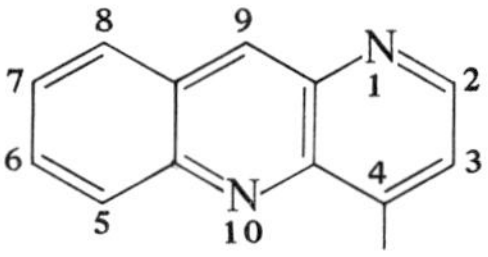

1,10-Diaza-4-anthryl

Rule B-6. Cationic Hetero Atoms

6.1—According to the "a" nomenclature, heterocyclic compounds containing cationic hetero atoms are named in conformity with the preceding rules by replacing "oxa-", "thia-", "aza-", *etc.*, by "oxonia-", "thionia-", "azonia-" *etc.*, the anion being designated in the usual way. Cationic hetero atoms follow immediately after the corresponding noncationic atoms: oxonia follows oxa, thionia follows thia, azonia follows aza, etc. (cf. Table I of Rule **B-1.1**).

Examples:

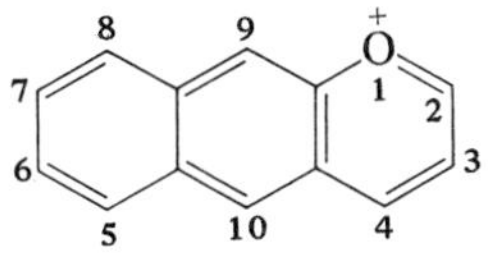

Cl⁻   1-Oxoniaanthracene chloride

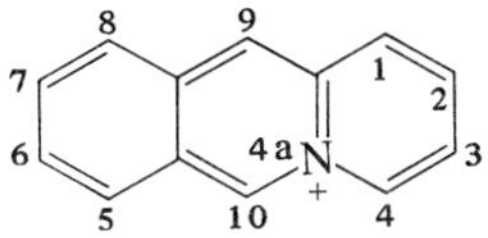

Cl⁻   4a-Azoniaanthracene chloride

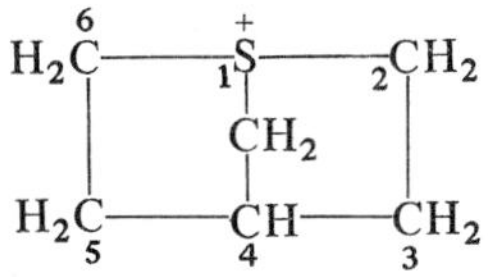

Cl⁻   1-Thioniabicyclo[2.2.1]heptane chloride

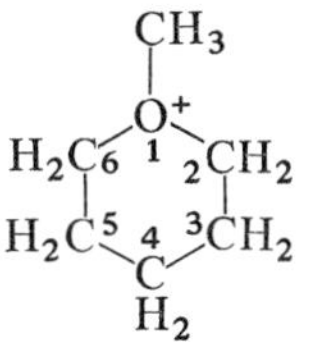

Cl⁻   1-Methyl-1-oxoniacyclohexane chloride

HETEROCYCLIC SPIRO COMPOUNDS

Rule B-10. Compounds: Method 1

10.1—Heterocyclic spiro compounds containing single-ring units only may be named by prefixing "a" terms (see Table I, Rule **B-1.1**) to the names of the spiro hydrocarbons formed according to Rules **A-41.1, A-41.2, A-41.3** and **A-41.6**. The numbering of the spiro hydrocarbon is retained and the hetero atoms in the order of Table I are given as low numbers as are consistent with the fixed numbering of the ring. When there is a choice, hetero atoms are given lower numbers than double bonds.

Examples:

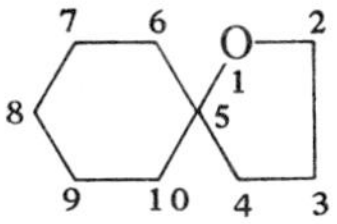

1-Oxaspiro[4.5]decane

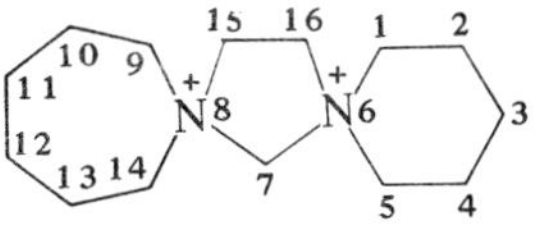

6,8-Diazoniadispiro[5.1.6.2]hexadecane dichloride

308　　　　　　　　　　IUPAC RULES　　　　　　　[Appendix

10.2—If at least one component of a mono- or poly-spiro compound is a fused polycyclic system, the spiro compound is named according to Rule **A-41.4** or **A-41.7**, giving the spiro atom as low a number as possible consistent with the fixed numberings of the component systems.

Examples:

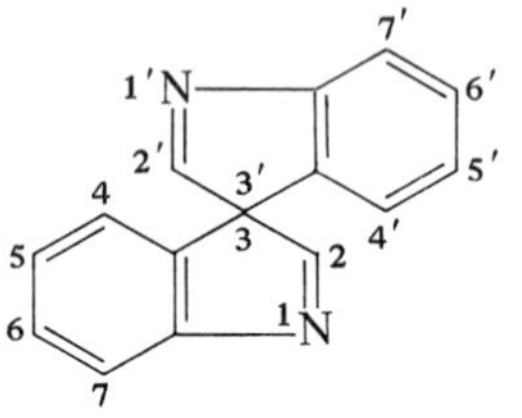

3,3'-Spirobi(3H-indole) 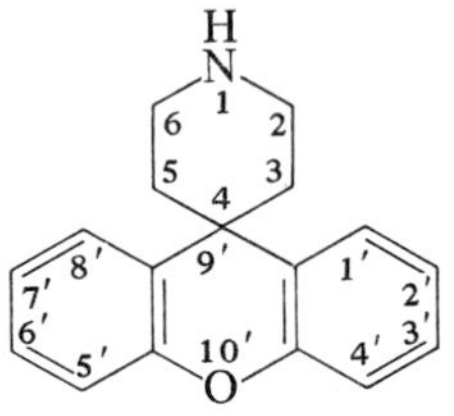　　　　　Spiro[piperidine-4,9'-xanthene]

Rule B-11. Compounds: Method 2

11.1—Heterocyclic spiro compounds are named according to Rule **A-42**, the following criteria being applied where necessary: (*a*) spiro atoms have numbers as low as consistent with the numbering of the individual component systems; (*b*) heterocyclic components have priority over homocyclic components of the same size; (*c*) priority of heterocyclic components is decided according to Rule **B-3**. Parentheses are used where necessary for clarity in complex expressions.

Examples:

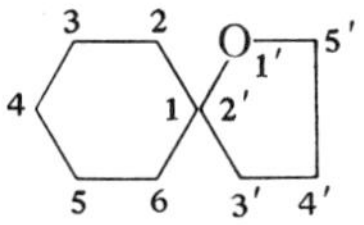

Cyclohexanespiro-2'-(tetra-　　Tetrahydropyran-2-spiro-　　3,3'-Spirobi(3H-indole)
hydrofuran)　　　　　　　　　cyclohexane

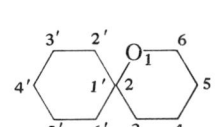

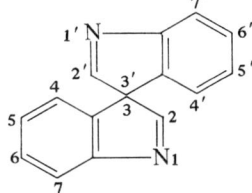

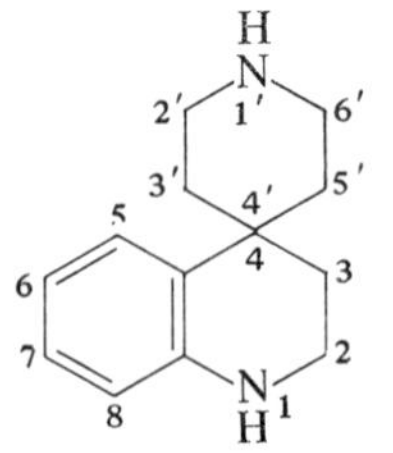

1,2,3,4-Tetrahydroquinoline-4-　　　Hexahydroazepinium-1-spiro-1'-
spiro-4'-piperidine　　　　　　　　imidazolidine-3'-spiro-1"-
　　　　　　　　　　　　　　　　piperidinium dibromide

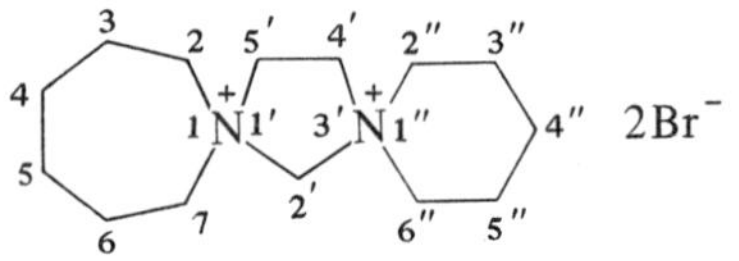

Rule B-12. Radicals

12.1—Radicals derived from heterocyclic spiro compounds named by Rule **B-10.1** are named according to the principles set forth in Rules **A-11** and **B-5.21**. Radicals derived from other heterocyclic spiro compounds are named by adding "-yl", "-diyl", *etc.*, to the name of the spiro compound (with elision of final "e", if present, before a vowel). The numbering of the spiro compound is retained and the point or points of attachment are given numbers as low as is consistent with any fixed numbering of the heterocyclic spiro compound.

Examples:

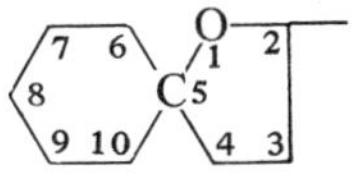

1-Oxaspiro[4.5]dec-2-yl
(cf. Rule **B-10.1**)
or Cyclohexanespiro-2'-(tetrahydrofuran)-5'-yl
(cf. Rule **B-11.1**)

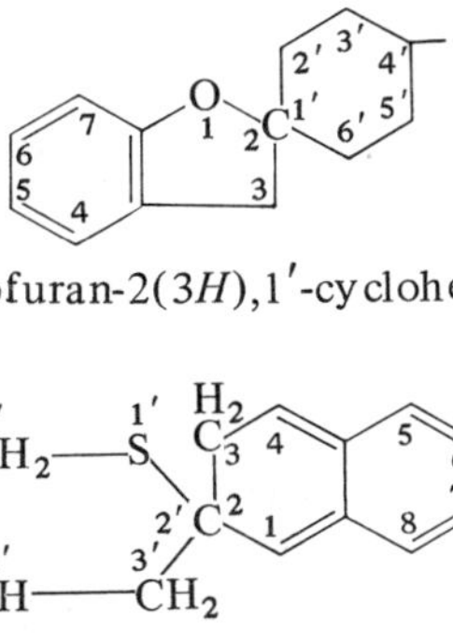

Spiro[benzofuran-2(3*H*),1'-cyclohexan]-4'-yl

Spiro[naphthalene-2(3*H*),2'-thian]-4'-yl

HETEROCYCLIC RING ASSEMBLIES

Rule B-13

13.1—Assemblies of two or more identical heterocyclic systems are named by placing the prefix "bi-", "ter-", "quater-", etc., before the name of the heterocyclic system or radical. The numbering of the assembly is that of the corresponding heterocyclic systems, one component being assigned unprimed and the others primed, doubly primed, *etc.*, numbers. The points of attachment are indicated by appropriate locants before the name. Other structural features are described as recorded for hydrocarbon ring assemblies in Rules **A-52.1** (double bond between two components), **A-52.3** (substituents), **A-53.3** (hydrogenation), **A-54.1** (numerical prefixes), and **A-55.1** and **A-56.1** (radicals), insofar as fixed numbering of the heterocyclic system allows.

Examples:

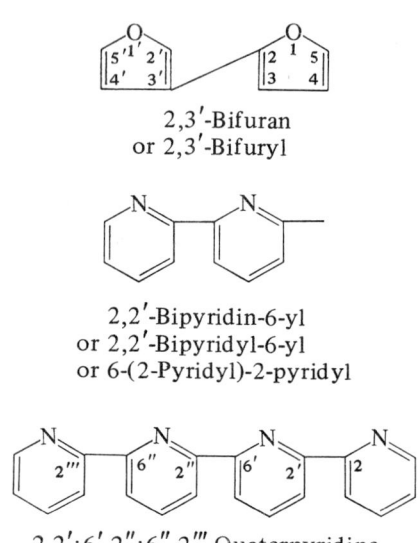

2,3'-Bifuran
or 2,3'-Bifuryl

2,2'-Bipyridin-6-yl
or 2,2'-Bipyridyl-6-yl
or 6-(2-Pyridyl)-2-pyridyl

2,2':6',2":6",2'''-Quaterpyridine

BRIDGED HETEROCYCLIC SYSTEMS

Rule B-14. Extension of the von Baeyer System

14.1—Bridged heterocyclic systems are named according to the principles of Rules **A-31** and **A-32**, the hetero atoms being indicated according to Rule **B-4.2** and derived radicals by the principles set forth in Rule **A-31.4**.

Examples:

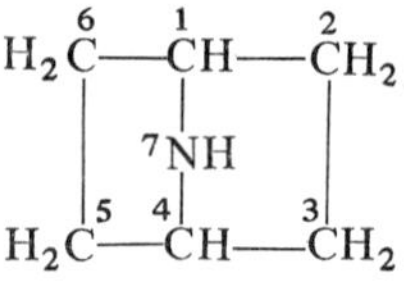

$$\begin{array}{c} \overset{6}{H_2C}\!-\!\overset{1}{CH}\!-\!\overset{2}{CH_2} \\ || \\ {}^7NH \\ |\overset{5}{}\overset{4}{}\overset{3}{}| \\ H_2C\!-\!CH\!-\!CH_2 \end{array}$$

7-Azabicyclo[2.2.1]heptane

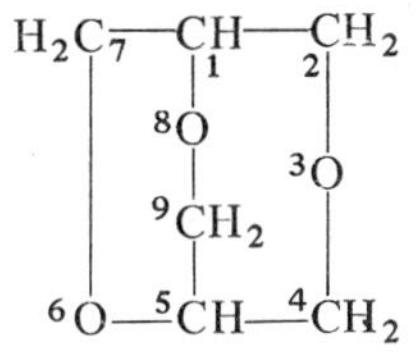

Appendix] BRIDGED HETEROCYCLIC SYSTEMS 311

$$\begin{array}{c} H_2C_7\!\!-\!\!\overset{1}{C}H\!\!-\!\!\overset{2}{C}H_2 \\ |\quad \overset{8}{O}\quad \overset{3}{O} \\ |\quad \overset{9}{C}H_2 \\ ^6O\!-\!\overset{5}{C}H\!\!-\!\!\overset{4}{C}H_2 \end{array}$$

3.6.8-Trioxabicyclo[3.2.2]nonane

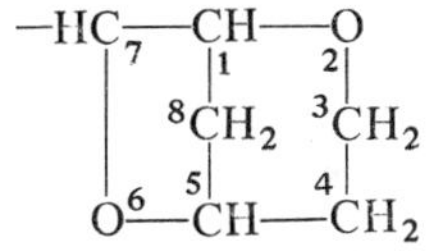

$$\begin{array}{c} -HC_7\!\!-\!\!\overset{1}{C}H\!\!-\!\!\overset{2}{O} \\ |\quad \overset{8}{C}H_2 \quad \overset{3}{C}H_2 \\ O^6\!-\!\overset{5}{C}H\!\!-\!\!\overset{4}{C}H_2 \end{array}$$

2.6-Dioxabicyclo[3.2.1]oct-7-yl

Rule B-15. Hetero Bridges

B-15.1—A hetero polycyclic system that contains an "*ortho*-fused" or an "*ortho*- and *peri*-fused" system according to Rule **A-21** or **B-3** and has one or more atomic bridges is named as an "*ortho*-fused" or "*ortho*- and *peri*-fused" system. The atomic bridges are then indicated by prefixes as exemplified in the annexed Table or by Rule **A-34.1**. The name of a bridge containing hetero atoms is constructed from units beginning with the terminal atom that occurs first in Table I of Rule **B-1.1**, the final "o" of a prefix being elided before a vowel in a following prefix; to illustrate this, the formulae in the annexed Table are arranged from left to right in the same order as the prefixes. If bridges of different types are present, they are cited in alphabetical order. For examples see Rule **B-15.2**.

Azimino	—N=N—NH—
Azo	—N=N—
Biimino	—NH—NH—
Epidioxy	—O—O—
Epidithio	—S—S—
Epithio	—S—
Epithioximino	—S—O—NH—
Epoxy (see also Rule **C-212.2**)	—O—
Epoxyimino	—O—NH—
Epoxynitrilo	—O—N=
Epoxythio	—O—S—
Epoxythioxy	—O—S—O—
Furano (usually 3,4-)	—C_4H_2O—
Imino (see also Rule **C-815.2**)	—NH—
Nitrilo	—N=

15.2—Systems described in Rule **B-15.1** are numbered according to the principles set forth in Rules **A-34.1** and **A-34.2** for compounds containing hydrocarbon bridges. In the name of the complete compound, the name of the bridge is preceded by two locants, that for the unit cited first in a composite bridge preceding that for the other end of the bridge. Radicals derived from the polycyclic systems described in Rule **B-15.1** are formed by the principles set forth in Rule **B-5**.

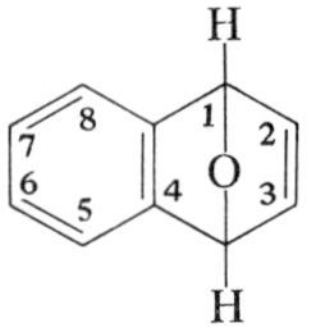

1,4-Dihydro-1,4-epoxynaphthalene

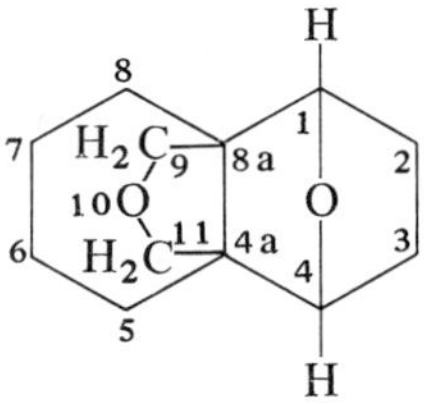

Perhydro-1,4-epoxy-4a,8a-(methanoxy-methano)naphthalene

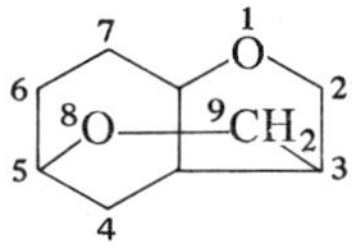

Perhydro-5,3-(epoxymethano)benzofuran

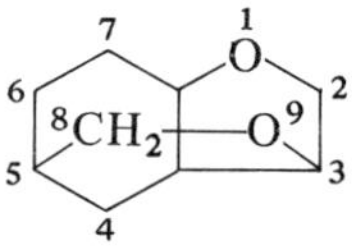

Perhydro-3,5-(epoxymethano)benzofuran

SUBSTITUTIVE NOMENCLATURE

C-0.14. SENIORITY OF RING SYSTEMS*

Rule C-14.1

14.11—Seniority of ring systems is decided by applying the following criteria, successively in the order given, until a decision is reached:

* Compare Rule **C-12.5**.

(a) All heterocycles are senior to all carbocycles.

Example:

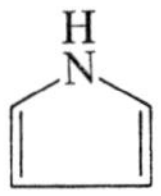

[pyrrole] senior to [benz[a]anthracene]

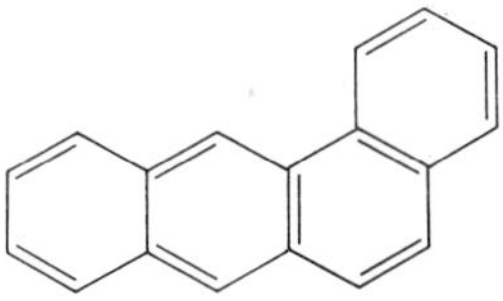

(b) For heterocycles the criteria based on the nature and position of the hetero atoms set out in Rule **B-3.1** (a)–(h).*

Examples:

See Rule **B-3.1**, such as:

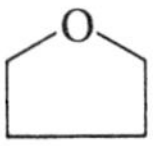

[tetrahydrofuran] senior to [tetrahydrothiophene]

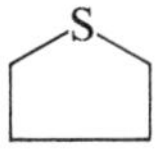

(c) Largest number of rings.

Example:

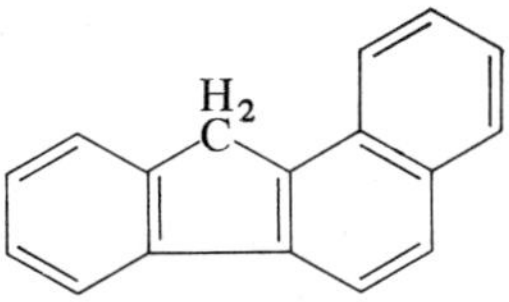

[fluorene-like tricyclic with CH₂] senior to [dibenzocyclooctene]

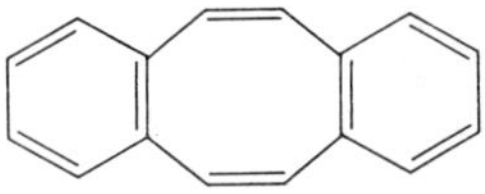

(d) Largest individual ring at first point of difference.

Examples:

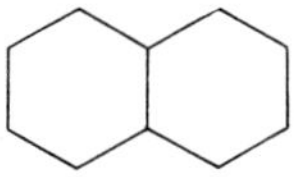

[decalin] senior to [hydrindane]

[cyclooctane fused to cyclopentane] senior to [cycloheptane fused to cyclohexane]

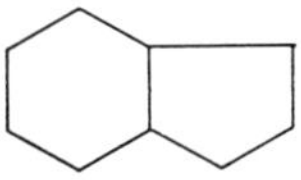

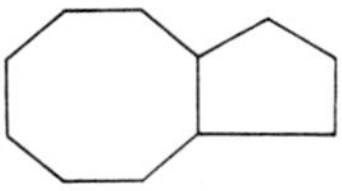

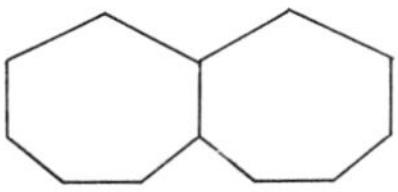

* In Rule **B-3.1** (b) "other than nitrogen" means "in the absence of nitrogen".

314   IUPAC RULES   [Appendix

(e) Largest number of atoms in common among rings.

Examples:

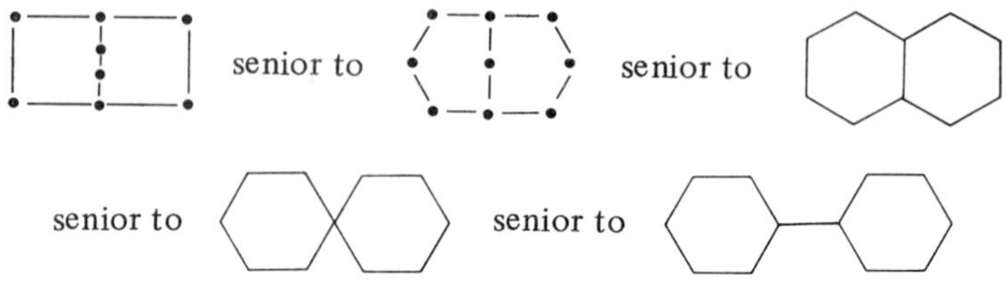

[structure] senior to [structure] senior to [structure]

senior to [structure] senior to [structure]

Note: Rings joined by a link (single or double) are included in this choice only when identical and named by the bi-, ter-, quater-, *etc.*, system (see Rule **A-54.1**).

(f) Lowest letters* (*a, b, etc.*, see Rule **A-21.5**) in the expression for ring junctions.

Example:

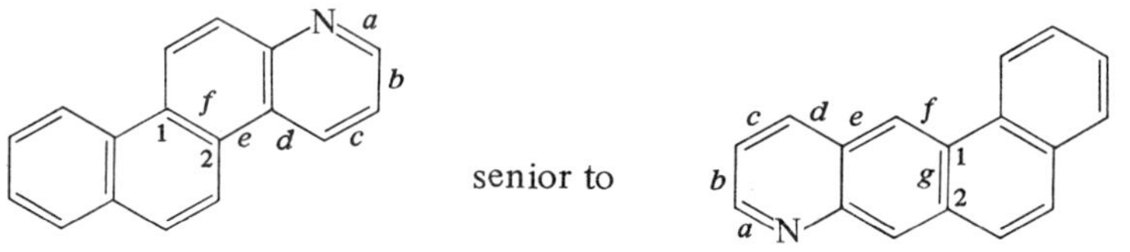

[structure] senior to [structure]

Naphtho[2,1-*f*]quinoline   Naphtho[1,2-*g*]quinoline

(g) Lowest numbers at the first point of difference in the expression for ring junctions. (See Rules: **A-21.5** for *ortho*-fusion and *ortho-peri*-fusion; **A-32** for tricyclo, *etc.*, systems; **A-41**, **A-42**, **B-10**, and **B-11** for spirans; **A-52** for assemblies of identical units.)

Examples:

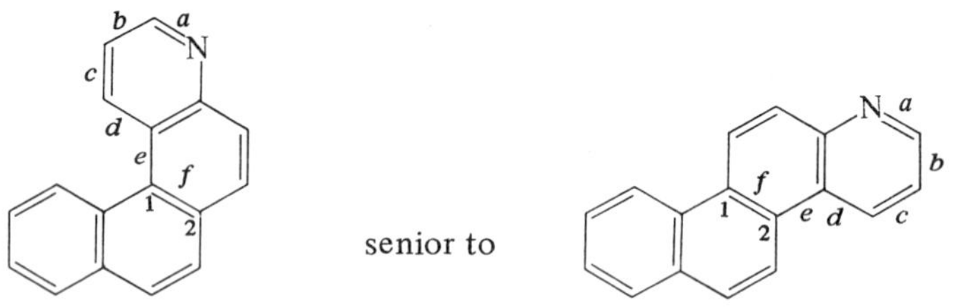

[structure] senior to [structure]

Naphtho[1,2-*f*]quinoline   Naphtho[2,1-*f*]quinoline

* Lowest means *a* before *b* before *c*, etc.

Appendix]   SUBSTITUTIVE NOMENCLATURE   315

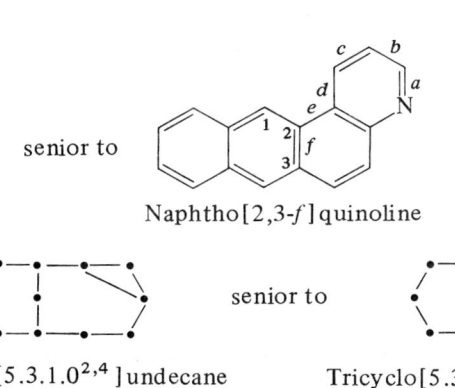

senior to

Naphtho[2,3-*f*]quinoline

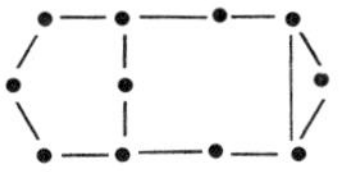

Tricyclo[5.3.1.0^{2,4}]undecane   senior to   Tricyclo[5.3.1.0^{3,5}]undecane

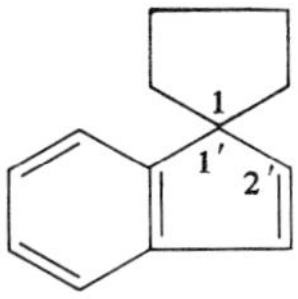

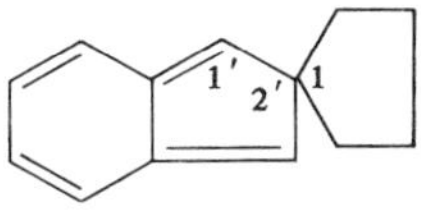

Spiro[cyclopentane-1,1′-indene]   senior to   Spiro[cyclopentane-1,2′-2′*H*-indene]
or Indene-1-spiro-1′-cyclopentane   or 2*H*-Indene-2-spiro-1′-cyclopentane

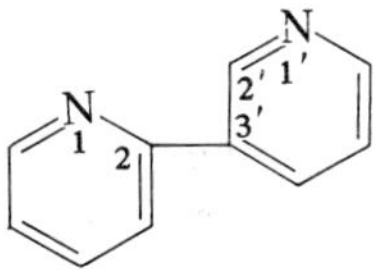

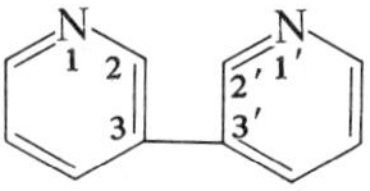

2,3′-Bipyridine   senior to   3,3′-Bipyridine

(h) **Lowest state of hydrogenation.**

Example:

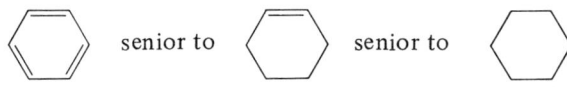

senior to   senior to

(i) **Lowest locant for indicated hydrogen.**

Example:

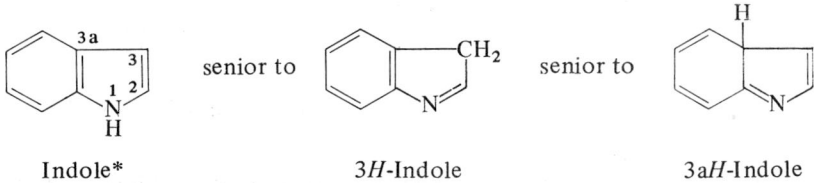

Indole*   senior to   3*H*-Indole   senior to   3a*H*-Indole

** 1*H*- is understood to be present.

(j) Lowest locant for point of attachment (if a radical).
Example:

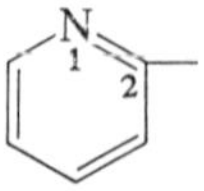

2-Pyridyl   senior to   3-Pyridyl

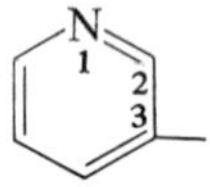

(k) Lowest locant for an attached group expressed as suffix.
Example:

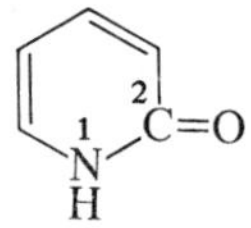

2(1H)-Pyridone   senior to   4(1H)-Pyridone

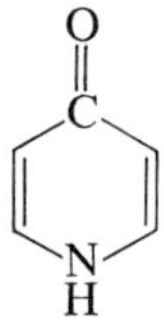

(l) Lowest locant for substituents named as prefixes, hydroprefixes, -ene, and -yne, all considered together in one series in ascending numerical order independently of their nature.
Examples:

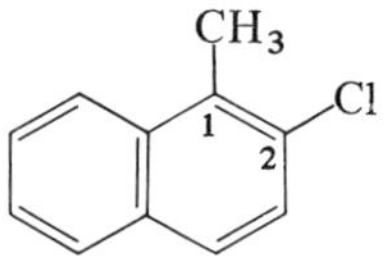

1,2
2-Chloro-1-methyl   senior to   2,3
2-Chloro-3-methyl

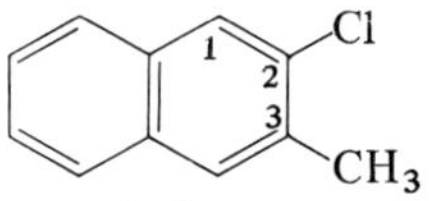

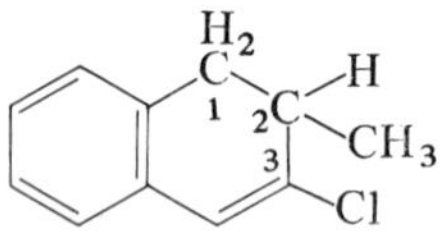

1,2,2,3
3-Chloro-1,2-dihydro-2-methyl   senior to   2,2,3,3
2-Chloro-2,3-dihydro-3-methyl

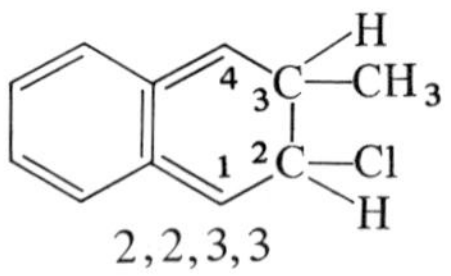

(m) Lowest locant for that substituent named as prefix which is cited first in the name (see Subsection C-0.16).

Appendix] SUBSTITUTIVE NOMENCLATURE 317

Example:

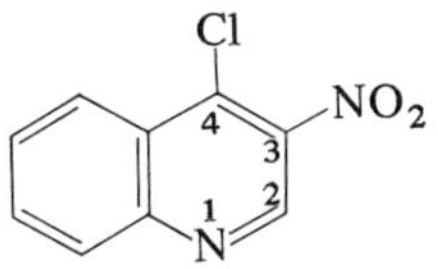

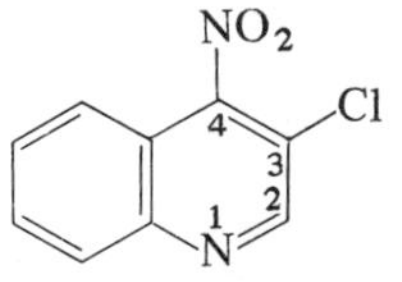

3-Chloro-4-nitro senior to 4-Chloro-3-nitro

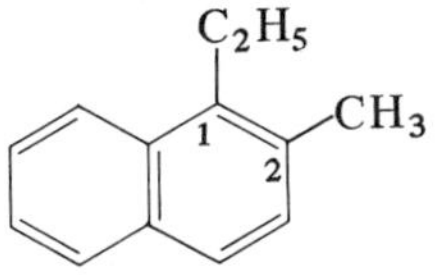

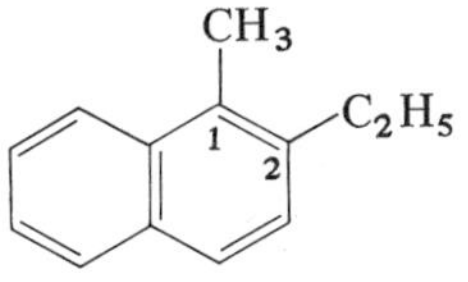

1-Ethyl-2-methyl senior to 2-Ethyl-1-methyl

Note: Hydro and dehydro prefixes, if treated as detachable (see Rule C-16.11), are considered along with prefixes for substituents when this criterion is applied.

C-0.15. NUMBERING OF COMPOUNDS*

Rule C-15.1

15.11—Insofar as Sections A and B of the IUPAC 1957 rules leave a choice, the starting point and direction of numbering of a compound are chosen so as to give lowest locants to the following structural factors (if present), considered successively in the order listed until a decision is reached:

(*a*) Indicated hydrogen (whether cited in the name or omitted as being conventional).

Examples:

1*H*-Phenalene-4-carboxylic acid

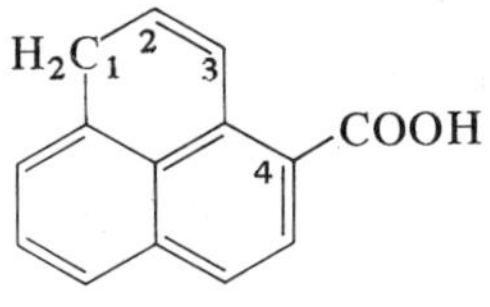

2*H*-Pyran-6-carboxylic acid

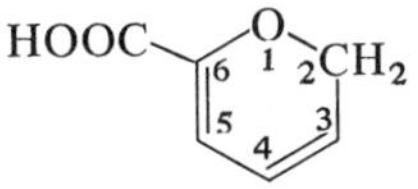

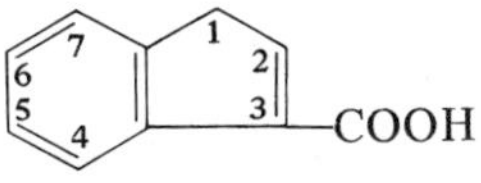

Indene-3-carboxylic acid

* *Cf.* Rule **C-12.8**.

(b) Principal groups named as suffix.

Examples:

3,4-Dichloro-1,6-naphthalene-dicarboxylic acid

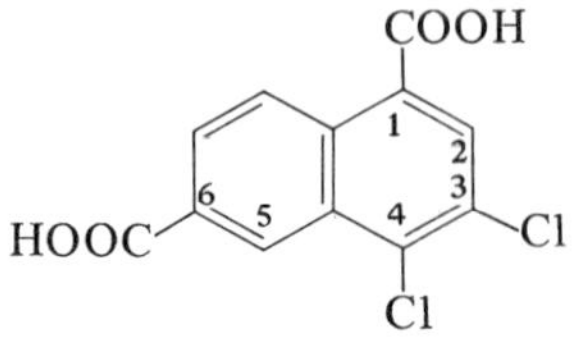

2-Cyclohexan-1-ol

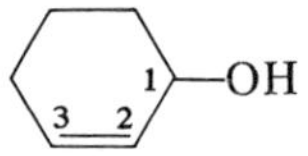

(c) Multiple bonds in acyclic compounds (see Rules **A-3.1** to **A-3.4**), in cycloalkanes (see Rule **A-11**), and in bi-, tri-, and poly-cycloalkanes (see Rules **A-31** and **A-32**), double having priority over triple bonds, and in heterocyclic systems whose names end in -etine, -oline, or -olene (see Rules **B-1.1** and **B-1.2**).

Examples:

3,4-Dichloro-1-cyclohexene

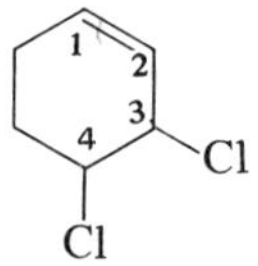

5-Chloro-2-pyrroline

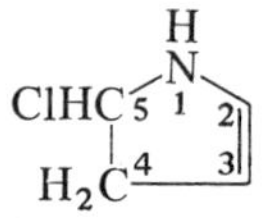

(d) Lowest locant for substituents named as prefixes, hydroprefixes, -ene and -yne, all considered together in one series in ascending numerical order.

Examples:

5,6-Dichloro-1,2,3,4-tetrahydronaphthalene

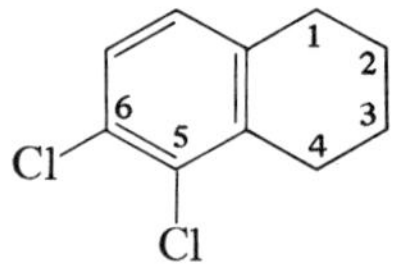

8-Hydroxy-4,5-dimethyl-2-azulenecarboxylic acid

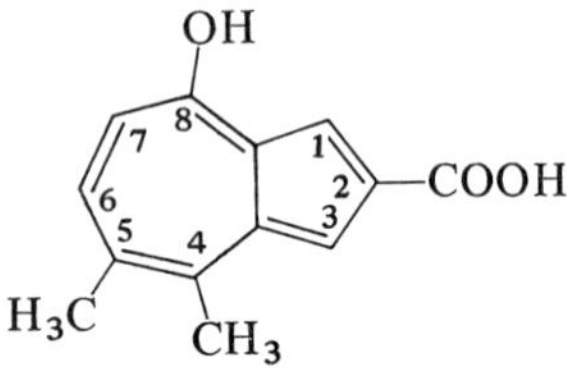

(e) Lowest locant for that substituent named as prefix which is cited first in the name (see Subsection C-0.16).

Examples:

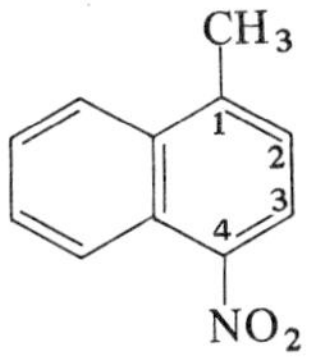

1-Methyl-4-nitronaphthalene

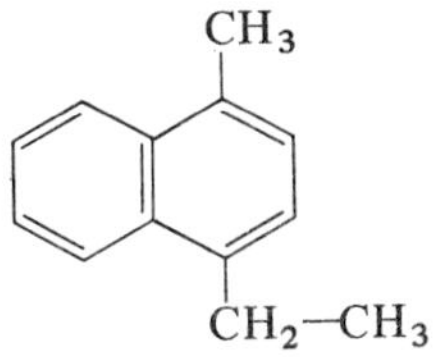

1-Ethyl-4-methylnaphthalene

Note: Hydro and dehydro prefixes, if treated as detachable (see Rule C-16.11), are considered along with prefixes for substituents when this criterion is applied.

Rule C-15.2

15.21—For cyclic radicals, indicated hydrogen and thereafter the point of attachment (free valency) have priority for lowest available number, the criteria of Rule **C-15.1** being applied only if a choice remains after the requirements of the rules in Sections A and B have been satisfied. For acyclic radicals see Rules **A-2.25, A-2.3,** and **A-3.5** to **A-4.4**.

Examples:

7(1H)-Phenalenyl

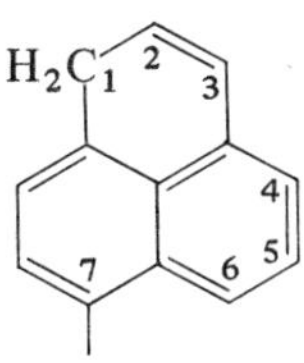

5-Carboxy-2-chlorophenyl

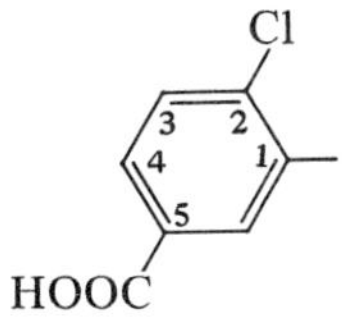

3,5,8-Trichloro-2-naphthyl

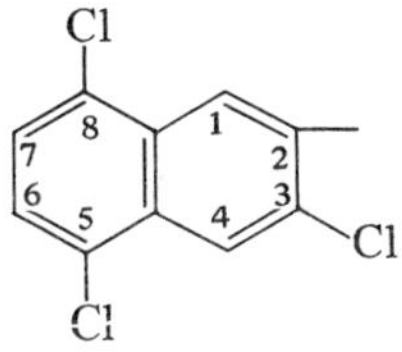

Cumulative Index of Titles

A

Acetylenecarboxylic acids and esters, reactions with N-heterocyclic compounds, **1**, 125
Acetylenic esters, synthesis of heterocycles through nucleophilic additions to, **19**, 297
Acid-catalyzed polymerization of pyrroles and indoles, **2**, 287
t-Amino effect, **14**, 211
Aminochromes, **5**, 205
Anthracen-1,4-imines, **16**, 87
Anthranils, **8**, 277
Applications of NMR spectroscopy to indole and its derivatives, **15**, 277
Applications of the Hammett equation to heterocyclic compounds, **3**, 209; **20**, 1
Aromatic quinolizines, **5**, 291
Aromaticity of heterocycles, **17**, 255
Aza analogs, of pyrimidine and purine bases, **1**, 189
7-Azabicyclo[2.2.1]hepta-2,5-dienes, **16**, 87
Azines, reactivity with nucleophiles, **4**, 145
Azines, theoretical studies of, physicochemical properties of reactivity of, **5**, 69
Azinoazines, reactivity with nucleophiles, **4**, 145
1-Azirines, synthesis and reactions of, **13**, 45

B

Base-catalyzed hydrogen exchange, **16**, 1
1-, 2-, and 3-Benzazepines, **17**, 45
Benzisothiazoles, **14**, 43
Benzisoxazoles, **8**, 277
Benzoazines, reactivity with nucleophiles, **4**, 145
1,5-Benzodiazepines, **17**, 27
Benzo[b]furan and derivatives, recent advances in chemistry of, Part I, occurrence and synthesis, **18**, 337
Benzofuroxans, **10**, 1
2H-Benzopyrans (chrom-3-enes), **18**, 159
Benzo[b]thiophene chemistry, recent advances in, **11**, 177
Benzo[c]thiophenes, **14**, 331
1,2,3-(Benzo)triazines, **19**, 215
Biological pyrimidines, tautomerism and electronic structure of, **18**, 199

C

Carbenes, reactions with heterocyclic compounds, **3**, 57
Carbolines, **3**, 79
Cationic polar cycloaddition, **16**, 289 (**19**, xi)
Chemistry
    of benzo[b]furan, Part I, occurrence and synthesis, **18**, 337
    of benzo[b]thiophenes, **11**, 178
    of chrom-3-enes, **18**, 159
    of diazepines, **8**, 21
    of dibenzothiophenes, **16**, 181
    of furans, **7**, 377
    of isatin, **18**, 1
    of lactim ethers, **12**, 185
    of mononuclear isothiazoles, **14**, 1
    of 4-oxy- and 4-keto-1,2,3,4-tetrahydroisoquinolines, **15**, 99
    of phenanthridines, **13**, 315
    of phenothiazines, **9**, 321
    of 1-pyrindines, **15**, 197
    of 1,3,4-thiadiazoles, **9**, 165
    of thienothiophenes, **19**, 123
    of thiophenes, **1**, 1
Chrom-3-ene chemistry, advances in, **18**, 159
Claisen rearrangements, in nitrogen heterocyclic systems, **8**, 143
Complex metal hydrides, reduction of nitrogen heterocycles with, **6**, 45
Covalent hydration
    in heteroaromatic compounds, **4**, 1, 43
    in nitrogen heterocycles, **20**, 117
Cyclic enamines and imines, **6**, 147
Cyclic hydroxamic acids, **10**, 199
Cyclic peroxides, **8**, 165
Cycloaddition, cationic polar, **16**, 289 (**19**, xi)

D

Development of the chemistry of furans (1952–1963), **7**, 377
2,4-Dialkoxypyrimidines, Hilbert–Johnson reaction of, **8**, 115
Diazepines, chemistry of, **8**, 21

1,4-Diazepines, 2,3-dihydro-, **17**, 1
Diazo compounds, heterocyclic, **8**, 1
Diazomethane, reactions with heterocyclic compounds, **2**, 245
Dibenzothiophenes, chemistry of, **16**, 181
2,3-Dihydro-1,4-diazepines, **17**, 1
1,2-Dihydroisoquinolines, **14**, 279
Diquinolylmethane and its analogs, **7**, 153
1,2- and 1,3-Dithiolium ions, **7**, 39

E

Electrolysis of N-heterocyclic compounds, **12**, 213
Electronic aspects of purine tautomerism, **13**, 77
Electronic structure of biological pyrimidines, tautomerism and, **18**, 199
Electronic structure of heterocyclic sulfur compounds, **5**, 1
Electrophilic substitutions of five-membered rings, **13**, 235

F

Ferrocenes, heterocyclic, **13**, 1
Five-membered rings, electrophilic substitutions of, **13**, 235
Free radical substitutions of heteroaromatic compounds, **2**, 131
Furan chemistry, development of the chemistry of (1952–1963), **7**, 377

G

Grignard reagents, indole, **10**, 43

H

Halogenation of heterocyclic compounds, **7**, 1
Hammett equation, applications to heterocyclic compounds, **3**, 209; **20**, 1
Hetarynes, **4**, 121
Heteroaromatic compounds
  free-radical substitutions of, **2**, 131
  homolytic substitution of, **16**, 123
  nitrogen, covalent hydration in, **4**, 1, 43
  prototropic tautomerism of, **1**, 311, 339; **2**, 1, 27; Suppl. 1
Heteroaromatic N-imines, **17**, 213

Heteroaromatic substitution, nucleophilic, **3**, 285
Heterocycles
  aromaticity of, **17**, 255
  nomenclature of, **20**, 175
  photochemistry of, **11**, 1
  by ring closure of ortho-substituted t-anilines, **14**, 211
  synthesis of, through nucleophilic additions to acetylenic esters, **19**, 279
  thioureas in synthesis of, **18**, 99
Heterocyclic chemistry
  applications of NMR spectroscopy to, **15**, 277
  literature of, **7**, 225
Heterocyclic compounds
  application of Hammett equation to, **3**, 209; **20**, 1
  halogenation of, **7**, 1
  isotopic hydrogen labeling of, **15**, 137
  mass spectrometry of, **7**, 301
  quaternization of, **3**, 1
  reaction of acetylenecarboxylic acids with, **1**, 125
  reactions of, with carbenes, **3**, 57
  reactions of diazomethane with, **2**, 245
N-Heterocyclic compounds, electrolysis of, **12**, 213
Heterocyclic diazo compounds, **8**, 1
Heterocyclic ferrocenes, **13**, 1
Heterocyclic oligomers, **15**, 1
Heterocyclic pseudo bases, **1**, 167
Heterocyclic sulfur compounds, electronic structure of, **5**, 1
Heterocyclic syntheses, from nitrilium salts under acidic conditions, **6**, 95
Hilbert–Johnson reaction of 2,4-dialkoxypyrimidines, **8**, 115
Homolytic substitution of heteroaromatic compounds, **16**, 123
Hydrogen exchange
  base-catalyzed, **16**, 1
  one-step (labeling) methods, **15**, 137
Hydroxamic acids, cyclic, **10**, 199

I

Imidazole chemistry, advances in, **12**, 103
N-Imines, heteroaromatic, **17**, 213
Indole Grignard reagents, **10**, 43
Indoles
  acid-catalyzed polymerization, **2**, 287
  and derivatives, application of NMR spectroscopy to, **15**, 277

Indoxazenes, **8**, 277
Isatin, chemistry of, **18**, 1
Isoindoles, **10**, 113
Isoquinolines
  1,2-dihydro-, **14**, 279
  4-oxy- and 4-keto-1,2,3,4-tetrahydro-, **15**, 99
Isothiazoles, **4**, 107
  recent advances in the chemistry of monocyclic, **14**, 1
Isotopic hydrogen labeling of heterocyclic compounds, one-step methods, **15**, 137
Isoxazole chemistry, recent developments in, **2**, 365

L

Lactim ethers, chemistry of, **12**, 185
Literature of heterocyclic chemistry, **7**, 225

M

Mass spectrometry of heterocyclic compounds, **7**, 301
Meso-ionic compounds, **19**, 1
Metal catalysts, action on pyridines, **2**, 179
Monoazaindoles, **9**, 27
Monocyclic pyrroles, oxidation of, **15**, 67
Monocyclic sulfur-containing pyrones, **8**, 219
Mononuclear isothiazoles, recent advances in chemistry of, **14**, 1

N

Naphthalen-1,4-imines, **16**, 87
Naphthyridines, **11**, 124
Nitriles and nitrilium salts, heterocyclic syntheses involving, **6**, 95
Nitrogen-bridged six-membered ring systems, **16**, 87
Nitrogen heterocycles
  covalent hydration in, **20**, 117
  reduction of, with complex metal hydrides, **6**, 45
Nitrogen heterocyclic systems, Claisen rearrangements in, **8**, 143
Nomenclature of heterocycles, **20**, 175
Nuclear magnetic resonance spectroscopy, application to indoles, **15**, 277
Nucleophiles, reactivity of azine derivatives with, **4**, 145
Nucleophilic additions to acetylenic esters, synthesis of heterocycles through, **19**, 299
Nucleophilic heteroaromatic substitution, **3**, 285

O

Oligomers, heterocyclic, **15**, 1
1,2,4-Oxadiazoles, **20**, 65
1,3,4-Oxadiazole chemistry, recent advances in, **7**, 183
1,3-Oxazine derivatives, **2**, 311
Oxazole chemistry, advances in, **17**, 99
Oxazolone chemistry, recent advances in, **4**, 75
Oxidation of monocyclic pyrroles, **15**, 67
3-Oxo-2,3-dihydrobenz[d]isothiazole-1,1-dioxide (Saccharin) and derivatives, **15**, 233
4-Oxy- and 4-keto-1,2,3,4-tetrahydroisoquinolines, chemistry of, **15**, 99

P

Pentazoles, **3**, 373
Peroxides, cyclic, **8**, 165
Phenanthridine chemistry, recent developments in, **13**, 315
Phenothiazines, chemistry of, **9**, 321
Phenoxazines, **8**, 83
Photochemistry of heterocycles, **11**, 1
Physicochemical aspects of purines, **6**, 1
Physicochemical properties
  of azines, **5**, 69
  of pyrroles, **11**, 383
3-Piperideines, **12**, 43
Polymerisation of pyrroles and indoles, acid-catalyzed, **2**, 287
Prototropic tautomerism of heteroaromatic compounds, **1**, 311, 339; **2**, 1, 27; Suppl. 1
Pseudo bases, heterocyclic, **1**, 167
Purine bases, aza analogs, **1**, 189
Purines
  physicochemical aspects of, **6**, 1
  tautomerism, electronic aspects of, **13**, 77
Pyrazine chemistry, recent advances in, **14**, 99
Pyrazole chemistry, progress in, **6**, 347
Pyridazines, **9**, 211
Pyridine(s)
  action of metal catalysts on, **2**, 179
  effect of substituents on substitution in, **6**, 229
  1,2,3,6-tetrahydro-, **12**, 43
Pyridoindoles (the carbolines), **3**, 79
Pyridopyrimidines, **10**, 149
Pyrimidine bases, aza analogs of, **1**, 189
Pyrimidines, tautomerism and electronic structure of biological, **18**, 199

1-Pyrindines, chemistry of, **15**, 197
Pyrones, monocyclic sulfur-containing, **8**, 219
Pyrroles
  acid-catalyzed polymerization of, **2**, 287
  oxidation of monocyclic, **15**, 67
  physicochemical properties of, **11**, 383
Pyrrolizidine chemistry, **5**, 315
Pyrrolopyridines, **9**, 27
Pyrylium salts, syntheses, **10**, 241

Q

Quarternization of heterocyclic compounds, **3**, 1
Quinazolines, **1**, 253
Quinolizines, aromatic, **5**, 291
Quinoxaline chemistry, recent advances in, **2**, 203
Quinuclidine chemistry, **11**, 473

R

Reduction of nitrogen heterocycles with complex metal hydrides, **6**, 45
Reissert compounds, **9**, 1
Ring closure of ortho-substituted t-anilines, for heterocycles, **14**, 211

S

Saccharin and derivatives, **15**, 233
Selenazole chemistry, present state of, **2**, 343
Selenophene chemistry, advances in, **12**, 1
Six membered ring systems, nitrogen bridged, **16**, 87
Substitution(s),
  electrophilic, of five-membered rings, **13**, 235
  homolytic, of heteroaromatic compounds, **16**, 123

nucleophilic heteroaromatic, **3**, 285
  in pyridines, effect of substituents, **6**, 229
Sulfur compounds, electronic structure of heterocyclic, **5**, 1
Synthesis and reactions of 1-azirines, **13**, 45

T

Tautomerism
  electronic aspects of purine, **13**, 77
  and electronic structure of biological pyrimidines, **18**, 199
  prototropic, of heteroaromatic compounds, **1**, 311, 339; **2**, 1, 27; Suppl. 1
1,2,3,4-Tetrahydroisoquinolines, 4-oxy- and 4-keto-, **15**, 99
1,2,3,6-Tetrahydropyridines, **12**, 43
Theoretical studies of physicochemical properties and reactivity of azines, **5**, 69
1,2,4-Thiadiazoles, **5**, 119
1,2,5-Thiadiazoles, **9**, 107
1,3,4-Thiadiazoles, chemistry of, **9**, 165
Thiathiophthenes (1,6,6aS^{IV}-Trithiapentalenes), **13**, 161
1,2,3,4-Thiatriazoles, **3**, 263; **20**, 145
Thienothiophenes and related systems, chemistry of, **19**, 123
Thiochromanones and related compounds, **18**, 59
Thiophenes, chemistry of, recent advances in, **1**, 1
Thiopyrones (monocyclic sulfur-containing pyrones), **8**, 219
Thioureas in synthesis of heterocycles, **18**, 99
Three-membered rings with two heteroatoms, **2**, 83
1,3,5-, 1,3,6-, 1,3,7-, and 1,3,8-Triazanaphthalenes, **10**, 149
1,2,3-Triazines, **19**, 215
1,2,3-Triazoles, **16**, 33
1,6,6aS^{IV}-Trithiapentalenes, **13**, 161